145 Topics in Current Chemistry

Synchrotron Radiation in Chemistry and Biology I

Editor: E. Mandelkow

With Contributions by
D. Bazin, M. Benfatto, A. Bianconi, R. Clement,
H. Dexpert, J. Galy, J. Garcia, P. Lagarde,
P. Laggner, Y. Mathey, A. Michalowicz, A. Mosset,
Z. Sayers, T.-K. Sham, H. B. Stuhrmann,
M. Verdaguer

With 128 Figures and 23 Tables

Springer-Verlag
Berlin Heidelberg GmbH

This series presents critical reviews of the present position and future trends in modern chemical research. It is addressed to all research and industrial chemists who wish to keep abreast of advances in their subject.

As a rule, contributions are specially commissioned. The editors and publishers will, however, always be pleased to receive suggestions and supplementary information. Papers are accepted for "Topics in Current Chemistry" in English.

ISBN 978-3-662-15109-9

Library of Congress Cataloging-in-Publication Data
Synchrotron radiation in chemistry and biology I / editor, E. Mandelkow;
with contributions by D. Bazi . . . [et al.].
p. cm. — (Topics in current chemistry; 145)
ISBN 978-3-662-15109-9   ISBN 978-3-540-47935-2 (eBook)
DOI 10.1007/978-3-540-47935-2
1. Radiation chemistry. 2. Radiobiology. 3. Synchrotron radiation. I. Mandelkow, E.
(Eckhard), 1943 —. II. Bazin, D. III. Series.
QD1.F58 no. 145   [QD641]   540 s—dc 19   [541.3′8]   87-28455

Preface

When synchrotron radiation research started in the 1960's as a small offshoot from high energy physics laboratories, few people would have predicted the rapid expansion we have withnessed over the past few years. The most visible evidence of this growth is the increasing number of synchrotrons dedicated largely or exclusively to the radiation particle physicists used to consider as an annoying by-product and a waste of energy. The advantages of synchrotron light become plain when considering its properties: High brightness, wide range of wavelengths, excellent collimation, polarization, and a pulsed time structure.

Physicists provided the driving force behind many of the new developments in synchrotron radiation research, but other fields are catching up rapidly. In fact biologists interested in the structure and contraction of muscle were among the early users of synchrotron X-rays because this enabled them to follow the scattering pattern in real time. Similarly, chemists noted quickly that interatomic distances could be determined to an unprecedented accuracy using X-ray spectroscopy. Scattering and spectroscopy techniques still represent the majority of synchrotron radiation experiments in biology and chemistry, and this is reflected in the articles in this book and its companion valume to follow shortly.

A variety of reviews on the uses of synchrotron radiation in chemistry and biology have appeared over the past few years. Many of these were published as books or in specialist journals that are often not easily accessible in an average library. It was therefore a timely decision by Springer Verlag to cover this area within the series of "Topics in Current Chemistry". The hope is to provide outsiders or newcomers to the field with an overview of current activities, written by scientists who are themselves engaged in synchrotron radiation projects. Their topics include theory, technical and methodological aspects, as well as the results that can be obtained. The first part of this book concerns applications in chemistry, most of which deal with X-ray spectroscopy (EXAFS, XANES); the second part covers biological aspects based mainly on X-ray scattering techniques.

The articles were written in late 1986 and in 1987 and thus represent the state of the art. During that time many of the authors were involved in the planning or construction of new synchrotron radiation facilities; it was not easy for them to take time off for writing, and I am grateful for their willingness to contribute in spite of their busy schedules. Finally I would like to thank Dr. Stumpe from Springer Verlag for the smooth collaboration and my secretary, Ms. Elke Spader, whose organizational talent was invaluable in putting this book together.

Hamburg, August 1987   Eckhard Mandelkow

Table of Contents

X-Ray Synchrotron Radiation
and Inorganic Structural Chemistry

Alain Mosset, Jean Galy

Laboratoire de Chimie de Coordination du C.N.R.S. 205 Route de Narbonne, 31400 Toulouse, France

Table of Contents

Alain Mosset and Jean Galy

This paper aims at showing the interest of synchrotron radiation for various applications in the field of Inorganic Chemistry. After a short survey of the main properties of this light source and some characteristics of the storage rings, two main applications are reviewed. The first one concerns X-ray scattering techniques. The study of very tiny single crystals and the improvements in the field of powder structure investigations are underlined. The second field of applications deals with X-ray absorption spectroscopies. After a short recall of the theory, remarkably enlightened during the last years, the discussion is mainly centred on the EXAFS applications in the various fields of inorganic chemistry. Some examples of XANES results are also given.

1 Introduction

Electromagnetic radiation is a universal probe for studying the chemical and structural properties of matter. This is quite evident when it is considered the amount of information brought by the different spectroscopies in the infrared, visible and ultra-violet wavelength regions. This is perhaps even more obvious in the field of X-ray diffraction, nowadays an unvaluable analytical tool for the Chemist, the Physicist or the Biologist.

Each improvement or discovery in the field of radiation sources (lasers for exemple) or in the field of detectors (monodimensional and area detectors) directly brings up new steps forwards in our knowledge of matter.

One of the most important improvements concerning radiation sources is the availability of synchrotron radiation. This light is emitted when relativistic charged particles move along a curved path under the guidance of a magnetic field. There are many synchrotron sources throughout the world providing a continuous spectrum whose range extends from infra-red to X-ray wavelengths. The properties of this radiation (intensity, brightness, spatial definition) have raised up a large interest in the scientific community.

The scientific activity of synchrotron radiation centres is growing very rapidly and covers a great number of applications and branches of knowledge.

As an example, the scientific report of the LURE-CNRS Laboratory at Orsay (France) for 1985[1] relates that over 300 projects have been developed, two-thirds of which in the X-ray field and the other third in the ultra-violet and soft X-ray regions. 50% of these experiments correspond to applications submitted by laboratories of Physics, viz. 12% and 38% from Biology and Chemistry laboratories.

The aim of this paper is to bring out the interest of synchrotron radiation, in the region of X-ray wavelengths, for inorganic structural chemistry. Two main types of applications will be presented and illustrated with recent results:
— X-ray scattering: structural studies of very small single crystals, phase transition in powders, local order in amorphous solids, glasses . . .
— X-ray absorption spectroscopy: particularly, structural studies by EXAFS and XANES spectroscopy of coordination compounds in the solid state or in solution, glasses . . .

2 Synchrotron Radiation Properties

2.1 Wide Spectral Range

Figure 1 shows the "universal" spectral curve of an electron moving in a curved path.

The variation of intensity with wavelength depends only on the 'critical wavelength" (λ_c), the vertical scale of intensity being simply defined by the electron current and energy. The value of λ_c is given by the following formula:

$$\lambda_c = 5.6 \, R/E^3 = 18.6/BE^2$$

1 L.U.R.E. Laboratory for Utilization of Electromagnetic Radiation. LURE-CNRS, University Paris-Sud, Bat. 209C, 91407 ORSAY Cedex. FRANCE.

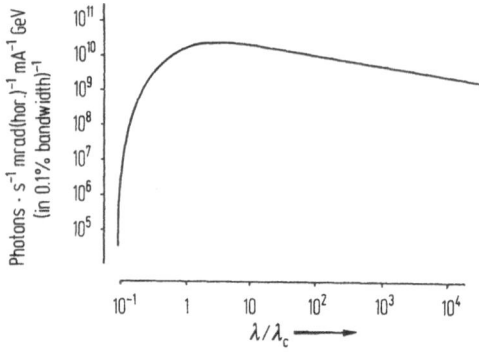

Fig. 1. Spectral curve of the synchrotron radiation

Where R is the magnet radius (meter)

E is the electron energy (GeV)

B is the bending field (Tesla)

Taking a magnetic field of 1.2 T, a wavelength of $\lambda_c = 1$ Å can be achieved with an electron energy E = 3.94 GeV. The shape of the spectrum indicates that, in such a case, a radiation extending from the X-ray region to the far infra-red can be obtained.

2.2 Excellent Spatial Coherence and Directional Properties

The radiation emitted by a single electron, for $\lambda = \lambda_c$, emerges at a mean angle $\theta = 0.511/E$ (MeV), E (MeV) being the electron rest mass energy. For the previous example (E = 3.94 GeV), $\theta \sim 1.3$ mrad.

2.3 Well-Defined Polarization

The light is linearly polarized in the plane of the electron orbit and elliptically polarized above and below this plane.

2.4 Fast-Time Structure

Because the electrons in a storage ring circulate precisely in the form of a bunch, the synchrotron light produced is very stable, with subnanosecond pulses repeated on a microsecond time scale.

2.5 High Intensity and Brightness

Figure 2 shows the vertically integrated flux (number of photons emitted per seconds) in a 0.1 % band pass for the de DCl storage ring (Orsay — France) compared with other synchrotron facilities [2].

The flux for 10 keV photons (1.24 Å) is equal to $1.5 \cdot 10^{12}$; this must be compared with 10^5 photons produced by a standard molybdenum target X-ray tube operating at 20 kV and 200 mA [3].

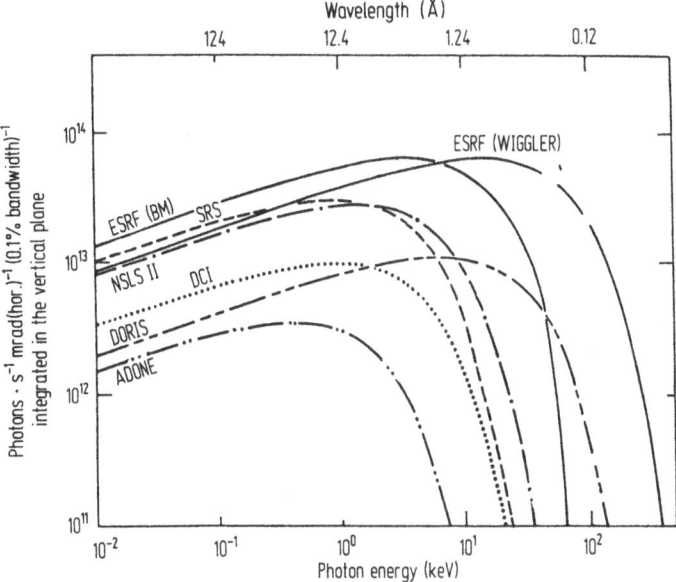

Fig. 2. Spectral curve for DCl and other synchrotron centers

The spectral brillance (or brightness), defined as the number of photons per second within a 0.1 % band pass per mrad² and mm², is equal to 3.3×10^{11}.

3 Storage Rings as Light Sources

In France, two synchrotron radiation sources are operational:
— ACO is working in the ultra-violet and soft X-ray regions. The installation of two new monochromators will allow the study of the S, C, O, N edges, very important elements in chemistry as well as in the field of biology.
— DCl working in the X-ray region, is mainly devoted to absorption experiments, EXAFS and XANES (2/3 of all experiments) and diffraction. This ring is the only one in the world to be equipped with an EXAFS apparatus working in the dispersive mode.

The super ACO ring, under construction at Orsay, will work in the energy regions extending from infra-red to 5 keV, in replacement of the ACO ring. The machine will be based on the use of positrons leading to a higher brightness [4].

The European Synchrotron Radiation Facility (ESRF), which will be settled at Grenoble (France), will have characteristics more advanced than those of existing machines built for high energy physics and even more advanced than machines under construction for synchrotron radiation purposes [2]. This machine will work in the X-ray region, i.e. $\lambda_c = 1$ Å for the normal bending magnet and $\lambda_c = 0.25$ Å for the strong field wigglers (a device which forces the beam in a trajectory with a shorter local radius of curvature).

Table 1. Synchrotron radiation facilities in use or under construction

NAME — LOCATION	E (GeY)	R (m)	l (ma)	λ_c (Å)
EUROPE				
ACO — Orsay, France	0.55	1.1	150	3.7
DCI — Orsay, France	1.8	3.8	250–400	3.9
Super ACO — Orsay, France	0.8	1.75	500	19
DORIS — Hambourg, FRG	5	12.1	100	0.54
BESSY — Berlin, FRG	0.8	1.8	500	20
SRS — Daresbury, UK	2.0	5.5	680	3.9
ADONE — Frascati, Italy	1.5	5.0	100	8.3
ESRF — Grenoble, France	5	22.4	560	1.0
USSR				
YEPP-2M — Novosibirsk	0.67	2	100	37
YEPP-3 — Novosibirsk	2.2	6.15	80–500	3.2
YEPP-4 — Novosibirsk	7	10	10	0.27
JAPAN				
INS-SOR — Tokyo	0.3	1.1	50	288
Photon Factory — Tsukuba	2.5	8.3	500	3.0
USA				
TANTALUS — Stoughton	0.24	0.65	100	263
SURF II — Washington	0.24	0.83	30	336
SPEAR — Standford	4	12.7	60	1.1
CESR — Cornell	8	32.5	100	0.35
NSLS I — Brookhaven	0.7	1.90	500	31
NSLS II — Brookhaven	2.5	6.95	500	2.5
ALLADIN — Stoughton	1.0	2.1	500	7.3

For 10 keV photons, the total emitted flux will be equal to $5 \cdot 10^{13}$ and the average brillance to $2.4 \cdot 10^{15}$, compared with $1.5 \cdot 10^{12}$ and $3.3 \cdot 10^{11}$ in DCl.

4 Applications of Synchrotron Radiation in Structural Inorganic Chemistry

The aim of this paper is not to give an exhaustive description of the applications in this field but rather to underline the main possibilities for the chemist involved in structural studies on bulk phases.

This means, for example, the study of the three dimensional order on very small single crystals, phase transitions or transformations under physical or chemical factors of crystalline powders but also vitreous transition of glasses, large angle and small angle X-ray scattering (LAXS—SAXS) of amorphous phases, liquids or co-ordination complexes in solution.

All these topics will be gathered under the heading "X-ray Scattering".

The second part of this illustration will concern the "X-ray Absorption Spectroscopies", mainly XANES and EXAFS the later becoming a routine structural investigation technique of the short range order.

4.1 X-Ray Scattering

4.1.1 Single Crystal Studies

Synchrotron radiation provides a number of advantages over the best high-power rotating anode X-ray generators: larger number of photons per second and larger spectral brillance. This allows to carry out experiments on a small number of scattering atoms or to shorten the measuring time. These conditions are particularly interesting in biology when sufficiently large crystals of protein molecules cannot be grown. Mineralogy and geochemistry are also potential fields for these applications because a number of minerals crystallize as tiny crystals (zeolithes, clay particles . . .) and of course the wide field of chemistry.

A quite nice illustration of such a type of study has been published in 1985 by Bachman et al. on a 6 µm CaF_2 crystal [5]. This was the smallest single crystal ever used for X-ray diffraction experiment. For an ideally imperfect crystal, the scattering power S may be defined as:

$$S = (\dot{F}_{000}/V_e)^2 \; V_c \lambda^3$$

where F_{000} and V_e are respectively the number of electrons and the volume in the elementary cell, V_c is the volume of the crystal and λ the wavelength. Usually, crystals with $S = 10^{17}$ are used for standard structure determinations.

The CaF_2 crystal, studied with $\lambda = 0.91$ Å, had $S = 1.3 \cdot 10^{14}$.

The structure refinements, carried out as a function of the degree of polarization K of the synchrotron beam (K ∼ 0.93), show that the diffraction data can be considered as quasi free of absorption, extinction and thermal diffuse scattering. The authors of this work conclude that this kind of experiment will allow the study of crystal structures on a level of accuracy unknown up to now. Moreover, with a slightly focussed beam and a vertical diffraction geometry it should be possible to study single crystals with a scattering power as low as $S = 10^{13}$. This corresponds to a crystal of 3 µm edge length, i.e. the size scale of a large number of mineral species such as clay materials.

The high flux of synchrotron radiation allows higher data acquisition rates. This can be of fundamental interest when the diffraction measurements must be very accurate. For example, the collection of X-ray data for *charge density investigations* by the X — n method, which takes more than a month with a classical source, could be performed within a few days. This high speed of data collection is also of tremendous importance in the study of single crystals such as those containing short-lived radio-element or corresponding to unstable phases.

4.1.2 Powder Studies

The advantages of the synchrotron radiation already mentioned permit excellent work on crystalline powders, especially accurate determination of unit cell. Moreover,

Table 2. Results of Rietveld refinement of powder data for Bi_2O_3 [6]
Previously reported Cu Kα single crystal results are also shown for comparison.

Atom	Powder			Single crystal		
	x	y	z	x	y	z
Bi(1)	0.526(1)	0.181(1)	0.363(1)	0.5240(1)	0.183(2)	0.3613(1)
Bi(2)	0.042(1)	0.046(1)	0.779(1)	0.0409(2)	0.0425(1)	0.7762(1)
O(1)	0.798(13)	0.286(7)	0.700(10)	0.780(4)	0.300(3)	0.710(3)
O(2)	0.275(13)	0.050(8)	0.155(10)	0.242(5)	0.044(4)	0.134(4)
O(3)	0.288(2)	0.043(9)	0.541(10)	0.271(4)	0.024(3)	0.513(3)
a		5.8480(1) Å			5.8486(5) Å	
b		8.1681(1) Å			8.1661(10) Å	
c		7.5111(1) Å			7.5097(8) Å	
β		112.983(<1) °			113.00(1) °	

it should be theoretically feasible to apply the Rietveld method to solve more complex structures than currently possible with existing neutron or classical X-ray sources. Cox et al. have described such experiments performed at the Cornell CESR storage ring [6]. The configuration adopted consists of a souble crystal Si(220) monochromator scattering in the vertical plane and a Philips diffractometer modified to accept a Si(111) analyser which is mounted before the detector and acts as a high angular resolution slit. Cox et al. have studied the structure of Bi_2O_3 with a 1,5 Å wavelength, obtaining a resolution roughly 2–4 times better than that of a conventional focusing diffractometer using Cu Kα radiation. The results of the Rietveld refinement are given in Table 2.

Anyhow we believe that structural studies on small crystals, when available, should be preferred as indicated previously.

Any kind of *phase transition* or *transformation* under the influence of physical (T, P, . . .) or chemical factors can be accurately studied with synchrotron radiation.

Bastie et al. have studied a new phase in quartz which lies between the classical α and β phases [7]. This phase exists in a very narrow range of temperature around 846 ± 1.3 K and has been proved to be incommensurate. Using a high temperature furnace especially designed for such a study, the α and β transitions were investigated and the evolution of the satellite reflections due to the incommensurate phase has been observed during the heating and the cooling of the sample.

Another example is provided by Tolocko et al., who took advantage of the high speed of data collection to study the recrystallization process in deformed silver by time-resolved X-ray diffraction [8]. The sample deformation was obtained by cutting and the first diffraction curve has been obtained 400 m sec after the deformation. Another 59 diffraction curves were stored with 50 m sec time interval. These curves allow to follow the structural changes during recrystallization.

4.1.3 Large Angle X-Ray Scattering (LAXS)

This method provides unvaluable informations on the short and medium range order in disordered atomic architectures: amorphous solids, glasses but also liquids or complexes in solution. The fast development and the economical importance of "modern" amorphous solids, like metallic glasses or glasses prepared by sol-gel process, the gel themselves, have raised a renewed interest for this technique.

In order to have accurate informations from the radial distribution function (RDF), it is necessary to have a good counting statistics and to extend the measurement as far as possible in the k space (k = $4\pi \sin\theta/\lambda$). This leads of course to time-consuming data collections. The improvement of linear detector permits to shorten the collection time; anyhow, with a classical X-ray tube and Mo Kα wavelength, the data collection takes a few hours to go up to 17 Å^{-1}. The synchrotron beam allows faster experiments and the possibility to work in the hard X-ray part of the spectrum extending considerably the k region available. A realistic possibility for the future ESRF machine seems to be k_{max} = 100 Å^{-1} (2).

Such a method has been used to elucidate the gold-sulfur framework of an impor-tant antiarthritic drug, sodium gold (1) thiomalate (Myochrisin). Elder [9] measured the X-ray scattering data on a mineral oil mull of solid Myochrisin at Stanford storage ring, with a 1.04 Å wavelength. Figure 3 shows the corresponding radial distribution function (RDF).

The first peak, at 2.30 Å, can be assigned as the Au—S vector. The major peak 2 corresponds to a second neighbour distance of 3.35 Å and a gold—gold interaction. This Au—Au distance leads to an Au—S—Au angle equal to 94°. The third major peak 3, at 5.8 Å, corresponds to the next Au ... Au interaction. Assuming the S—Au—S angle equal to 180°, the chain conformation depends only on the torsional

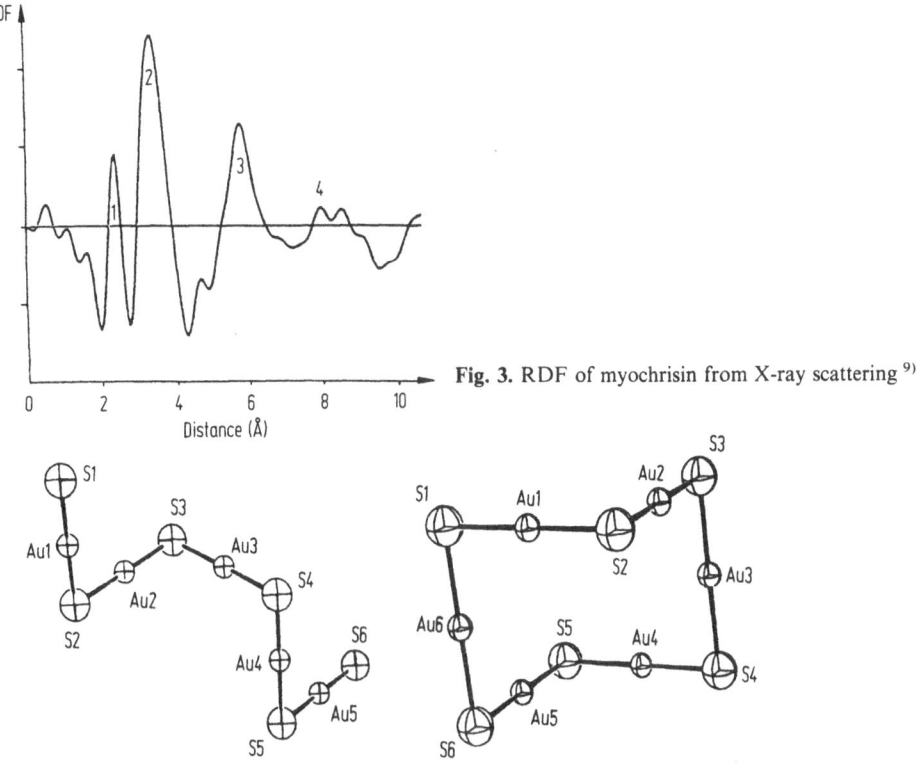

Fig. 3. RDF of myochrisin from X-ray scattering [9]

Fig. 4. Two possible Au—S frameworks for myochrisin [9]

angle S(1)-S(2)-S(3)-S(4). The Au(1)-Au(3) distance is equal to 5.8 Å when this angle is 90°. Figure 4 shows two possible frameworks: a hexameric ring or an open-chain structure.

An interesting modification of the LAXS method has been used by Sadoc [10] to study the local order in amorphous metallic alloys. The scattering data of Cu_5Y and Cu_2Y were collected successively with four wavelengths (the inflection point of the two absorption edges and some hundred of eV below). This allows to extract the partial structure factors. In the case of Cu_5Y, the Y—Y contribution is very low and it was not possible to calculate the Y—Y pair distribution. The RDF, corresponding to the Cu—Cu pair, shows a great similarity with the distribution in the crystalline phase up to 10 Å. But the Cu—Y pair distribution is very different in the crystalline and in the amorphous solids. The model proposed can be described in terms of disclination lines in an icosahedral packing.

4.1.4 Small Angle X-Ray Scattering (SAXS)

Among all the techniques listed in this heading "X-ray Scattering", the SAXS method is probably one of the most often used. The studies of phase separation and recrystallization in organic polymers or polymer mixtures very often requires SAXS experiments. In the inorganic chemistry field too, this method is used to tackle a variety of scientific purposes:
— fractal structures and ageing kinetics of $Al(OH)_x$ solutions [11];
— medium range order in amorphous hydrogenated Cu—Ti alloys [12];
— phase separation in AlZnAg solid solutions [13];
— physico-chemical studies of $AlCl_3$, 6 H_2O aqueous solutions [14];
— phase separation in B_2O_3—PbO—Al_2O_3 glasses [15];
— ... .

An interesting application, performed at LURE-DCl storage ring concerns the study of the swelling properties of layered minerals [16]. Pons et al. have studied the swelling of ornithine-exchanged and Li-exchanged vermiculites. The SAXS experiments allow to extract three order parameters: the first moment $d_{moy.}$ of the interlayer distance distribution; the ratio $\Delta^2/(d_{moy.})^2$ with Δ^2, the distribution variance and the ratio $d_{max.}/d_{moy.}$ with d_{max} the most probable interlayer distance.

The SAXS diagrams show a very intricate transition: hydrated solid → gel in the Li-exchanged solids. Three phases are coexisting: a hydrated solid with two water layer, a gel and a very disordered solid. This study should allow to elucidate the formation process of a colloidal structure in such systems.

Pons et al. have also followed the dessication-rehydration cycle in the montmorillonite clays by the SAXS method. This is an interesting study because the swelling properties of clays play an important role in the structural organization of soils. Two main conclusions arise from this work:
— during the first dessication, the clay network is distorted and, as the loss of water is going on, the structural units gather to form thicker particles. Water is essentially extracted from the interparticle porosity.
— during the rehydration process, it is shown that the clay mineral partly loses its property to absorb water. This loss is directly proportional to the level of dessication. This is due to the irreversible deformation of the network.

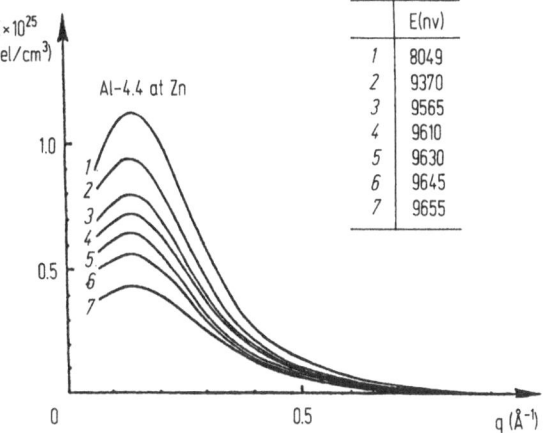

Fig. 5. ASAX curves of Al — 4 % Zn alloy [17]

As in the case of LAXS measurements, *the anomalous dispersion effects*, arising when the energy is closed to the absorption edge, can be used with the SAXS technique. FONTAINE did ASAXS experiments, at LURE-DCI, to probe the distribution of Zn atoms in phase separated Al—Zn alloys [17]. He measured the variation of SAXS from Guinier-Preston zones when the energy is scanned below the Zn absorption edge (Fig. 5).

The absolute integrated intensities, scattered by an aluminum alloy at 4.4% Zn, allows to deduce the variation of f'_{Zn} with a good accuracy.

SAXS data collections on these Al—Zn alloys were also performed at Stanford [18] to study the kinetics of the early stages of decomposition (unmixing) at the ageing temperature (400 K < T < 470 K). In this experiment, the quenching and ageing were performed in situ and the scattering curves recorded every 10 seconds (Fig. 6).

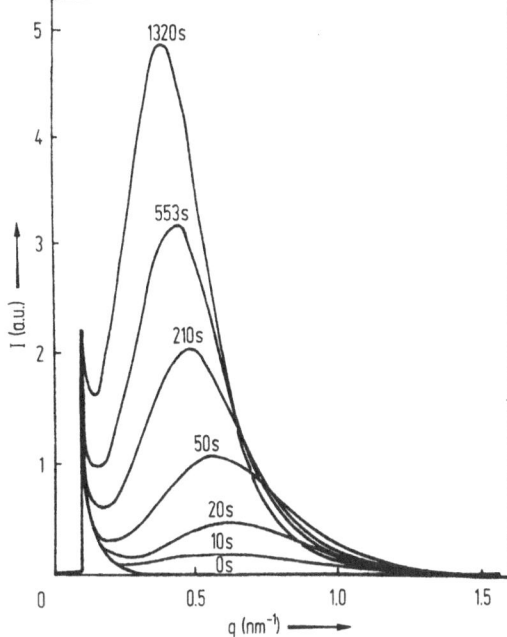

Fig. 6. Time — resolved SAXS curves for Al—Zn alloy [18]

4.2 X-Ray Absorption Spectroscopies

4.2.1 EXAFS: Theory and Applications

X-ray absorption spectroscopy is a technique that depends on the absorption of the radiation by a selected type of atom within the material. A typical X-ray absorption spectrum is pictured in figure 7. The abrupt change in the absorption coefficient μ, called absorption edge, occurs when the incident X-ray photon energy is sufficient to ionize an inner electron shell from the absorbing atom (K-edge when the ejected electron is a 1 s electron).

Above this threshold, the absorption coefficient shows a modulation known as Extended X-ray Absorption Fine Structure (EXAFS). The origin of this modulation can be qualitatively explained as follows:
— the ejected photoelectron can be viewed as a spherical wave centered on the absorbing atom (Fig. 8);
— this wave can be backscattered by the neighbouring atoms producing an incoming wave;
— the interferences of the outgoing and incoming waves result in the oscillatory behaviour of the absorption coefficient.

This modulation depends on the chemical nature of the neighbouring atoms, the structural environment of these atoms and their distances to the absorber.

Details of EXAFS theory and analysis are given in numerous reviews and books [19-22]. The main features only will be recalled in this paper.

For moderate thermal or static disorders, the modulation of the absorption coefficient is given by:

$\chi(E) = (\mu(E) - \mu_0(E))/\mu_0(E)$; ($\mu x = Ln(I_0/I)$, where I_0 and I are the incident and transmitted X-rays intensities and x the sample thickness)

$$\chi(k) = \sum_j N_j S_j(k) F_j(k) \exp(-2\sigma_j^2 k^2)$$

$$\times \exp(-2r_j/\lambda(k))(\sin(2kr_j + \varphi_j(k)))/kr_j^2$$

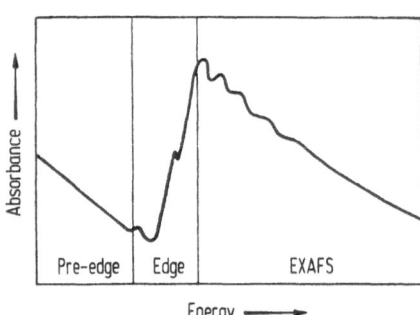

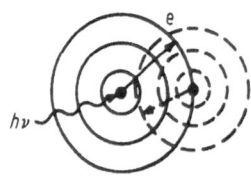

Fig. 7. General shape of an X-ray absorption edge

Fig. 8. Schematic picture of the extended electron wave function. The dotted circles correspond to the backscattering from the neighbours

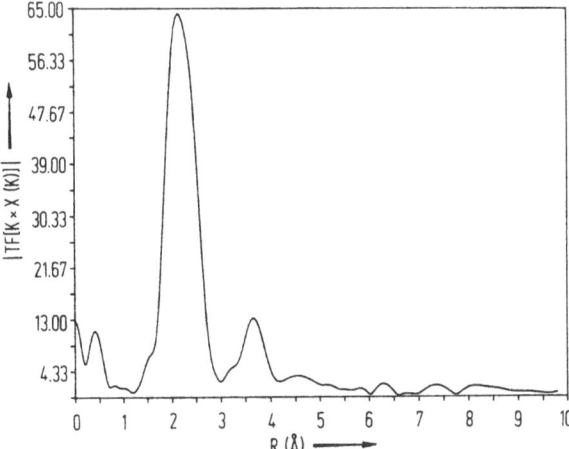

Fig. 9. Typical radial distribution function in EXAFS

where $k = \sqrt{(2\,m/h^2)\,(E - E_0)}$, E is the incident photon energy, E_0 is the threshold energy. $F_j(k)$ is the backscattering amplitude from each of the N_j neighbouring atoms with a Debye-Waller factor σ_j at a distance r_j from the absorber. $\varphi_j(k)$ is the total phase shift, depending on both the absorber and backscatterer nature of the atoms.

In strongly disordered systems (liquids-amorphous solids), the previous treatment is no longer appropriate. The EXAFS expression can be written:

$$\chi(k) = F(k)/k \int_0^\infty g(r) \exp(-2r/\lambda(k)) (\sin(2kr + \varphi(k)))/r^2 \, dr$$

$g(r)$ is a continuous radial distribution function characteristic of the system.

There are two main methods of analysis for the EXAFS data: the Fourier transform and the curve fitting. The first method gives a radial distribution function $\varrho(r)$ versus interatomic distances (Fig. 9):

$$\varrho(r) = 1/\sqrt{2\pi} \int_{k\,min.}^{k\,max.} k^3 X(k) \exp(12kr) \, dk$$

By comparison with model compounds, this function predicts the interatomic distances and the number of neighbouring atoms in the studied compound.

The curve fitting method attempts to fit by least-squares refinements the k^3. X(k) curves with some calculated models (Fig. 10).

EXAFS oscillations extend from ~ 50 eV to ~ 800 eV apart from the absorption threshold. In this energy range, the ejected photoelectron is weakly scattered by the neighbouring atoms and EXAFS can be analyzed by the simple scattering theory [23].

On the contrary, the structure which extends about 10 eV on each side of the edge, called X-ray Absorption Near Edge Structure (XANES), is strongly affected by multiple scattering effects. From this structure, it is possible to obtain information on the coordination geometry and the effective charge on the absorber.

Several theoretical methods to calculate the XANES spectra have been proposed [24–26]. One of the most efficient method is that proposed by Durham et al. which

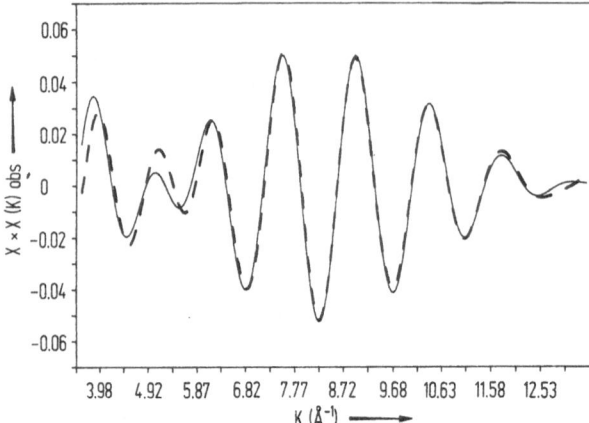

Fig. 10. Curve fitting in the k space (experimental: full line; theoretical: dotted line)

allows to calculate the near edge structure with the help of the multiple scattering theory and a muffin-tin form for the potential of each atom. The atoms are classified into shells surrounding the absorbing atoms and the whole multiple scattering within and between the atomic shells is taken into account.

4.2.2 Coordination Chemistry

There has been a growing interest for the structural study of disordered or amorphous coordination complexes in the past few years. Thus, particular situations account for this interest:
— it is sometimes very difficult or impossible to grow suitable single crystals;
— the properties of the amorphous phase are different from the properties of the crystalline phase.

An example of the previous situation is enlightened by the case of the $Cu^{II}Ni^{II}\mu$-oxalato chain studied by Verdaguer et al. [28]. The aim of this study is to establish a structural model of this ordered bimetallic chain in order to explain the magnetic properties. As it has been impossible to obtain single crystals, the EXAFS technique has been used at both the copper and the nickel edges.

EXAFS data were recorded at LURE-DCl at room temperature and 30 K, for the (CuNi) and (CuZn) bimetallic chains and for the (Cu), (Ni) and (Zn) homometallic compounds. The Fourier transform spectra are shown in Fig. 11.

The least-squares fitting on the first shell leads to the following conclusion: there are four neighbours around copper and six around nickel (or zinc). The authors assign the three first peaks to the M—O, M—C and M—O' shells. Taking the stoichiometry into account, this bridging network is only compatible with a chain structure. The magnetic properties allow to rule out the stacking of alternating ribbons of $Cu(C_2O_4)$ and $Ni(C_2O_4)$. So, the most probable model is the bimetallic chain. The strong peak around 4 Å, observed at 30 K in the "CuOx" spectrum, can give a clue to build up a model of the chain stacking.

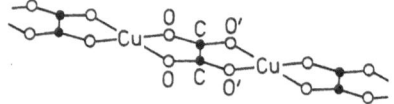

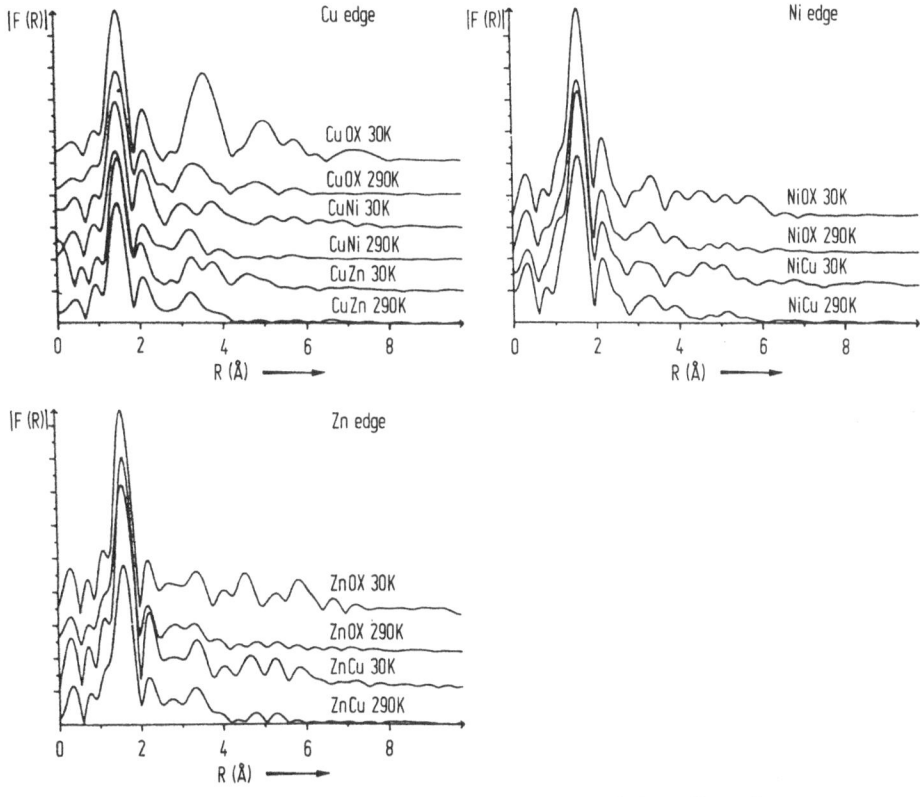

Fig. 11. Fourier transforms at 30 and 290 K of various homo- and bimetallic oxalates.
Cuox = $Cu(C_2O_4)$, $1/3 H_2O$; CuNi = NiCu = $CuNi(C_2O_4)_2$, $4 H_2O$;
CuZn = ZnCu = $CuZn(C_2O_4)_2$, $4 H_2O$; Niox = $Ni(C_2O_4)$, $2 H_2O$;
Znox = $Zn(C_2O_4) \cdot 2 H_2O$ [28]

Indeed, this peak corresponds to four copper neighbours. This peak is also observed, but less intense (two neighbours), in the (CuNi) complex at the copper edge but not at the nickel edge. The authors conclude that copper "sees" two other copper atoms situated at 4 Å in neighbouring chains and propose the model shown in Fig. 12. This structural model is quite similar to that observed by Gleizes et al. in the compound $CuMn(C_2O_2S_2)_2$, $7.5 H_2O$, the structure of which has been determined by X-ray single crystal diffraction methods [29].

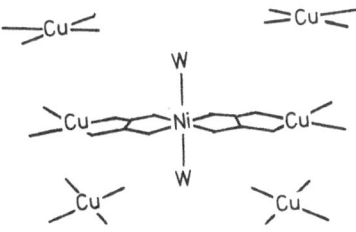

Fig. 12. Proposed structure for bimetallic $CuNi(C_2O_4)_2$ $\times 4 H_2O$ oxalate [28]

15

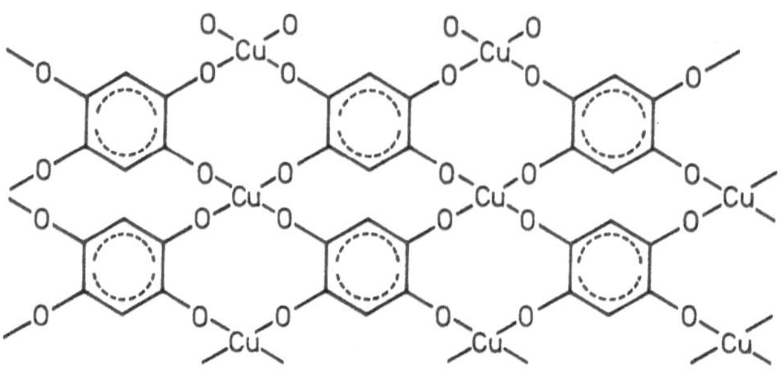

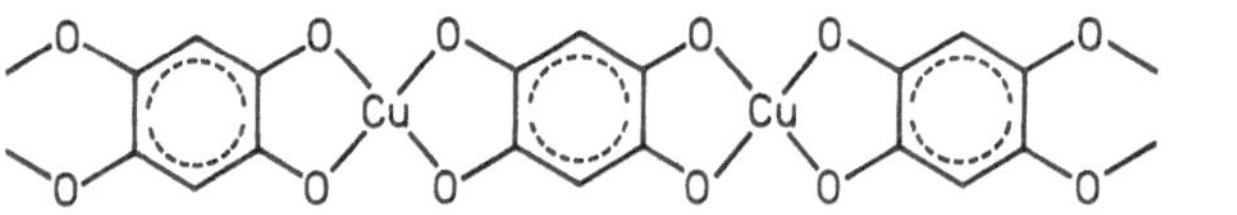

Fig. 13. Proposed structures for $Cu(C_6O_4H_2)$ [30]

A very similar study, with the same motivation, has been performed by Verdaguer et al. on copper (II) chloranilato and bromanilato complexes [30]. Two structures were proposed for the compound $Cu(C_6O_4X_2)$, as shown in Fig. 13 for X = H.

EXAFS data were collected at the Cu and Br edges. The fitting results give the following distances: Cu—O = 1.95 Å; Cu—C = 2.6 Å; Br — C = 1.86 Å; Br—Cu = 5.06 Å; Br—Br = 6.55 Å. These results are quite compatible with a planar ribbon structure. Moreover, the authors demonstrate that a planar layer structure or a bent ribbon structure can be ruled out as incompatible with the Cu—O and Cu—C distances.

Synthetic zeolithes, exchanged with transition metals or rare earth ions, are of extreme importance in fundamental and applied catalysis. However, it is generally difficult to obtain suitable single crystals and, moreover, X-ray diffraction studies often show long-range disorder in the tunnels. EXAFS spectroscopy can be used to probe the local environment of metal ions or even to study coordination complexes formed inside the cages of the zeolithes. Morrison et al. have collected EXAFS data on $Cu(en)^{2+}$ exchanged Y-zeolithe at Stanford Synchrotron Radiation Laboratory (Fig. 14) [31]. The first shell analysis of their spectra gives: Cu—N = 1.98 Å and 4.2 neighbours. This result shows unambiguously that the bis(ethylenediamine)-copper(II) complex remains unchanged in the zeolithe.

Another important application of the EXAFS technique is the study of transition metal ions involved in biological macromolecules. The analysis of EXAFS data from gold metallothionein suggests that the gold atoms are twofold coordinated whereas the zinc atoms, which gold replaces, are fourfold coordinated. Eidsness and Elder [32] have studied several Au—S and Au—P type structures as models for the biological systems. The results of these studies are summarized in Table 3.

The refinement of gold sulfur complexes is quite good. Bond distances agree within 0.01 Å and the error on the coordination numbers is less than 10%. On the contrary, the refinement of gold-phosphorus complexes yields correct bond distances but some of the coordination numbers (CN) are quite wrong. The authors show that the ratio

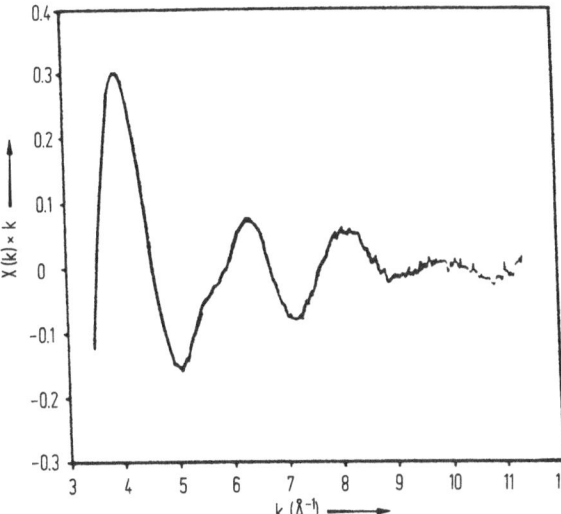

Fig. 14. EXAFS of $Cu(en)_2^{2+}$-exchanged zeolithe [31]

Table 3. Bond distances and coordination numbers for Au—S and Au—P structures [32]

Structure	xtal		Room Temp Data		Low Temp Data	
	dist (Å)	n	R (Å)	n	R (Å)	n
$[Au(PPh_2CH_3)_4]^+$	2.449	4	2.44	2.0	2.44	2.8
$[Au(dpe)_2]^+$	2.389	4	2.40	2.7	2.39	3.3
$[Au(dpp)Br]_2$	2.309	2	2.31	1.8	2.31	2.0
$[Au(dte)_2]^-$	2.289	4	2.29	3.9	2.29	4.1
$[Au(dtt)_2]^-$	2.310	4	2.30	4.1	2.30	4.4
$[Au(etu)_2]^+$	2.279	2	2.28	2.8	2.28	2.1
Models:	$[Au(S_2O_3)_2]^{3-}$		$Au-S_{xtal} = 2.276$ Å			
	$[Au(PPh_2CH_3)_2]^+$		$Au-P_{xtal} = 2.316$ Å			

dpe   dpp   dte   dtt   etu

EXAFS(CN)/Crystal(CN) is linearly related to the bond distance in the crystal (Fig. 15).

Their conclusion is that a rescaling of the CN is necessary with increasing bond distances as indicated in Fig. 15. This situation may be encountered in cases where metal-coordinating atom distances can take values very different from those in the model structure. It is to be noticed that the results for CN are not improved either by variations of the disorder parameter in the curve fitting or by measurements at low temperature.

17

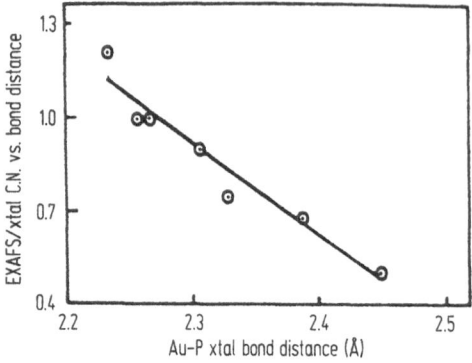

Fig. 15. Plot of the ratio EXAFS(CN)/ crystal (CN) versus Au−P bond distances [32]

This study shows that the transferability of results obtained on model compounds to more intricate systems must be done very carefully, especially concerning the appreciation of the coordination numbers.

4.2.3 Solution Chemistry

In recent years, many studies have been devoted to the transition metal halides in aqueous solution by X-ray and neutron diffraction and by EXAFS [33−36]. The local environment of the metal ion differs greatly for the different elements and valence states. Thus, for the chloride solutions, Fe^{3+} and Cu^{2+} show an important bonding to Cl^- whereas Ni^{2+} and Co^{2+} exhibit little or no bonding.

Sadoc et al. [37] have studied the aqueous solution of $CdBr_2$ by EXAFS measurements on the Cd and Br K-edges at the LURE-DCI storage ring. The prospect of this work was to confirm the conclusion of the Raman study [38]: viz. the absence of monomeric $Cd−Br_2$ species and correlatively the existence of $CdBr_3^-$ and $CdBr_4^{2-}$.

EXAFS data have been collected for six solutions whose concentration varies from the saturation (concentration $= C_s$) to 0.05 C_s. The first important result deals with the radial distribution functions which are identical for all concentrations on both edges. This means a local environment of both Cd^{++} and Br^- ions which do not change over this concentration range. The refinement, on the cadmium edge, gives 2.0(5) Br ions at 2.55(5) Å and 3.0(5) oxygen atoms at 2.18 Å, for the cadmium environment. The refinement on the bromine edge yields to 1.70(25) neighbouring cadmium at 2.58 Å.

In isolated $CdBr_n$ complex ions, the bromine atom "sees" only one cadmium; this is not in agreement with the EXAFS result which indicates 1.7. Local order consisting in $CdBr_3$ or $CdBr_4$ isolated units is no more correct as it leads to disorder parameters anomalously large. Thus, the authors propose a model, which is consistent with Raman results, and built up with chains of $CdBr_3$ pyramids or $CdBr_4$ tetrahedra (Fig. 16). Such a local structure may give a correct average coordination number, the exact value depending on the extent of these chains. However, the chains of $CdBr_3$ units seem more probable as the best fit was obtained with a second subshell of three oxygen atoms. So, the local order should be made of corner-shared octahedra chains.

A similar study has been performed by Lagarde et al. [39] on several strong II−I electrolytic solutions: $CuBr_2$, $CuCl_2$, $NiBr_2$, $NiCl_2$, $ZnBr_2$, $SrBr_2$. These solutions show very different local orders.

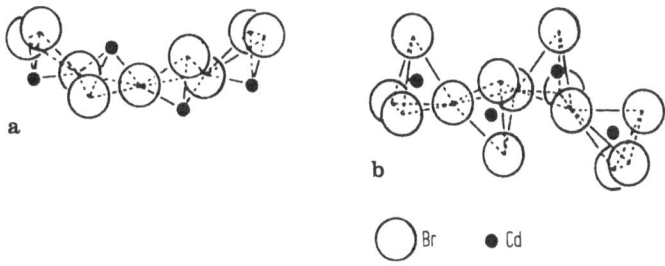

Br   ● Cd

Fig. 16. Two possible models for the local order of CdBr$_2$ solutions: a) chain of CdBr$_3^-$ pyramids; b) chain of CdBr$_4^{2-}$ tetrahedron [37]

In the case of CuBr$_2$ and ZnBr$_2$ concentrated solutions, the authors suggest a model built up with MBr$_4$ tetrahedral units together in a solid like structure, ressembling that of the corresponding crystals. In the case of CuBr$_2$, these extended structures involve 50 % of the metal ions and 85 % in the case of ZnBr$_2$. The surrounding of the metal ion and of the bromine atom are given in Table 4.

When the concentration decreases, the percentage of metal atoms involved in these structures decreases too but, even at low concentrations (0.1 M), there is still evidence of metal-halogen bonding. The local order in the CuCl$_2$ solutions seems to be similar but it is rather difficult to obtain quantitative informations.

On the opposite, the study of a SrBr$_2$ saturated solution does not show any Sr—Br bonding. EXAFS data can be analyzed with a shell of four oxygen atoms at 3.40 Å.

An other example of the interest of EXAFS experiments for solution chemists is the determination (see Yamaguchi et al. [40]) of the hydration structure of silver ions in aqueous perchlorate and nitrate concentrated solutions.

3 M and 9 M solutions have been studied, at Stanford, on the Ag K-edge. The data analysis reveals that the hydration structure of the silver ions consists of three to four oxygen atoms provided by water molecules, the bond distance being around 2.31–2.36 Å (Table 5).

In the case of 9 M AgNO$_3$ solution, the Fourier transform shows a small peak around 4 Å which is ascribed, by the authors, to Ag ... Ag interactions. This hypothesis is supported by an X-ray scattering study [41] of the AgNO$_3$ melt which shows such metal-metal interactions.

4.2.4 Oxides, Chalcogenides and Metallic Glasses

The number of studies in this field has been growing very rapidly in the past few years since there is a great demand from industry for improved "classical" glasses

Table 4. Coordination numbers CN and Bond distances extracted from EXAFS experiments performed on CuBr$_2$ and ZnBr$_2$ solutions [39]

	CN				Distances (Å)		
	M—Br	M—O	Br—M	Br—O	M—Br	M—O	Br—O
ZnBr$_2$	3.5	2.5	1.75	1.75	2.37	1.94	3.26
CuBr$_2$	1.5	3.5	1–1.5	3	2.39	1.93	3.3

Table 5. EXAFS results issued about silver salt solutions [40]

		3M AgClO₄	3M AgNO₃	9M AgNO₃
Ag—O	r (Å)	2.31	2.36	2.34
	N	2.9	3.9	2.6
Ag ... Ag				3.84 (Å)

and for new materials such as glasses prepared by sol-gel process or metallic glasses.

X-ray scattering studies often lead to ambiguous answers concerning the local structure, especially in the case of multicomponent glasses. As EXAFS provides a partial radial distribution function, this method becomes very powerful when the distribution relative to each atomic species can be joined up. This situation is often encountered in metallic glasses.

Maeda et al. [42] have prepared glassy samples of $Fe_{90}Zr_{10}$ alloy by the melt-quenching technique. EXAFS data were collected on both edges at Tsukuba Photon Factory. Results of the curve fitting are given in Table 6.

The main conclusions of this study are:
— the nearest neighbour distances are 0.2 to 0.3 Å shorter
— the coordination numbers are lower, in the amorphous state, than those in the corresponding crystals.

However, according to Raoux and Frank [43] these underestimated coordination numbers and too short bond distances are caused by an intrinsic limitation of EXAFS, i.e. the spectrum is not analysed below $k = 6 Å^{-1}$. These authors propose to overcome this limitation by a simultaneous analysis of EXAFS and X-ray scattering data. They have applied this method to the Cu_2Y alloy.

The EXAFS on Y-edge is very well fitted by a two shell model: 3.3 Cu at 2.83 Å and 1.5 Cu at 3.03 Å. But, in this second shell, the coordination number is extremely ambiguous as it can be raised to 6 by changing the Debye-Waller factor from 0.1 Å to 0.15 Å. The study of the X-ray scattering data allows to solve this problem. The fitting of the radial distribution functions gives evidence of Cu—Cu interactions. And, finally, the only model which leads to a good reconstruction of the X-ray RDF *and* of the EXAFS spectrum implies four different subshells, two of them for Cu—Cu correlations and the other ones for Cu—Y correlations. Interatomic distances and coordination numbers appear in Table 7.

The study of oxide glasses by EXAFS is not so favourable as it is very difficult to collect extended spectrum of the oxygen K-edge. EXAFS information is therefore limited to the cation environments [44].

For the network-forming elements, EXAFS will be used to detect some occasional change in coordination numbers and bonding distances, as a function of chemical composition or thermal treatment. For network-modifying or doping elements, two main problems have to be solved: (i) the nature of the coordination shell and (ii) the description of the second shell which characterizes the degree of local order.

Petiau and Calas [44] have studied the zirconium surrounding in various silicate and borosilicate glasses where Zr is a doping element (1 %). The Fourier transform of the EXAFS signal clearly indicates the existence of a two-shell distribution: (i) a first shell of six oxygen atoms and (ii) a second one corresponding to 2.5 silicon atoms.

Table 6. Results of least square refinements on EXAFS data from Fe—Zr alloys [42]

(a) Zr K-edge EXAFS

Sample	Shell	R (Å)*	N*
Fe_2Zr	Zr—Fe	(2.931)	(12.0)
(crystal)	Zr—Zr	(3.061)	(4.0)
$Fe_{90}Zr_{10}$	Zr—Fe	2.62(1)	5.2(1)
(as-quenched)	Zr—Fe	2.72(1)	3.2(1)
	Zr—Zr	3.04(1)	0.9(1)
	Zr—Zr	3.29(1)	0.3(1)
$Fe_{90}Zr_{10}$	Zr—Fe	2.61(1)	2.7(1)
(annealed)	Zr—Fe	2.73(1)	2.2(1)
	Zr—Zr	3.02(1)	0.8(1)
	Zr—Zr	3.26(1)	0.3(1)

(b) Fe K-edge EXAFS

Sample	Shell	R (Å)*	N*
Fe_2Zr	Fe—Fe	(2.500)	(6.0)
(crystal)	Fe—Zr	(2.931)	(6.0)
$Fe_{90}Zr_{10}$	Fe—Fe	2.30(1)	1.8(1)
(as-quenched)	Fe—Fe	2.39(1)	2.0(1)
	Fe—Zr	2.64(1)	0.2(1)
	Fe—Zr	2.73(1)	1.4(1)
$Fe_{90}Zr_{10}$	Fe—Fe	2.28(1)	2.2(1)
(annealed)	Fe—Fe	2.38(1)	4.3(1)
	Fe—Zr	2.64(1)	0.4(1)
	Fe—Zr	2.76(1)	0.4(1)

* R and N values in brackets are from crystallographic study.

Thus, the local order is made by ZrO_6 octahedra sharing corners with SiO_4 tetrahedra with a non random angular distribution.

The same type of analysis on glasses with titanium as doping element shows a connection between SiO_4 and TiO_4 tetrahedra and a Ti—O—Si angle (130°) significantly different from the usual value for Si—O—Si angle (152°) in silica glasses.

Another study of the environment of network modifiers in oxide glasses has been made by Greaves et al. [45] at Daresbury. This work is interesting in the way that the

Table 7. Comparison between EXAFS and X-ray scattering results for Cu_2Y alloy [43]

		R	N (±0.5)	σ	R	N (±0.5)	σ
Cu—Cu	Exafs	2.47 ± 0.02	4	0.11	3.05 ± 0.05	2	0.11
	Scattering	2.47 ± 0.05	4.5	0.15	3.03	2	0.15
Cu—Y	Exafs	2.83 ± 0.03	3.3	0.13	3.03 ± 0.05	1.5	0.13
	Scattering	2.84 ± 0.05	3.3	0.15	3.20	1.3	0.15
Y—Y	Scattering	>3.58	>5	0.15			

used modifiers are more sensitive to structural changes in the network than the network-forming atoms and because few data are available on these elements for experimental limitations. The authors measured the EXAFS of various silicate and borate glasses at Na and Ca K-edges. In $Na_2Si_2O_5$, the first coordination shell of Na is well defined with a mean-square relative displacement in Na—O distances of 0.07 Å. By contrast, the O shell in $Na_2CaSi_5O_{12}$ is split into two subshells with bonding oxygen atoms (Na—O ~ 2.3 Å) and non-bonding oxygen atoms (Na—O ~ 3 Å). This situation has already been observed in crystal structures of ternary silicates such as Na_2CaSiO_4 where the sodium environment is very irregular.

In crystalline calcium silicates, the alkaline earth environments, with 8 or 9 oxygen atoms in the first coordination sphere, are extremely distorted. The same situation is encountered in the corresponding glasses. EXAFS analysis of anorthite ($CaAl_2SiO_8$) and diopside ($CaMgSi_2O_6$) leads to a distorted surrounding with a mean displacement of around 0.2 Å in Ca—O distances. The anorthite glass shows exactly the same pair distribution function as the crystalline phase. On the contrary, both glassy and crystalline diopside have different calcium sites, with Ca—O distances significantly longer in the glass.

4.2.5 Catalysis

EXAFS is a powerful technique for the structural studies of heterogeneous catalysts. In these solids, the metallic atoms are often arranged in small clusters difficult to study by X-ray scattering. EXAFS allows to study the structural properties of these metallic sites, the coordination of adsorbed molecules and, moreover, the eventual modification of the catalyst during a chemical reaction.

This subject will be treated in details elsewhere in this issue and Lagarde and Dexpert have published an excellent review about this subject [46]. The wish here is to mention an EXAFS characterization of intermetallic compounds before and after various chemical reactions: carbon oxide hydrogenation, ethane hydrogenolysis and propane hydrogenation [47]. $LaNi_5$, $LaNi_4Fe$ and $LaNi_3Mn_2$ have been measured by comparison with pure Ni. The Fourier transforms for these compounds are given in Fig. 17.

It is clearly seen that, after Co,H_2 conversion, $LaNi_5$ has a different behaviour from substituted compounds. There is a decrease of the first shell coordination number instead of an increase in the other compounds. The authors explain this phenomenon by the contribution of small nickel particles. In the other cases, there is an increase of the main peak and a growth of two new peaks resembling to those which appear in the pure Ni spectrum.

4.2.6 Mineralogy

A large variety of geological materials can be studied by EXAFS: poorly organized solids — natural glasses or gels — natural solutions — coals ... Apart from the major constituents (Si, Al, . . .), minerals contain minor ($\sim 1\%$) or trace (1–100 ppm) elements which often are transition metal ions. The study of these elements can give informations on the physico-chemical conditions during the formation of minerals.

For example, Manceau et al. [48] have collected EXAFS data on Ni-containing phyllosilicates from New Caledonia. In this series of layered minerals, nickel atoms can

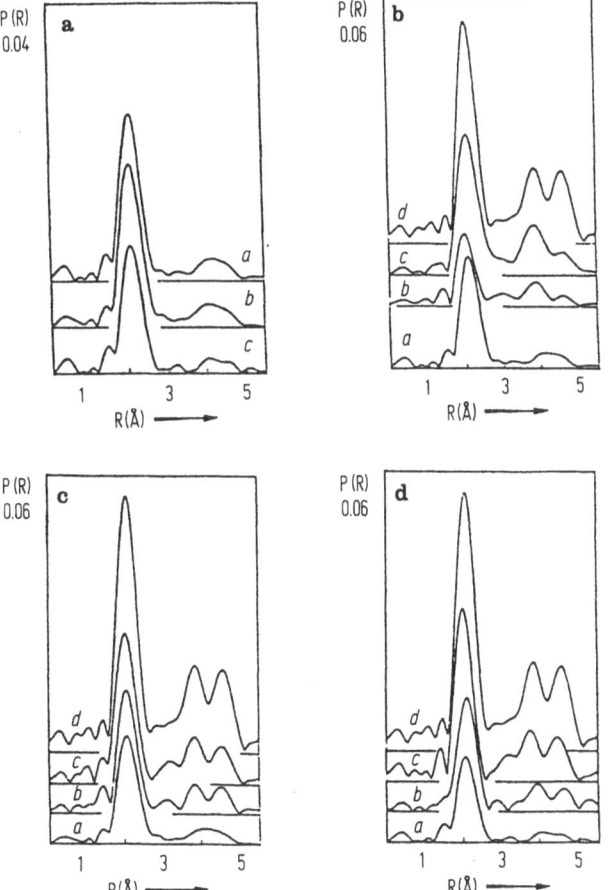

Fig. 17 a–d. Fourier transforms of intermetallics:
a — a = LaNi$_5$; b = LaNi$_4$Fe; c = LaNi$_3$Mn$_2$.
b — a = LaNi$_5$ before reaction; b = after CO, H$_2$; c = after C$_2$H$_6$; d = Ni reference.
c — idem for LaNi$_4$Fe;
d — idem for LaNi$_3$Mn$_2$ [47]

substitute magnesium atoms. Two kinds of informations have been provided by EXAFS experiments:

1) Ni is substituted to Mg in the octahedral sheet of these phyllosilicates, as seen from Ni—O distances (2.07(1) Å) and coordination numbers.
2) The analysis of the second atomic shell permits discussion of the Mg—Ni cation ordering. The studied phyllosilicates can be divided in two series: a serpentine series and a talc series. In the later, nickel atoms are segregated into discrete domains whose minimal size can be estimated roughly around 30 Å. In the former, on the contrary, EXAFS results are more consistent with the existence of nickel-enriched areas without segregation. The authors conclude that the Ni—Mg cation ordering is a memory of the formation mechanism of the ore mineral. Indeed, the minerals

of the serpentine series result from the transformation of primary serpentines by a Ni enrichment and a subsequent Mg loss whereas the minerals of the talc series are neoformed products.

4.3 Xanes: Applications

We have detailed the different fields of application in the case of EXAFS because this technique is now routinely used. This is not yet the case of XANES though there is a growing interest in this field. The reason for this delay may be that, until recently, there was no general theory to explain the phenomenon and to allow accurate modelization. Anyhow, the literature gives some examples in the same fields as EXAFS. We selected three of them to illustrate the potentialities of the method.

Marcelli et al. [49] have studied the XANES spectra, at the oxygen K-edge, of well characterized silica glasses. The aim of this experiment, performed at Frascati, was to show that oxygen XANES can be a probe of oxygen site geometry in glasses. Silica glasses are built up with SiO_4 tetrahedral clusters and the disorder leads to variation of the Si—O—Si bridging angle. The various methods, used to study this disorder (EXAFS, MAS NMR, Raman Spectroscopy) give contradictory results. A recent review of this problem has been made by Kreidl [50].

Three samples were investigated by Marcelli et al.: — an α-quartz crystal, — a "natural" glass prepared by melting — and a commercial silica glass Suprasil prepared by oxidation of $SiCl_4$. Figure 18 gives the oxygen K-XANES of these compounds.

The differences observed in the spectra can be explained in terms of Si—O—Si: angle variations. Indead, theoretical spectra (Fig. 19) show the sensitivity of XANES to bonding angles. However, the agreement with experiment is poor, and the authors conclude that it is necessary to use bigger clusters of atoms.

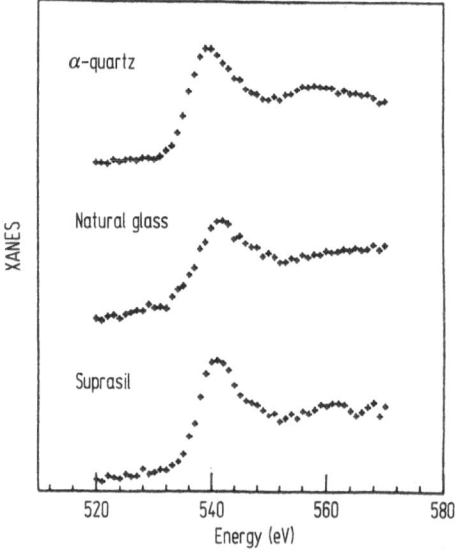

Fig. 18. XANES on oxygen K-edge for α-quartz and silica glasses [49]

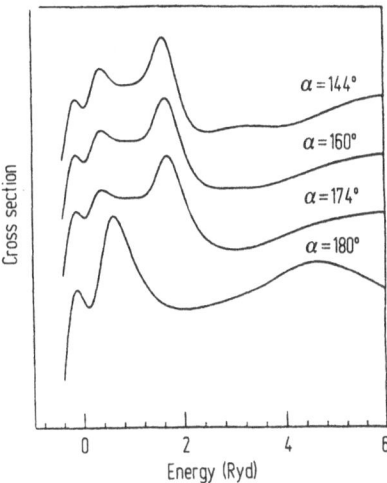

Fig. 19. Total cross section as a function of the Si—O—Si angle (α) (from 49)

Table 8. Peak position in the XANES spectra of f.c.c. Cu and α-Mn [51]

peak position (eV)	1st peak	2nd peak	3rd peak	4th peak
f.c.c. Cu	2.4	14.0	24.8	45.6
α-Mn	3.6	17.0	28.2	45.8

The application of the XANES method to metallic elements is even more demonstrative. Greaves et al. [51] have collected K-edge absorption spectra for f.c.c. Cu and α-Mn. The two spectra show four strong peaks in the first 60 eV of fine structure. But, although both metals have close-packed structures of similar size atoms, the positions and amplitude of the peaks are different (Table 8).

The reconstruction of the f.c.c. Cu spectrum has been made by theoretical multiple scattering calculations. A quite good agreement with experiment is obtained with a four-shell 54-atom cluster, organized with 12 atoms at 2.55 Å, 6 at 3.61 Å, 24 at 4.42 Å and 12 at 5.09 Å. The four peaks of the spectrum are only reproduced when the third shell, at 4.42 Å, is included. The reconstruction of α-Mn spectrum was not realized as either a b.c.c. or a f.c.c. packing lead to a good fit.

Bianconi et al. [52] report a quantitative interpretation of the XANES of Fe(II) and Fe(III) hexacyanide complexes. XANES data show two strong and sharp peaks, A and B, for the two compounds with slight differences in the precise positions. This peak shift can be associated with different Fe—C and C—N distances as illustrated in Fig. 20.

The more probable effect is the contraction of the Fe—C distance in the Fe(III) complex by 0.04 Å. This result would be in agreement with the diffraction data giving a 0.026 Å contraction of the Fe—C distance. The energy separation between peaks A and B are respectively 16.5 eV and 17 eV in Fe(II) and Fe(III) complex. This increase can be attributed to a 0.02 Å contraction of the C—N distance, in agreement with the 0.028 Å contraction observed by X-ray diffraction. Finally, the differences of line shapes in the two complexes can be explained, according to the authors, by small distortions of the $Fe(CN)_6$ group.

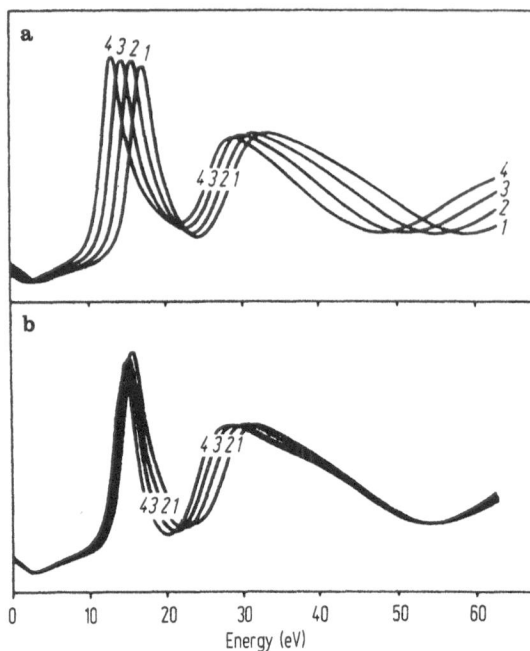

Fig. 20. Calculated K-edge XANES of Fe(CN)$_6$ clusters: **a)** C—N distance fixed at 1.134 Å, Fe—C distances of 1.852 Å (curve 1), 1.903 Å (curve 2), 1.958 Å (curve 3), and 2.011 Å (curve 4). **b)** Fe—C distance fixed at 1.903 Å, C—N distances of 1.134 Å (curve 1), 1.164 Å (curve 2), 1.191 Å (curve 3), and 1.217 Å (curve 4). (from [52])

5 Conclusion

The development of new storage rings, among which ESRF is an enligthened example, is directly related to the remarkable and successful contribution in Science to the study of new phenomena and especially to the understanding of the fine structure of the matter.

Since synchrotron radition requires heavy investments, a precise and safe organization, it cannot be used as a tool for daily laboratory investigations but it will be increasingly used for thorough understanding of carefully prepared structural studies of various materials, chemical reactions or phase transitions.

A wide spectrum of scattering studies is opened on single crystals or microcrystalline powders: (i) on the former, through the possibility of using small crystals, of collecting quickly accurate data or of playing with the wavelengths in order to use, at its best, the anomalous scattering phenomenon; (ii) on the later, the intensity of the X-ray beam allows to follow fast phenomena occuring to materials placed under specific pressure, temperature or chemical factors, and also, thanks to its fine wavelength definition, to obtain accurate patterns extremely well defined permitting, eventually, structural analysis.

In the field of spectroscopies, joint EXAFS and LAXS investigations (especially for the former at different wavelengths) give a powerful tool to tackle the difficult problem of the structural organization of amorphous or glassy materials, complexes in solution, metallic glasses, melts ... EXAFS is also a good technique to handle the determination of the surroundings of a doping element used as a probe in a material. SAXS and XANES further provide valuable informations, even if the later is not yet an easy facility. SEXAFS should know a great development since the chemistry on surfaces (molecular catalysis on metal surfaces, for example) tremendously needs original ways for investigation of the structural features in order to understand the chemical mechanisms and to master them.

The use of synchrotron radiation has a remarkably positive impact in the field of the structural inorganic chemistry. Further, it will be a reference, which is not the least paradox, to improve current laboratory apparatus and to create new ones (EXAFS, LAXS, SAXS . . .) in order to assume daily chemical researches and to prepare the experiments at their best level (scientifically and technically) prior to carrying them out in the storage ring centres.

Aknowledgements: The authors are grateful to Mrs. Helene Morelle for her help in reviewing the english of the manuscript.

6 References

1. Activity Report (1985) LURE-CNRS Laboratory, Université Paris Sud, Orsay (France)
2. Farge Y, Duke PJ, (1979) European Synchrotron Radiation Facility. The scientific case, European Science Foundation, Strasbourg (France)
3. Cohen GG, Deslattes RD, (1982) Nucl Instr Methods, *193*; 33
4. Activity Report (1983) LURE-CNRS Laboratory, Université Paris Sud, Orsay (France)
5. Bachmann R, Kohler H, Schulz H, Weber HP (1985) Acta Cryst *A41*: 35
6. Cox DE, Hastings JB, Thomlinson W, Prewitt CT, (1983) Nucl Inst Methods *208*: 573
7. Bastie P, Capelle B, Dolino G, Zarka A (1985) Progr X-rays stud synchrotron rad, Strasbourg, (France), 3-3(A)4
8. Tolochko BP, Masliv AJ, Lyakhov NT (1985), Prog X-ray stud synchrotron rad. Strasbourg (France). 3-4(P)9
9. Elder RC, Ludwig K, Cooper JN, Eidsness MK, (1985) J. Am. Chem. Soc *107*: 5024
10. Frank AM, Raoux D, Naudon A, Sadoc JF (1984) J. Non-Crystall Solids *61–62*: 445
11. Axelos M, Tchoubar D, Bottero JY, (1985) Activity Report, 226, LURE-CNRS Laboratory, Université Paris Sud, Orsay (France)
12. Goudeau P, Maudon A (1985) Activity Report, 253, LURE-CNRS Laboratory, Université Paris Sud, Orsay (France)
13. Simon JP, Lyon O (1985) Activity Report, 254, LURE-CNRS Laboratory, Université Paris Sud, Orsay (France)
14. Bottero JY, Tchoubar D (1983) Activity Report, 116, LURE-CNRS Laboratory, Université Paris Sud, Orsay (France)
15. Craievich A, (1985) Progr X-ray stud synchrotron rad. Strasbourg (France). 3-1(C)3
16. Pons CH, Rausell-Colon JA, Ben Rhaiem H, Tessier D, (1983) Activity Report, 118, LURE-CNRS Laboratory, Université Paris Sud, Orsay (France)
17. Goudeau P, Maudon A, Fontaine A, Williams CE (1985) Activity Report, 252, LURE-CNRS Laboratory, Université Paris Sud, Orsay (France)
18. Simon JP, Lyon O, DeFontaine D, (1985) Progr X-ray stud synchrotron rad. Strasbourg (France). 3-4(A)5
19. Stern EA, (1974) Phys. Rev. B *10(8)*: 3027
20. Lee PA, Pendry JB (1975) ibid. *11(8)*: 2795
21. Bianconi A, Incoccia L. Stipchich S (Eds), (1983) EXAFS and Near-Edge Structure 11 Springer Berlin
22. Teo BK, Joy DC (Eds), (1981) EXAFS spectroscopy — Techniques and Applications, Plenum Press New York
23. Lee PA, Citrin PH, Eisenberger P, Kincaid BM, (1981) Rev. Mod. Phys. *53*: 769
24. Muller JE, Jepson O, Wilkins JW (1982) Solid State Comm. *42*: 365
25. Kutzler FW, Scott RA, Berg JM, Hodgson KO, Doniach S, Cramer SP, Chang CH (1981) J. Am. Chem. Soc. *103*: 6083
26. Dill D, Dehmer JL (1974) J. Chem. Phys. *61*: 692

27. Durham PJ, Pendry JB, Hodges CH (1981) Solid State Comm. *38*: 159
28. Verdaguer M, Julve M, Michalowicz A, Kahn O (1983) Inorg. Chem. *22*: 2624
29. Gleizes A, Verdaguer M (1981) J. Am. Chem. Soc. *103*: 7373
30. Verdaguer M, Michalovicz A, Girerd JJ, Alberding N, Kahn O (1980) Inorg. Chem. *19*: 3271
31. Morrison TJ, Shenoy GK, Iton LE, Stucky GD, Suib SL (1982) J. Chem. Phys. *76*: 5665
32. Eidsness MK, Elder RC, (1984) EXAFS and Near-Edge Structure III. 83 Hodgson KO, Hedman B Penner-Hahn JE (Eds) Springer Berlin
33. Sandstrom DR, (1979) J. Chem. Phys. *71*: 2381
34. Licheri G, Paschina G, Piccaluga G, Pinna G, Vlaic G (1981) Chem. Phys. Lett *83*: 384
35. Fontaine A, Lagarde P, Raoux D, Fontana MP, Maisano G, Migliardo P, Wanderlingh F, (1978) Phys. Rev. Lett *41(7)*: 504
36. Sandstrom DR, (1989) EXAFS and Near-Edge Structure III, 409 Hodgson KO, Hedman B Penner-Hahn JE (Eds), Springer Berlin
37. Sadoc A, Lagarde P, Vlaic G (1985) J. Phys. C: Solid State Phys. *18*: 23
38. Macklin JW, Plane PA (1970) Inorg. Chem. *9*: 821
39. Lagarde P, Fontaine A, Raoux D, Sadoc A, Migliardo PJ. (1980) Chem. Phys. *72*: 3061
40. Yamaguchi T, Lindqvist O, Boyce JB, Claeson T (1984) Acta Chem. Scand *A38*: 423
41. Holmberg B, Johansson G (1983) ibid. *A37*: 367
42. Maeda H, Terauchi H, Hida M, Kamijo N, Osamura K (1984) EXAFS and Near-Edge Structure III, 328 Hodgson KO, Hedman B Penner-Hahn JE (Eds) Springer Berlin
43. Raoux D, Frank AM, EXAFS and Near-Edge Structure III, 321 Hodgson KO, Hedman B Penner-Hahn JE (Eds) Springer Berlin
44. Petiau J, Calas G (1985) J. Physique *46*: C8, 41
45. Greaves GN, Binsted N, Henderson CMB, EXAFS and Near-Edge Structure III, 297 Hodgson KO, Hedman B Penner-Hahn JE (Eds), Springer Berlin
46. Lagarde P, Dexpert H (1984) Adv. in Physics *33(6)*: 567
47. Paul-Boncour V, Percheron-Guegan A, Achard JC, Barrault J, Guilleminot A, Dexpert H, Lagarde P, (1984) EXAFS and Near-Edge Structure III, 199 Hodgson KO, Hedman B Penner-Hahn JE (Eds) Springer Berlin
48. Manceau A, Calas G, Petiau J, (1984) EXAFS and Near-Edge Structure III, 358 Hodgson KO, Hedman B Penner-Hahn JE (Eds), Springer Berlin
49. Marcelli A, Davoli I, Bianconi A, Garcia J, Gargano A, Natoli CR, Benfatto M, Chiaradia P, Fanfoni M, Fritsch E, Calas G, Petiau J (1985) J. Physique *46*: C8, 107
50. Kreidl MJ (1983) Glass Sci Technol Vol 1, 107 Kreidl NJ Uhlmann DR (Eds)
51. Greaves GN, Durham PJ, Diakun G, Quinn P, (1981) Nature *294*: 139
52. Bianconi A, Dell'Ariccia M, Durham PJ, Pendry JB (1982) Phys. Rev. B *26*: 6502

XANES in Condensed Systems

Antonio Bianconi[1], J. Garcia[2]*, M. Benfatto[2]

1 Dipartimento di Fisica, Università di Roma "La Sapienza", 00185 Roma, Italy
2 INFN Laboratori Nazionali di Frascati, 00044 Frascati, Italy

Table of Contents

---

* Permanent adress: Departamento de Fisica de la Materia Condensada, Universidad de Zaragoza, 50009-Zaragoza, Spain.

1 Introduction

X-ray absorption spectroscopy concerns the study of electronic transitions from atomic inner shells to unoccupied states. Electronic and structural information on selected sites can be obtained in complex systems.

The characteristic of x-ray spectroscopy is that by tuning the x-ray energy it is possible to select the atomic species in which the excitation takes place and by selecting different core levels in the same atom the symmetry of the unoccupied states (the final electronic states) can be selected.

The spectra of atoms are characterized by a set of bound Rydberg states below the photoionization threshold E_0 for excitations in the continuum [1-3]. The Rydberg states become very weak in molecules [4-7] and disappear in condensed systems. Review papers on x-ray absorption spectra for atoms [1-3] and molecules [4-8] are available and these spectra will not be discussed here.

The spectra of condensed systems show structures above the continuum threshold. The study of the spectra within ~ 10 eV, of the absorption edge called Kroning structure [9], was first performed using standard x-ray sources [10, 11]. Recently the use of synchrotron radiation sources has produced a rapid development of x-ray absorption near edge spectroscopy [12-15].

Kroning [9] first pointed out that the fine structure above the absorption threshold can be assigned to unoccupied molecular orbitals. According to the Fermi golden rule for optical transitions the transition rate from a core level to an unoccupied state at energy E can be calculated in **k**-space, and the measured absorption coefficient is given by $\alpha(E) \sim P(E) \cdot D_{l'}(E)$, where $P(E)$ is the dipole matrix element and $D_{l'}(E)$ is the density of states of selected angular momentum l'. The dipole selection rule for an electronic transition from a core level of well defined angular momentum l selects the unoccupied density of states of angular momentum $l' = l \pm 1$.

In a set of early experiments with synchrotron radiation [12, 16] the important role of the atomic cross section also in condensed systems was pointed out. In fact, large peaks of atomic origin have been observed at the absorption edges of condensed systems where the atomic cross section has strong resonances, e.g. for $3p \rightarrow \varepsilon d$ and $2p \rightarrow \varepsilon d$ transitions in transition elements [16, 17], and for $3d \rightarrow nf$ transitions in rare earths [18-20]. In fact the total absorption cross section can be factorized into an atomic part and a modulating part due to the crystal structure $\alpha(E) = \alpha_a(E) \cdot \alpha_s(E)$.

Where the atomic factor $\alpha_a(E)$ has no resonances $\alpha_s(E)$ is the main term. α_s is determined by scattering of the photoelectron by the atoms of the condensed system surrounding the absorbing atom.

In the interpretation of the x-ray absorption spectra the inelastic interaction between the photoelectron and the valence electrons is an important process. The mean free paths for Au and Si are shown in Fig. 1 [21-23]. Therefore, in x-ray absorption near edge spectra extending up to several tens of eV above the Fermi level, inelastic losses become highly probable. Because of inelastic scattering of the photoelectron and finite core hole life time only a cluster of finite size is relevant to the determination of the final state wave function of the photoelectron.

It has been shown in a series of papers [24-26] that the description of the near edge spectra depends mainly on the geometrical structure of a finite cluster of atoms around the absorbing atom. The size of the cluster changes in different systems,

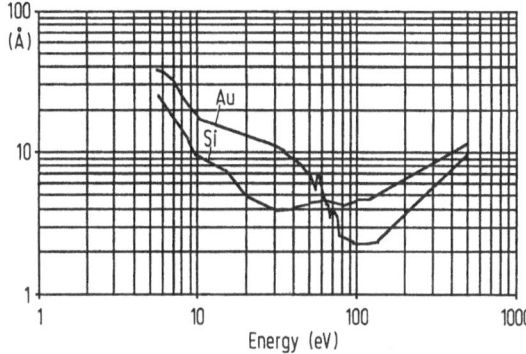

Fig. 1. Photoelectron mean free path for inelastic scattering in Si [22] and Au [23] in the energy range of interest for XANES spectra. The data were obtained from the measurement of electron escape depth in photoemission experiments

ranging from a single shell to several shells. Therefore, neither the description of the final states considering an infinite crystal, without including inelastic effects, nor simple molecular orbital calculations could explain the experimental data. An interpretation of the x-ray absorption near edge structure (called XANES) of condensed systems was proposed in terms of multiple scattering resonances of the photoelectron in a cluster of finite size [25].

The absorption cross section for core transitions can be solved in *real space* using the Green's function approach in the frame of multiple scattering theory [27-31]. In this approach the limited mean free path of the photoelectron determines the finite size of the cluster of atoms around the central absorbing atom to be considered in the calculations. An important feature of the solution of the absorption cross section in real space is that physical processes determining the unoccupied density of states of condensed matter appear explicitly. Moreover, the spectra of systems without long range order can also be solved.

Interest in determination of higher order correlation functions of local atomic distribution in complex systems has stimulated the growth of XANES. In fact XANES is sensitive via multiple scattering, to the geometrical arrangement of the environment surrounding the absorbing atom.

The generally strong scattering power of the atoms of the medium for low kinetic energy photoelectrons favors multiple scattering (MS) processes [24-33]. At higher energies, such that the atomic scattering power becomes small, a single scattering (SS) regime is found, where the modulation in the absorption coefficient is substantially due to interference of the outgoing photoelectron wave from the absorbing atom with the backscattered wave from each surrounding atom, giving the extended x-ray absorption fine structure (EXAFS) [34,35]. Hence EXAFS provides information about the pair correlation function. By decreasing the photoelectron kinetic energy a gradual transformation occurs from the EXAFS regime to the XANES regime [32,33].

1.1 XANES Energy Range

In the analysis of XANES spectra it is possible to define the absorption threshold in different ways.

i) The "absorption threshold" is at the energy of the transition to the lowest unoccupied energy state,

ii) The "absorption jump edge" is at the energy at which the absorption coefficient is at half the height of the atomic absorption jump. Because the dipole selection rule can suppress the absorption cross section over a large energy range, the energy of the absorption jump edge can be at much larger energy than the absorption threshold (In some cases the energy difference can be up to 100 eV [31]).

iii) The "continuum threshold" is at the energy at which the electron is ejected to the continuum, i.e. the vacuum level in atoms and molecules, the Fermi level in metals and the bottom of the conduction band in insulators.

In the XANES data analysis the first step is to remove the continuum absorption background due to transitions from levels of lower binding energy. This procedure is generally called pre-edge background subtraction. The relative absorption $\alpha(E)/\alpha_a$ is usually plotted, where α_a is the smooth atomic absorption above threshold determined by the usual EXAFS analysis. Fig. 2 shows two typical XANES spectra where the XANES are normalized to the atomic absorption.

In order to give experimental evidence of different EXAFS and XANES regions the absorption spectra of $[Mn(OH_2)_6]^{2+}$ and $[MnO_4]^-$ are compared in Fig. 2. The corresponding energy scales are in the ratio $(d_2^*/d_1^*)^2 = 0.47$, where d_2^* and d_1^* are the Mn—O distances in the two complexes corrected for the linear term coefficient of

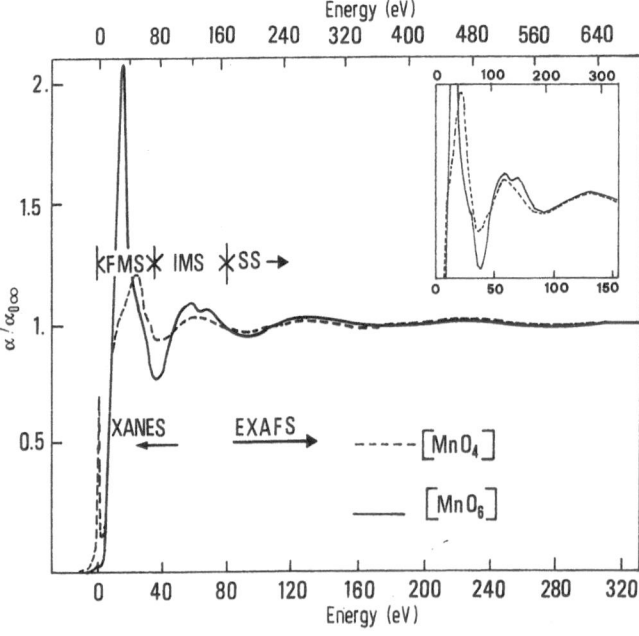

Fig. 2. Comparison of normalized Mn K-edge x-ray absorption spectra of $(MnO_4)^-$ and $(Mn(OH_2)_6)^{2+}$ ions in 50 mM aqueous solution. The respective energy scales are given in the upper and lower parts of the figure. A partition of the spectra in to FMS (full multiple scattering) IMS (intermediate multiple scattering) and SS (single scattering) regions is sketched. The insert shows a comparison of the two spectra after a rescaling of the oscillating amplitudes in the ratio 4:6

the scattering phase shift, to eliminate the effects in the spectra arising from the different bond length and from the dependence of the phase shifts on the energy. The zero of the energy has been set at the absorption threshold (the 1s-3d transition) in both spectra. In this way the two spectra (after further rescaling of the oscillating amplitude to take into account the different number of nearest neighbors, see insert in Fig. 2) show a superposed sinusoidal behaviour in that energy region that contains information only about the pair correlation function (EXAFS). Below 160 eV (75 eV with reference to the $[Mn(OH_2)_6]^{2+}$ scale) the absorption spectrum of $[MnO_4]^-$ clusters deviates from the spectrum of $[Mn(OH_2)_6]^{2+}$. This fact is a clear indication that below these energies multiple scattering contributes to the spectra, and that above, the single scattering theory is enough to predict the fine structure. Therefore, the energy range of multiple scattering is largely dependent on the interatomic distance which agrees with the physical idea that where the photoelectron wavelength is larger than the interatomic distance the single scattering approximation breaks down.

We can actually distinguish three parts in a x-ray absorption spectrum:

1) The first part extending over about 8 eV, is called "edge region", threshold region or low energy XANES region. The physical origin of the absorption features in the edge region is different in different classes of materials: Rydberg states in atoms, bound valence states (or bound multiple scattering resonances) in molecules, core excitions in ionic crystals, many-body singularities in metals, bound atom-like localized excitations in solids.

2) The region in the continuum where multiple scattering processes are relevant is called the XANES region.

3) The region of single scattering at higher energies is called the EXAFS region.

In this chapter we will discuss one electron excitations in XANES. The multiple scattering theory and its applications for simple octahedral and tetrahedral clusters will be discussed. Then some cases where larger clusters contribute to the XANES spectra are considered. In the second part, the band structure approach for the study of metals will be described.

2 Experimental Methods

The advances in x-ray spectroscopy are related to the development of x-ray sources. Limitations in energy range and intensity of radiation from conventional x-ray tubes were overcome by using synchrotron radiation sources for soft x-rays in the 1960s and high energy, E > 1 GeV, electron storage rings in the 1970s as stable, tunable, intense x-ray sources. A schematic view of an experimental setup for x-ray absorption spectroscopy is shown in Fig. 3. The experimental setup and detection methods depend on several factors [36]. First is the energy range of the x-rays. It is possible to distinguish different energy ranges requiring different types of monochromators; soft x-ray (50–700 eV) need glancing incidence grating mono-chromators; the range 700–3000 eV needs special crystals like InSb and beryl; for the x-ray range higher than 3000 eV mostly double reflection silicon crystals are used. Moreover, the soft x-ray range requires special beam lines and sample chambers because of ultra high vacuum conditions.

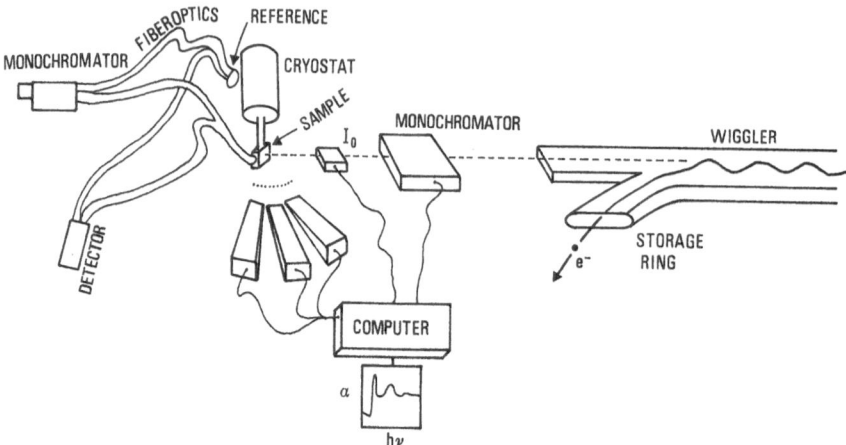

Fig. 3. Schematic view of a beam line in a synchrotron radiation facility for x-ray absorption experiments using a wiggler as x-ray source and a fluorescence detection system. The incident radiation flux I_0 is measured usually by a ionization chamber. The absorption is measured by the ratio of the x-ray fluorescence I detected by the photomultipliers and I_0
The optical monochromator using a fiber optics system is used for sample characterization.

The detection system depends mostly on the concentration of absorbing atoms in the material. For bulk x-ray absorption spectroscopy for concentration above 10^{-3} atomic ratio standard x-ray transmission techniques are used. Incident and transmitted flux are measured by photoionization chambers.

The absorption can also be measured by recording core-hole decay products in the case of diluted systems. The inner shell photoionization process can be described as a two-step process. In the first step the photon excites a core hole-electron pair, and in the second step the recombination process of the core hole takes place. There are many channels for the core hole recombination. These channels can produce the emission of photons, electrons or ions, which can be collected with special detectors. The recombination channel that is normally used to record bulk x-ray absorption spectra of dilute systems is the direct radiative core-hole decay producing x-ray fluorescence lines. In Fig. 3a beam line with an apparatus to record absorption spectra in the fluorescence mode is represented schematically.

The total x-ray reflection can be measured to obtain the XANES spectra [37, 38] and surface sensitivity can be increased by detecting the x-ray fluorescence in the total reflection geometry [39].

In the soft x-ray range, hv < 4000 eV Auger recombination has higher probability than the radiative recombination. As the energy of the Auger electrons is characteristic of a particular atom, it was suggested that the selective photoabsorption cross section of atomic species chemisorbed on a surface could be measured by monitoring the intensity of its Auger electrons as a function of photon energy. An intense Auger line is selected by an electron analyzer, operated in the constant final state (CFS) mode with a energy window of a few eV. A standard experimental setup for this type of surface x-ray absorption measurements is shown in Fig. 4 [25].

For bulk measurements the total electron yield, which has been found to be pro-

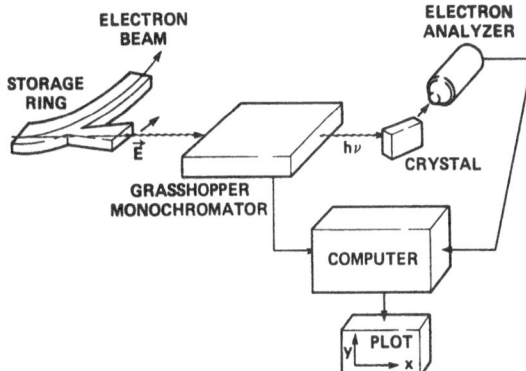

Fig. 4. Experimental setup for surface XANES detection. In the soft X-ray region a grasshopper monochromator is often used. The absorption is measured by detecting the flux of emitted electrons of energy selected by the electron analyzer

portional to the absorption coefficient, is used. This technique measures the integral over the entire energy range of the emitted electrons. The advantage of this method is that the maximum counting rate is obtained since emitted electrons over a large solid angle can be collected by applying a positive voltage to the detector. The other detection method is the low energy partial yield, where the secondary electrons within a kinetic energy window around the maximum in the inelastic part of the energy distribution curve (EDC) are detected. Because of the long escape depth for low energy electrons the bulk absorption is recorded by this method.

The surface absorption of clean surfaces can be detected by recording Auger electrons at kinetic energies in the range 50–100 eV [40] which have a very short escape depth.

The technique which uses the ionic emission is called photon-stimulated desorption (PSD) [41]. The ion current due to PSD is proportional to the number of created core holes., i.e., to the photoabsorption cross section of the absorbate. It is a measurement of the surface absorption with a very high surface contrast in comparison with the previously discussed detection methods. A comparison of different techniques is given by Stohr et al. [42].

Another technique used for the study of core level excitation is electron energy loss spectroscopy (EELS). Experimentally an electron beam is incident on a target and the energy loss of the reflected [43] or transmitted [44] electrons is analyzed.

High resolution is required in XANES spectroscopy, because important physical information can be extracted from small variations in intensity and/or from small energy shift of an absorption peak. For this reason, careful preparation of pin-hole free samples and suppression of high harmonics in the incident photon beam are required in XANES spectroscopy. The energy band width ΔE of the photon beam monochromatized by Bragg diffraction using a crystal monochromator is determined by the angular divergence and by the crystal monochromator rocking curve. The angular divergence in synchrotron radiation beam lines depends on the intrinsic vertical spread of the synchrotron radiation, which is determined by the energy of the electron beam in the storage ring and by the source size, i.e., the diameter of the electron beam and its divergence at the emission point determined by the electron optics.

The resolution can be improved by changing either the crystal or the reflection plane. In a classical two crystals monochromator, two parallel reflections produce a monochromatized photon beam parallel to the incident one. These two reflections reduce the tails of the rocking curve, and consequently they increase the resolution but leave the harmonic content like that of a single reflection. High resolution crystal monochromators have been described using antiparallel reflections [45] and high resolution XANES spectra have been measured using higher order reflections [46].

Harmonic rejection was achieved in devices with two crystals by detuning one crystal with respect to the other [47]. In fact, when the two crystals are misaligned the intensity of the harmonics drops off much more rapidly then the intensity of the fundamental, because the band width $\omega_n(\lambda)$ is much narrower for then harmonics than for the fundamental. The higher harmonic content in the synchrotron radiation beam is due to the intense continuum of the primary beam extending towards high energies. Therefore the harmonic contamination of the monochromatic beam can be drastically reduced by lowering the kinetic energy of the electrons in a bending magnet line or by lowering the wiggler magnetic field in a wiggler beam line. In future XANES experiments will be performed by using many pole wigglers or undulators insertion devices in the new high brilliance synchrotron radiation sources like the European 6 GeV storage ring. Recently the dispersive EXAFS system [38, 48, 49] has reached the same resolution as standard channel-cut monochromators and this opens up the field of time resolved XANES experiments.

3 Multiple Scattering in XANES

3.1 Multiple Scattering Theory

The absorption cross section for x-rays in the range 100–20000 eV is determined by photoexcitation of electrons from atomic core levels. In this energy range pair production is forbidden and the weakness of electromagnetic field is such that only first order processes are important.

To calculate the cross section for photoionization, multiple scattering theory within the framework of a single-particle is presented. The multiple scattering method has been developed in nuclear physics to calculate nuclear scattering cross sections and in solid state physics to compute the electronic structure of solids.

Here we show an extension of the bound state molecular scattering method of Johnson and co-workers [50] to determine the one-electron wave function for continuum states. The extension of the method, formulated for the first time by Dill and Dehmer [51], is one in which the continuum wave function is matched to the proper asymptotic solution of the Coulomb scattering states. In this way the multiple scattering problem is changed from a homogeneous eigenvalue problem (bound states) to an inhomogeneous one in which the continuum wave function is determined by an asymptotic T-matrix normalization condition [52]. In this scheme the total potential is represented by a cluster of nonoverlapping spherical potential centered on the atomic sites and the molecule as a whole is enveloped by an "outer sphere". Three regions can be identified in this partitioning: atomic regions (spheres centered upon nuclei, normally called region I); extramolecular region (the space beyond the outer sphere radius, region III) and an interstitial region of complicated geometry in which the molecular potential is approximated by a constant "muffin-tin" potential. The Coulomb and exchange part of the input potential are calculated on the basis of a total charge density obtained by superimposing the atomic charge densities, here calculated from Clementi and Roetti tables [53], of the individual atoms constituting the cluster. For the exchange potential it is possible to use both the usual energy independent Slater approximation

$$V_{ex} = -6\alpha((3/8\pi) \varrho(x))^{1/3} \tag{1}$$

where $\varrho(x)$ is the charge density and α, the Slater exchange factor, is obtained from Schwartz [54] and the energy dependent Hedin-Lundqvist [55] potential in order to incorporate the dynamical effect.

Following this theory, the expression of the photoabsorbing cross section for a cluster of atoms with light polarized in the vector e direction is given, in the dipole approximation [52], by

$$\sigma(E; \varepsilon) = k/\pi(4\pi^2 h\nu\alpha \sum_{L} |D_L(E; \varepsilon)|^2 \tag{2}$$

where $L = (l, m)$ is the angular momentum, **a** is the fine structure constant, k is the photoelectron wave number,

$$D_L(E; \varepsilon) = (\Psi_L(E) |r \cdot \varepsilon| \Psi_{in}) \tag{3}$$

and $\Psi_L(E)$ is the continuum final state wave function in T-matrix normalization with incoming wave boundary conditions [56], the solution of the Schrödinger equation associated with the potential described above and relative to exciting wave L. The initial state Ψ_{in} is the core state completely localized at the site of the absorbing atom. Within each atomic sphere and beyond the outer sphere the Schrodinger equation is solved with the usual regularity condition at the nuclear center and the scattering condition (unbounded states) at infinity. Hence, one function is obtained for each angular momentum value L of each atom and for each L value of the outer sphere. Each function carries the wave function amplitude as undetermined coefficient. Its magnitude will be found by imposing continuity of the logarithmic derivatives with the intersphere wave function, which, because of the constant potential in this region, is a linear combination of spherical Bessel and Neumann functions. The continuity requirement defines a set of secular inhomogeneous equations which depend of course, on the energy and geometry of the cluster [51]. From this ensemble of equations the total wave function in the interstitial region can be interpreted as a sum of the exciting wave plus the scattered wave coming from each sphere in the cluster. The model is therefore a multiple scattering model for several centers with free propagation in the interstitial region. The physical details of waves scattered by the various centers are contained in the phase shifts.

The scattering total solution $\Psi_L(E)$, for the T-matrix normalization, behaves asymptotically as

$$\Psi_L(E) = f_l(E) \, Y_L(\Omega) - i \sum_L T_{LL}\{f_l(E) + ig_l(E)\} \, Y_L(\Omega) \qquad (4)$$

where $f_l(E)$ and $g_l(E)$ are the regular and irregular solutions of the radial Coulomb wave equation of energy $E = h\nu - E_0$, where E_0 is the ionization potential. Because the initial state is a core state localized at the site of absorbing atom, just the expression for $\Psi_L(E)$ calculated at this site (referred to as 0) is needed. So Eq. (3) becomes

$$D_L(E; \varepsilon) = (k\varrho_0^2)^{-1} \sum_L W\{j_l, R_l^0\}^{-1} B_l^0(\underline{L}) \, (R_l^0 Y_L | \varepsilon_0 r \, |\Psi_{in}) \qquad (5)$$

Here R_l^0 is the radial part of the wave function calculated inside the sphere of the absorbing atom of radius ϱ_0, j_l is the Bessel function, $W\{A, B\}$ is the Wronskian calculated at $r = \varrho_0$, $B_l^0(\underline{L})$ is the vector solution of the multiple scattering equation for the particular exciting wave $\underline{L}$ and L is the final angular momentum selected by the dipole selection rule. All geometrical information is included in this coefficient, which for the case of an isolated atom is simply proportional to δ_{LL}.

To clarify the physical implication of the expression written above, it is better to make contact the Green's function approach by a generalized optical theorem [57] and to write for the cross section the expression

$$\sigma(E; \varepsilon) = -k/\pi(4\pi^2 h\nu\alpha \sum_{LL'} m_L(\varepsilon) \, \text{Im} \, \{(I + T_a H)^{-1} T_a\}_{LL'}^{00} \, m_{L'}(\varepsilon) \qquad (6)$$

where $m_L(\varepsilon)$ is the matrix element, between the function $\underline{R}_l^0 = R_l^0(k\varrho_0^2)^{-1} W\{j_l, R_l^0\}^{-1}$ and the corresponding function of the absorbing atom, which choses the final L by the dipole selection rule; $T_a = \delta_{ij}\delta_{LL'} t_l^i$ is the diagonal matrix describing the process of the spherical wave with angular momentum l being scattered by the atom located at site i through the atomic T-matrix element which has the usual dependence on atomic phase shifts, $H = H_{LL'}^{ij}(1 - \delta_{ij})$ is the free amplitude propagator of the photoelectron in the spherical wave state from site i with angular momentum L to site j with angular momentum L' and I is the unit matrix. As is apparent from this last equation the geometrical information about the medium around the photoabsorber is contained in the matrix inverse $(I + T_a H)^{-1}$. When the maximum eigenvalue $\varrho(T_a H)$ of the matrix $T_a H$ is less than one it is possible to expand the inverse in series [32, 33, 57], that is absolutely convergent relative to some matrix norm, so

$$\sigma(E; \varepsilon) = \sum_{n=0} \sigma_n(E; \varepsilon) \qquad (7)$$

where

$$\sigma_n(E; \varepsilon) = -4\pi h\nu\alpha k \sum_{LL'} m_L(\varepsilon) \, \text{Im} \, \{(-1)^n \, (T_a H)^n \, T_a\}_{LL'}^{00} \, m_{L'}(\varepsilon) \qquad (8)$$

$n = 0$ represents the smoothly varying "atomic" cross section while the generical n is the contribution to the photoabsorption cross section coming from processes in which the photoelectron has been scattered $n - 1$ times by the surrounding atoms before returning to the photoabsorbing site. Before showing the explicit dependence on the geometrical factor of the first two contributions ($n = 2$ and $n = 3$) (besides the atomic one) to the total cross section it is better to average over the polarization of the electric field. Therefore, the unpolarized absorption coefficient which is proportional to the total cross section is given by [58]

$$\alpha_F = Ahv \{(l + 1) M_{l, l+1} \varkappa_{l+1} + l M_{l, l-1} \varkappa_{l-1}\} \tag{9}$$

where l indicates the angular momentum of the core initial state ($l = 0$) for K excitation), $M_{l, l\pm 1}$ is now the atomic dipole transition matrix element for the photo-absorbing atom and [32]

$$\varkappa_l = \{(2l + 1) \sin^2 \delta_l^0\}^{-1} \sum_m \mathrm{Im} \{(1 + T_a H)^{-1} T \}_{1m, 1}^{00} \tag{10}$$

Here δ_l^0 is the phase shift of the absorbing atom. The total absorption coefficient can be expanded as a series $\alpha_F = \alpha_0 (1 + \sum_{n=2} \varkappa_n)$ where α_0 is the atomic absorption coefficient and because α_1 is always zero since $H_{1m, 1m} = 0$. For the K edge, neglecting the angular dependence of the Hankel function in the free propagator, the following expression can be obtained for $n = 2$:

$$\alpha_2 = \alpha_0 \varkappa_2 = \alpha_0 \sum_j \mathrm{Im} \{f_j(k, \pi) \exp (2i\delta_1^0 + kr_j)/kr_j^2\} \tag{11}$$

where r_j is the distance between the central atom and the neighboring atom j and $f_j(k \pi)$ is the usual backscattering amplitude. This contribution is the usual EXAFS signal times the atomic part and it contains information about the pair distribution function only. The first multiple scattering contribution is the α_3 term which can be written [59]

$$\alpha_3 = \alpha_0 \sum_{i \neq j} \mathrm{Im} \{P_1(\cos \varphi) f_i(\omega) f_j(\theta) \exp (2i(\delta_1^0 + kR_{tot}))/kr_i r_{ij} r_j\} \tag{12}$$

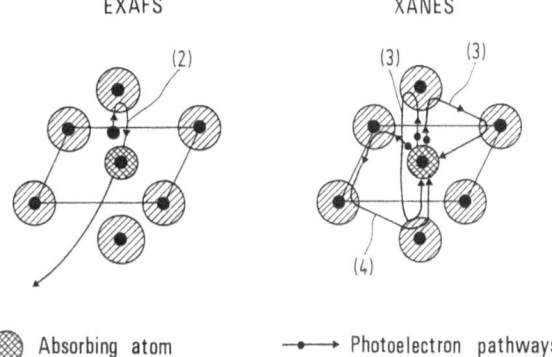

EXAFS   XANES

Absorbing atom   Photoelectron pathways

Fig. 5. Schematic view of the multiple scattering pathways in the EXAFS single scattering regime and in the XANES multiple scattering regime

Here r_{ij} is the distance between atoms i and j, $f_i(\omega)$ and $f_j(\theta)$ are the scattering amplitude, which now depend on the angles in the triangle which joins the absorbing atom to the neighboring atoms located at sites r_i and r_j and $R_{tot} = r_i + r_{ij} + r_j$; in this expression $\cos\varphi = -r_i \cdot r_j$, $\cos\varphi = -r_i \cdot r_{ij}$ and $\cos\theta = r_j \cdot r_{ij}$. So the terms with n higher than 2 clearly contain information about the n-th order correlation function. A pictorial view of this interpretation is shown in Fig. 5.

To conclude it is possible to observe that because $P_1(\cos\varphi) = \cos\varphi$ there is a selection rule in the pathways that contribute to the α_3 term, in fact all pathways where r_i is perpendicular to r_j do not contribute to this term since $\cos\varphi = 0$.

3.2 Tetrahedral Cluster

The x-ray absorption spectra of transition metals with tetrahedral coordination have attracted a great deal of attention since the beginning of x-ray spectroscopy using laboratory x-ray sources [10]. The spectrum of MnO_4^- in aqueous solution recently measured with synchrotron radiation is shown in Fig. 6. The K-XANES of transition metal ions MnO_4^-, CrO_4^{2-} and VO_4^{3-} show a characteristic narrow and strong absorption line at the K threshold which was called a white line because of its image on the photographic plates used in early experiments. Best [60] has interpreted these spectra by calculating the unoccupied molecular orbitals. The fact that these spectra can be used to probe the geometrical coordination of metal ions in solution was first

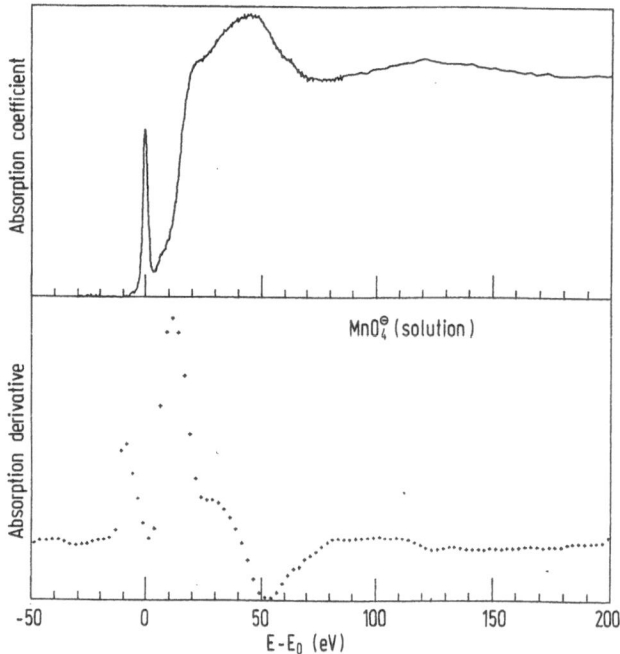

Fig. 6. Manganese K-edge XANES spectrum of $[MnO_4]^-$ in aqueous solution, and its derivative (lower panel)

pointed out by Lytle [61]. The occupied molecular orbitals of the MnO_4^- cluster was the subject of one of the first multiple scattering calculations in the $X-\alpha$ multiple scattering theory of Johnson [50].

Kutzler et al. [27] have extended the $X\alpha$ multiple scattering calculations in the continuum for the $[CrO_4]^{2-}$ cluster. It was demonstrated that the white line at threshold is due to a dipole allowed transition to the unoccupied antibonding orbital t_2. This final state is in an energy region at threshold where the final states are of d-like symmetry and therefore the cross section is quenched by dipole selection rules. The absorption jump edge is about 6 eV above the white line. It is determined by the threshold of transitions to continuum states of p-like symmetry. Recently this system has been chosen to establish the role of multiple scattering in the XANES in open structures, i.e., clusters with no atoms in collinear configurations. The tetrahedral cluster has the characteristic that there are not multiple scattering pathways with atoms in collinear configuration.

The spectrum of MnO_4^- ion in solution is determined mainly by the multiple scattering from the first neighbors because of structural disorder in futher shells. This point has been determined by measuring the spectrum of this ion in different solvents [62]. This is in agreement with the analysis of the EXAFS oscillations in the range of wave vector **k** above 4 Å^{-1} by Rabe et al. [63] where the contribution of only the first shell has been observed.

Tetrahedral clusters offer a good example to test the transition from the multiple scattering to the single scattering regime by increasing the kinetic energy of the photoelectron. In Fig. 7 the results of multiple scattering calculations using the theory discussed above are shown.

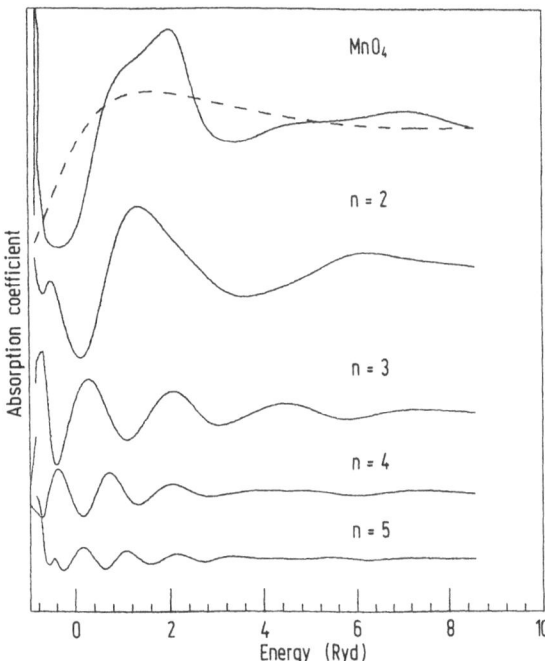

Fig. 7. Theoretical XANES absorption spectrum for tetrahedral MnO_4 cluster showing (as in Fig. 3) the total (upper curve), atomic (dashed line) and $\alpha_n = \alpha_0 \cdot \chi_n$ n = 2, 3, 4, 5 partial higher-order contribution

For the MnO_4 cluster the condition $\varrho(T_aH) < 1$ is verified for all energies greater then the white line, therefore the convergence of the series $\alpha_F(E) = \alpha_0(1 + \Sigma_{n=2} \varkappa_n(E))$ has been found over the full spectrum in the continuum. Hence the experimental spectrum can be analyzed in terms of contributions of successive orders of multiple scattering, from the EXAFS term $n = 2$ to the higher terms $n = 3, 4, 5, 6 \dots$ The experimental spectrum shown in Fig. 6 can be compared with the full multiple scattering calculations of α_F shown in Fig. 7 where the sum up to infinity is calculated by inversion of the scattering matrix.

Using for the exchange part of the potential the usual energy independent Slater approximation ($X\alpha$ potential) we must expand the energy scale by a factor of about 1.1 to obtain the agreement between theory and XANES experiments. Good agreement with the energy positions of experimental peaks has been obtained by considering the energy dependent exchange Hedin-Lundqvist potential.

The atomic absorption $\alpha_0(E)$ coefficient is shown in Fig. 7 as a dashed curve. The terms $\varkappa_2(E)$, $\varkappa_3(E)$, $\varkappa_4(E)$ and $\varkappa_5(E)$ times $\alpha_0(E)$ are reported in Fig. 7. The $\varkappa_2(E)$ term is the EXAFS signal, which is not enough to account for experimental features below 150 eV. The $\varkappa_3(E)$ term is due to multiple scattering paths involving the central atom (manganese) M_0 and two oxygens $M_0-O-O-M_0$. We observe a negative interference between the higher order terms $\varkappa_4(E)$ and $\varkappa_5(E)$. From the figure it is clear that the higher-order contributions attenuate faster with increasing kinetic energy of the photoelectron. The approach in terms of partial contributions to the absorption coefficient shows the continuous merging of the multiple scattering (MS) regime into the single scattering (SS) regime above 150 eV in $[MnO_4]^-$.

The general picture of an absorption spectrum that emerges from the preceding considerations consists of a full multiple scattering region where many or an infinite number of multiple scattering paths of high order contribute to the final shape of the spectrum. Sometimes in this region the series does not converge at all. These full multiple scattering resonances are followed by an intermediate region where only a few MS paths of low order are relevant (typically $n < 4$), this region merges continuously into the EXAFS regime (SS region). The energy extent of each region is obviously system dependent.

The experimental three-atom correlation [32,62] function for the $[MnO_4]$ cluster in the range 20–140 eV has been extracted using the following procedure. The experimental modulated part $\varkappa_m$ of the measured absorption coefficient α_m due to the photoelectron scattering is obtained as $\varkappa_m = (\alpha_m - \alpha_0)/\alpha_0$. Neglecting higher order contributions $\varkappa_4(E)$ and $\varkappa_5(E)$ it is possible to identify $\varkappa_m = \varkappa_2 + \varkappa_3$ in the intermediate region between the EXAFS and the full multiple scattering region. Therefore an experimental $\varkappa_3(E)$ can be obtained by subtracting from $\varkappa_m(E)$ the $\varkappa_2(E)$ term which can be determined by theory using spherical waves and a reliable phase shift down to very low kinetic energy of the photoelectron. In Fig. 8 the comparison between the theoretical and experimental $\varkappa_3$ is shown. The experimental spectrum has been corrected for a double electron excitation centered at 58 eV [62]. The Fourier transform of both theoretical and experimental signals are shown in the lower part of Fig. 8. The peaks correspond to the sum of the $M_0-O-O-M_0$ distances corrected for the phase shift of the $n = 3$ scattering processes.

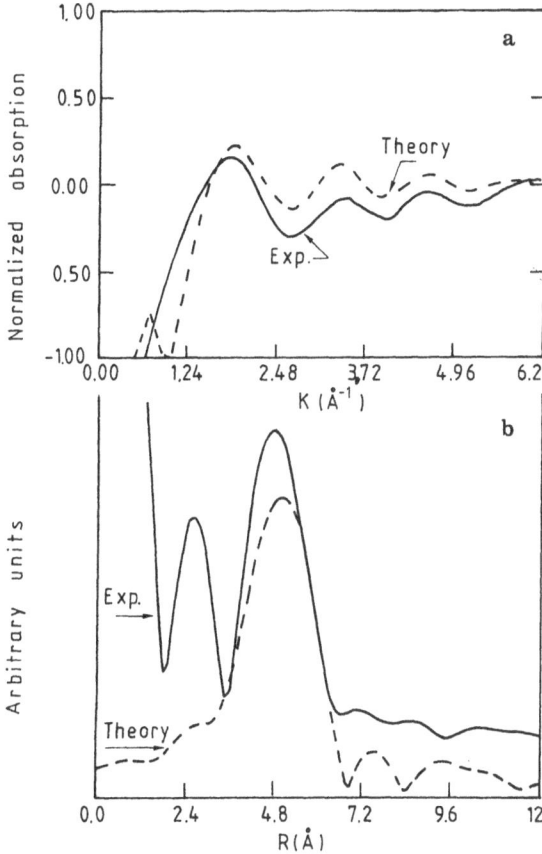

Fig. 8. The experimental $\varkappa_3(k)$ signal (solid line) obtained from the measured $\mu(k)$ as $\varkappa_3(k) = \{(\mu(k) - \mu_0)/\mu_0\} - \varkappa_2(k)$ is compared with the theoretical $\varkappa_3(k)$. In the lower panel the Fourier transform of the experimental (solid line) and theoretical $\varkappa_3(k)$ are shown

3.3 Octahedral Clusters

In the octahedral clusters there are $O-M_0-O$ atoms in a collinear configuration. When there are atoms in collinear configurations, the scattering amplitude is enhanced due to the high probability of forward scattering. This effect is known to enhance the amplitude and to modify the phase shift of the EXAFS signal of a distant neighboring atom; it is called the focusing effect [64-66]. The focusing effect determines the presence of multiple scattering contributions also at high kinetic energies of the photoelectron. In Fig. 9 the x-ray absorption spectrum of Mn^{2+} ions in aqueous solution at 50 mM concentration is shown.

The theoretical spectrum for an octahedral cluster formed by a central manganese and six neighboring oxygens has been calculated according to the multiple scattering theory described in the Sect. 3.1. The expansion of the total cross section $\alpha_t = \alpha_0(1 + \sum_n \varkappa_n)$ is demonstrated to be feasible above threshold. The results are shown in Fig. 10 where the atomic absorption α_0 and the successive multiple scattering contributions $\alpha_0\varkappa_n$ are reported. The multiple scattering contributions are classified according to the multiple scattering pathways within the octahedral cluster as shown in Fig. 5. The full multiple scattering calculation reported in the upper part

43

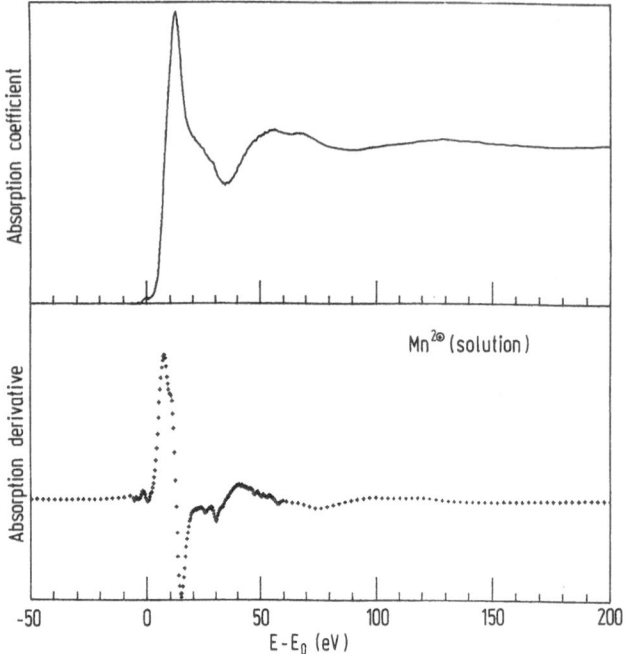

Fig. 9. X-ray absorption spectrum and its derivative (lower panel) of Mn^{2+} ions in solution. The zero of the energy scale is fixed at the weak 1s-3d excitation at threshold

of the figure is in good agreement with the experimental data of Fig. 9. No white line is observed at threshold because in the octahedral cluster the transition from the 1 s (1 = 0) level to the first unoccupied states is dipole forbidden because the first unoccupied states have t_{2g} and e_g symmetry, which are 1 = 2 symmetries.

The multiple scattering pathways where the angle between the outgoing wave and the reflected wave is 90° (for example the n = 3 contribution for the M—O—O—M pathway) has zero amplitude in the plane wave approximation as can be seen from formula (12) in Sect. 3.1. Therefore the pathways for atoms in collinear configurations provide the main contribution to multiple scattering.

This fact explains the results in Fig. 10 where $\varkappa_3(E)$ has as large an amplitude up to high energies as the EXAFS term $\varkappa_2(E)$. Also, the attenuation of the higher terms $\varkappa_4(E)$ and $\varkappa_5(E)$ with increasing kinetic energy is much less than that observed for tetrahedral clusters.

By comparing the total cross section obtained from a full multiple scattering calculation shown in the upper part of Fig. 10 with the EXAFS term $\alpha_0 \varkappa_2$ we observe that the two spectra are very similar above 3 Rydberg. This effect is determined by the destructive interference between the n ≥ 3 contributions. The negative interference between n = 3 and n = 4 contributions is clearly shown in Fig. 10. This destructive interference between higher order contributions is characteristic of octahedral coordination and disappears for distortion of the octahedra.

The main peak observed in the first 30 eV energy range is clearly a multiple scattering resonance. This peak has contributions from all multiple scattering pro-

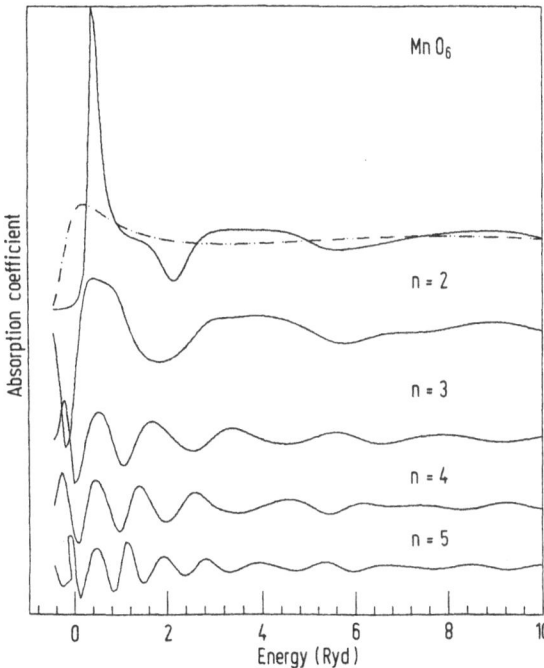

Mn O_6

n = 2

n = 3

n = 4

n = 5

Absorption coefficient

0   2   4   6   8   10

Energy (Ryd)

Fig. 10. Theoretical absorption coefficient for a cluster formed by the central Mn and six oxygens. Going from top to bottom the total cross section, the atomic contribution (dotted-dashed line), the single scattering contribution, n = 2, (EXAFS) and the contributions of successive orders of multiple scattering pathways of orders n = 3, 4 and 5 are shown

cesses, which, due to the particular geometrical coordination, happen to be all in phase at one particular energy, the resonance energy.

3.4 Contribution of Further Shells

In the analysis of XANES spectra of condensed systems the first step is to identify the size of the cluster of atoms relevant for the XANES spectrum [25, 26]. The cluster consists of the central absorbing atom and its local environment. The size of the cluster can range from the smallest one, including only the nearest neighbors, to clusters including up to many surrounding shells.

The finite size of the cluster is determined by the mean free path for elastic scattering of the photoelectron as shown in Fig. 1 and by the core hole life time. In the energy range 1–10 eV, where the mean free path becomes longer than 10 Å, the size limitation due to the core hole lifetime becomes the relevant one. The contribution of further shells can be cancelled by structural disorder as already discussed for transition metal ions in dilute solutions.

Here we discuss some cases where contributions from further shells are essential in order to interpret the XANES spectra. The iron K edge spectra of $K_4Fe(CN)_6$ and $K_3Fe(CN)_6$ crystals are reported in Fig. 11 (panel b) together with their derivative [67].

These spectra have been interpreted by the multiple scattering approach of XANES developed by Durham et al. [68, 69]. In this approach an independent particle approximation similar to the local density approximation (LDA) to the density functional theory has been used [70]. The effective one-particle Schrödinger equation using multiple

45

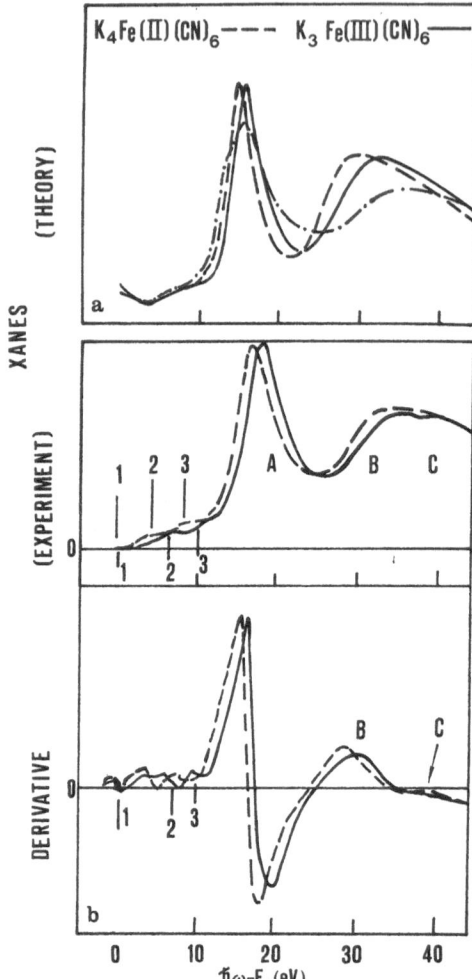

Fig. 11. X-ray absorption spectra of $K_3Fe(CN)_6$ and $K_4Fe(CN)_6$ (panel b) and their derivative (panel c). In panel a the calculated XANES of $Fe(CN)_6$ clusters are plotted. The solid line is obtained using the coordinates of diffraction experiments for the $K_3Fe(CN)_6$ crystal. The dashed line is obtained using the coordinates of diffraction experiments for the $K_4Fe(CN)_6$ crystal. The dashed-dotted line is the XANES spectrum using the coordinates of an alternative structure suggested by diffraction experiments

scattering theory in real space for a finite cluster of atoms surrounding the absorbing atom has been solved. Neither translational symmetry nor site symmetry of the cluster are required. This aspect of the theory makes it applicable for the extraction of structural information from XANES of complex systems such as proteins and disordered systems. This is a full multiple scattering theory because the key quantity that determines the interaction of the excited electron with the environment is the scattering path operator $\underline{\tau}$, which sums all scattering paths which begin and end on the central atom.

The cluster is divided into concentric shells of atoms and the multiple scattering equations are solved first within each shell and then recursively between the shells themselves and the central atom. Therefore the contribution of successive shells can be easily studied using this method.

The key role of the second shell in the spectra of ferro-cyanides is demonstrated in Fig. 12. The multiple scattering XANES calculation for a single shell of FeC_6 is

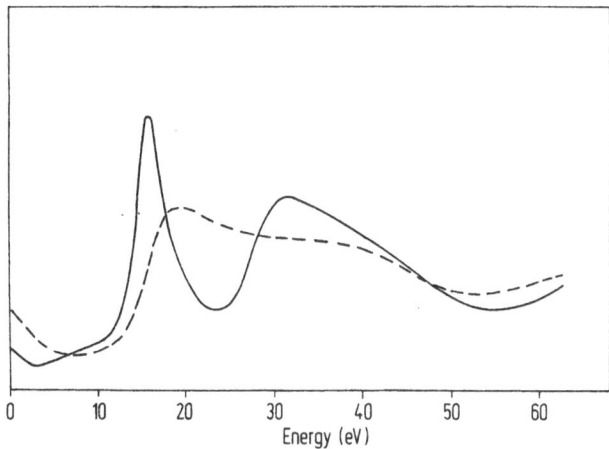

Energy (eV)

Fig. 12. Calculated Fe K-edge XANES for an $Fe(CN)_6$ cluster ($R_{Fe-C} = 1,903$ Å, $R_{C-N} = 1.134$ Å) with both C and N shells included (solid line) and with the C shell only included (dashed line)

compared with a full cluster calculation including the second shell. It is clear that it is only on including the second shell that the experimental features A, B and C in the spectra of Fig. 11 appear. For these systems where the C—N distance is short the full multiple scattering within two shells plays a major role in the XANES.

The XANES spectra of ferro- and ferricyanide complexes have been used to solve a structural problem concerning these systems. The structure of these complexes is approximately octahedral, with six CN groups surrounding a central Fe atom, with the FE—C—N bond being nearly linear. The difference between the structure of $K_4Fe(CN)_6$ and $K_3Fe(CN)_6$ crystals has been the object of long standing research using diffraction methods. In fact the Fe—C distances do not show large difference between the two complexes, and the valence electronic charge is delocalized in the $Fe(CN)_6$ complex. The difference between the two structures is caused by subtle distortions of the octahedral symmetry. Two structures [71,72] were published for $K_4Fe(CN)_6$. Iron K-edge XANES spectroscopy has been used to solve the controversy over the two proposed structures. The structure of $K_3Fe(CN)_6$ was well established.

The Fe—C interatomic distance was determined by EXAFS, which gives the same interatomic distance for both complexes. The EXAFS signal of Fe—N is strongly affected by multiple scattering due to the focusing effects. Because of these complications, EXAFS spectroscopy cannot solve the difference between the two samples because the spectra are very similar.

The parameters of the multiple scattering theory, such as the muffin tin radii, were fixed by reaching a good fir of the spectrum of $K_3Fe(CN)_6$ using the crystallographic coordinates. The results are shown by the curve (solid line) in panel a of Fig. 11. Then the spectrum of $K_4Fe(CN)_6$ was calculated by changing just the coordinates of the C and N atoms according to the two proposed structures. The results are plotted in panel a of Fig. 11 (dotted-dashed line and dashed line). Clearly one calculation (dashed line) is in agreement with the experimental XANES spectrum while the other disagrees. The fine distortions of the structure determine the variation of the line shape of the spectra and one of the two proposed structures predicts large distortions of the line shape (dotted-dashed curve). Therefore XANES spectra confirm the proposed structure of Ref. 72. In this way the structure of $K_4Fe(CN)_6$ can be solved.

An interesting feature of XANES for structure determination in condensed matter is that it can be used to study the local structure both in crystalline and disordered materials. Therefore once a local structure has been solved for a crystalline material also the similar structure in amorphous, liquid or complex materials, where diffraction methods cannot be applied, can be solved.

The size of the cluster for the XANES spectra of crystalline materials is expected to be determined mainly by the elastic mean free path, however, the focusing effect for crystal structures with collinear atoms enhances the contributions from further shells in the XANES region as well as in the EXAFS region. NiO is a good example of a simple cubic crystalline structure with important contribution from multiple scattering in collinear atoms. We show in Fig. 13 the oxygen K-edge of stoichiometric NiO measured by the electron yield method and a grazing incidence "grasshopper" monochromator [73]. This spectrum has also been obtained by electron energy loss measurements [74, 75]. The oxygen K-edge spectrum has been calculated by the full multiple scattering approach and is shown in Fig. 13. To obtain good agreement between theory and experiment a cluster out to a radial distance of 5 Å, including 30 atoms, was necessary [76,77]. The features **b** and **f** of the NiO spectrum in Fig. 13 have been assigned to many body effects due to configuration interaction in the ground state of NiO [73].

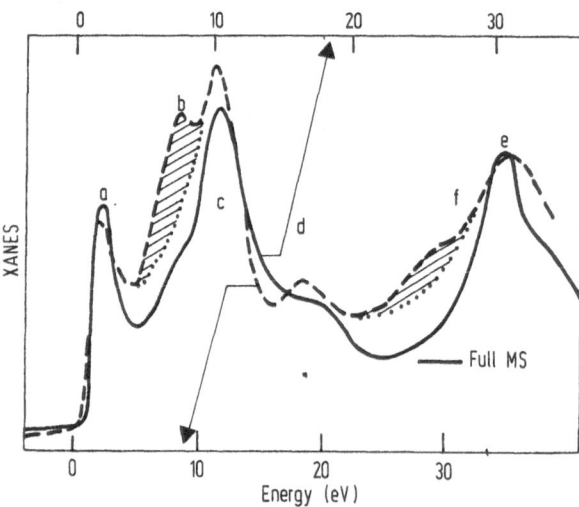

Fig. 13. Oxygen K-edge XANES spectrum (dashed line) and full multiple scattering theoretical spectrum (solid line). The theoretical curve has been obtained from a large cluster including up to five shells. The energy scale of the theoretical spectrum has been expanded to fit the experiment. The features b and f not predicted by one-electron theory can be assigned to many body configuration interactions

The single scattering EXAFS calculations disagree with the experimental data over a range of 30 eV above the peak **a** at threshold. Vvedensky and Peddry [77] have also divided the multiple scattering paths into two classes, those for which an intervening atom lies collinearly with the central atom and the backscattering atom (type 1), and all other types of events (type 2). The number of type 2 paths contributing to XANES is greatly reduced for a cubic crystal like NiO because there are several symmetry-inhibited paths where the angle between the outgoing wave and the incoming wave at the central atom is 90°, as discussed in Sect. 3.3 for octahedral clusters. However, the type 2 multiple scattering paths, involving intershell scattering, contribute to the full cross section. The Ni K-edge XANES spectrum of NiO and of the series of MO (M = transition metal) metals was measured by Knapp et al. [78] and by Norman et al. [79]. It was shown that the XANES spectra are similar and that their shape is determined by the crystalline structure, although they show a wide variety of electronic and magnetic properties. Multiple scattering calculations were performed for the series of MO crystals, obtaining good agreement with the experiments [79].

3.5 The Role of Interatomic Distance

Here we discuss the effect of bond length on XANES spectra. It is well known that EXAFS oscillations are determined by the interatomic distances and coordination numbers. Also a full multiple scattering resonance in the XANES range is strongly affected by interatomic distance.

The effect of bond distance on multiple scattering resonances was first observed in the carbon K-edge spectra of simple molecules like hydrocarbons [80]. In C_2H_2 and C_2H_4 the main resonance above the continuum photoionization threshold is determined by the scattering from the other carbon atom. A linear relation between k_r, the wave vector of the photoelectron at the multiple scattering resonance, and $1/R$, where R is the interatomic distance has been found. The rule $k_r.R = $ constant was justified theoretically by the multiple scattering theory which shows that this is a general feature of full multiple scattering resonances [81,82]. This rule is valid only for a range of 20% variation of the interatomic distance R where the energy dependence of scattering phase shifts is negligible. The rule $k_r.R = $ constant $= C_0$ has been verified to be valid in a large set of XANES spectra of simple molecules measured by energy loss method [83-85].

In the XANES spectra of condensed systems this rule has been verified for tetrahedral clusters [86]. In Fig. 14 the spectra of a series of transition metals with tetrahedral coordination are plotted. The similarity of the XANES spectra demonstrates the similar tetrahedral coordination. A good alignment of the peaks A and D was obtained by expanding the energy scales of each spectrum by a factor d_{sample}^2/d_{ref}^2, where the d_{ref} is the Cr—O distance (1.59 Å) in Na_2CrO_4, and d_{sample} is the corresponding one for each sample, fixing the origin of the energy scale at the energy of the weak 1 s to 3 d transition at threshold (often called prepeak). The peaks A and C, common to all spectra, are due to the first shell of neighboring atoms, as shown by XANES calculations reported in Sect. 3.2, while other small features are due to further shells.

In the framework of the Xα-multiple scattering theory, the absorption cross section is determined by the multiple scattering matrix M of the photoelectron with kinetic

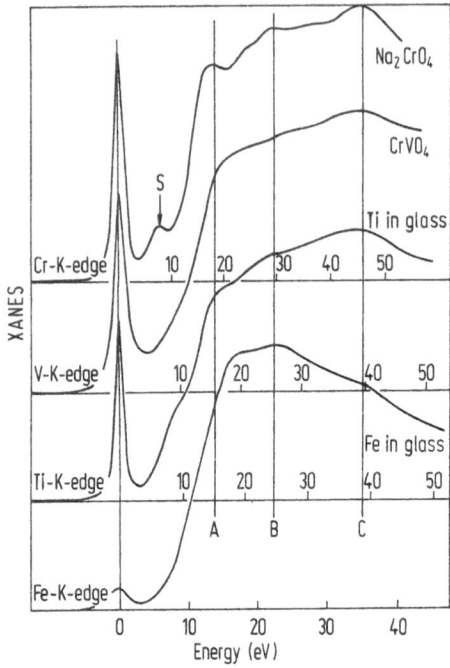

XANES

Cr-K-edge 10 20 30 40 50 | Na₂CrO₄

CrVO₄

Ti in glass

V-K-edge 10 20 30 40 50

Fe in glass

Ti-K-edge 10 20 30 40 50

A B C

Fe-K-edge

0 10 20 30 40

Energy (eV)

Fig. 14. The XANES spectra at the K edge of transition elements in tetrahedral coordination: (a) Na_2CrO_4. Cr edge; (b) $CrVO_4$, V edge; (c) Ti^{4+}-containing glass; Ti K-edge; (d) Fe^{3+}-containing silicate glass, Fe edge. The energy scale of each curve has been multiplied by a factor $1/d^2$, where d is the metal-oxygen distance. The zero of the energy scale has been fixed at the prepeak maximum

energy $(\hbar k)^2/2\,m = E_r - E_0 + V$, where E_r is the energy of the resonance, E_0 the absorption threshold and V is the average muffin tin interstitial potential. The maxima in the absorption correspond to the condition $\det M = 0$. This gives the relation $k_r \cdot R = C_0 = \text{constant}$.

In Fig. 15 the results of multiple scattering calculations of the Mn K-edge XANES spectra for a MnO_4 cluster with different Mn—O distances $R_1 = 1.73$ Å and $R_2 = 1.66$ Å. The white line at threshold is almost unaffected by this small change of distance and the general shape of the spectra is conserved. The multiple scattering peaks clearly moves toward higher energies with decreasing interatomic distance, following the theoretical rule given above. Weak variations of the lineshape with distance are observed, as is expected because there is a variation of the interference pattern between the different multiple scattering contributions. These changes become very important for large changes of interatomic distance, therefore this is another fact that limits the validity of the $k \cdot R = $ constant rule to distance variations smaller than 20%.

The extraction of the variation of the interatomic distances in unknown systems is complicated by the determination of $k_r = \{\hbar/2\,m(E_r - E_0 + V)\}^{1/2}$ because V, the average interstitial potential in the muffin tin approximation, is unknown. In order to overcome the problem of V, which cannot be determined experimentally, it has been pointed out that the energy separation of a multiple scattering resonance in the continuum from a bound state at threshold $\Delta E_r = (E_r - E_b)$ can be used to determine the variation of the interatomic distance.

A bound resonance at the threshold of a XANES spectrum can be determined by a resonance in the atomic cross section or by a molecular bound state, like the white line at threshold of the K-edge of tetrahedral clusters, shown in Fig. 14 and 15. Multiple

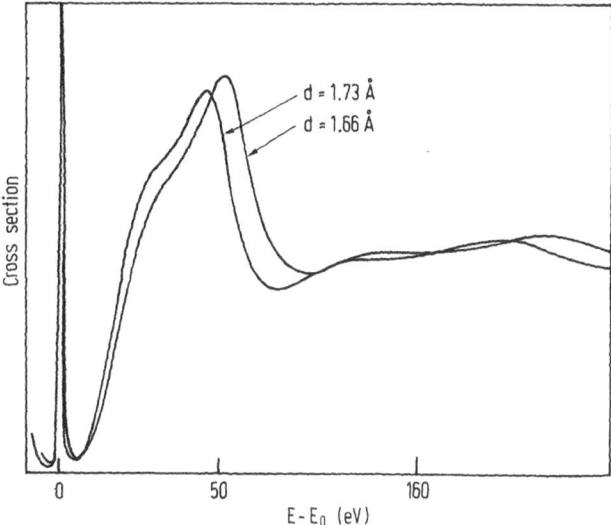

Fig. 15. The effect of the variation of the interatomic distance on the Mn K-edge XANES spectrum for a MnO_4 cluster

scattering theory [82] shows that a similar relation $k_b \cdot R = C_b$ is valid also for a bound state at k_b, but with a different constant C_b, if its variation with the distance is very small as shown in Fig. 15. Therefore the relation $(E_r - E_b) R^2 = C \approx C_0 - C_b$ is valid with a different constant C. Now this constant C can be derived from experimental data on a set of spectra of known compounds and the interatomic distance can be determined in unknown system.

This approach has been applied in a condensed system for the determination of the V—O distance in a silica glass. The overall shape of the vanadium k-edge spectrum

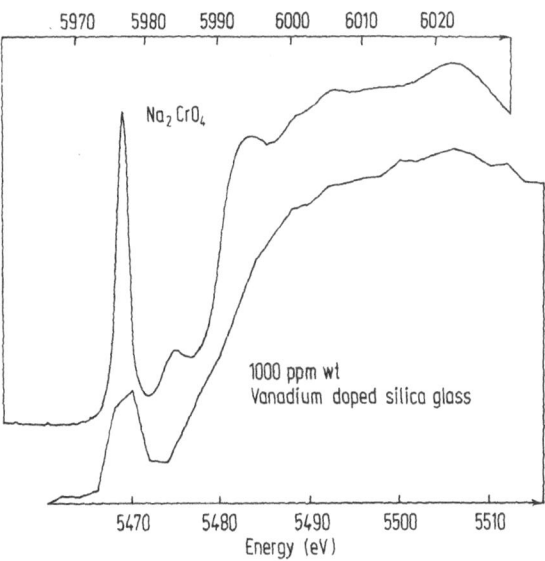

Fig. 16. Comparison of the K XANES of 0.1 wt.-% vanadium doped silica glass and of Na_2CrO_4 documents tetrahedral coordination of vanadium in the glass. This shows the application of XANES for local structure determination in disordered materials. The analogy of the overall shape of the XANES spectra is obtained by dilation of the energy scale with a factor $1/d^2 = 0.318$ ($d \approx 1.77$ Å) of the energy scale of the glass spectrum

shown in Fig. 16 of a vanadium doped silica glass at 1000 ppm wt shows that the metal ion V is tetrahedrally coordinated by comparison of the spectrum with a spectrum characteristic of tetrahedral clusters. The constant C was determined by a set of samples shown in Fig. 14. The interatomic V—O distance in the glass was determined by the $(E_r - E_b) R^2 = C$ rule to be $R = 1.77 \pm 0.05$ Å.

The effects due to the variation of interatomic distances in a spectrum determined by a cluster including several shells will be discussed for the case of ferrocyanides. Figure 17 shows the effects on full multiple scattering theoretical spectra of the variation of the interatomic distance d_1(Fe—C) between the central atom and its first neighbors leaving the C—N distance constant [67]. The XANES spectra show two main peaks due to the full cluster as shown in Fig. 12. Both multiple scattering resonances move toward higher energy with decreasing d_1 according to the $(E - V) \cdot d_1^2 =$ constant rule where the calculated average interstitial potential V was about 13 eV. The energy separation ΔE between the two peaks remains constant with constant C—N distance as shown in Fig. 17a. The energy separation ΔE depends on the C—N distance as shown in Fig. 17b. This shows that the two multiple scattering peaks can be considered as arising from multiple scattering within the C—N groups and therefore their separation is strongly dependent on the C—N distance.

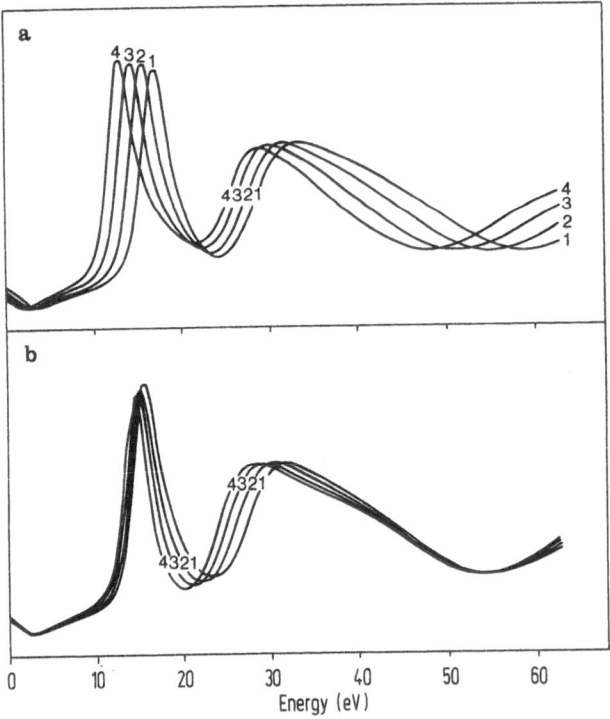

Fig. 17. Calculated K-edge XANES of Fe(CN)₆ clusters: panel a) C—N distance fixed at 1.134 Å, Fe—C distances of 1.852 Å (curve 1), 1.903 Å (curve 2), (1.958 Å (curve 3), and 1.217 Å (curve 4). In the panel b the Fe—C has been fixed and the C—N distance changes. The zero of the energy is chosen to be close to the 3d resonance on the Fe atom

In conclusion, when different shells contribute to the XANES spectra a detailed analysis of the spectra is required and no simple relations can be used. Moreover, the simple relation $k \cdot R = $ constant cannot be used when there are changes of the symmetry of the cluster due to structural distortions.

3.6 Angular Resolved XANES

The polarized x-ray absorption spectra show a strong dichroism for anisotropic sites. The strong dichroism of XANES spectra of single crystals was observed experimentally by several groups for II–VI layer compounds [87], uranyl nitrate [88], Cu complexes [89–92] and in proteins such as crystalline myoglobin [93] and plastocyanin [94] and in vanadium sites in sol-gel crystals [95,96] and in porphyrins [97]. The dichroism for electronic transitions is well known in the classical quantum theory of the absorption cross section. The system is described by N electrons in the positions r_n. Following the interaction with the photon beam the system is excited from the initial state i at energy E_i to the final state f at energy E_f in the dipole approximation, which is given by

$$\sigma(h\nu) \sim h\nu \sum_f |M_{if}|^2 \delta(E_i - E_f - h\nu)$$

where the sum is extended over all the possible final states f and M_{if} is the matrix element involving the many body radial wave functions:

$$M_{if} = \int \Psi_f^*(r_1, r_2, \dots, r_n)\, \Sigma_n(r_n \cdot e)\, \Psi_i(r_1, r_2, \dots, r_n)\, dr$$

where e is the unitary polarization vector of the electric field. Therefore, in anisotropic systems the matrix element is dependent on the direction of the polarization vector via the terms $(r_n \cdot e)$.

In the multiple scattering description of the absorption cross section in real space the angular dependence is predicted by formulas 2 and 3 of Sect. 3.1. The polarization effects are determined by the terms $|(r_i \cdot e)|^2$ where r_i are the positions of neighboring atoms with the central absorbing atom at the origin. The presence of this term remains in the single scattering approximation which describes the polarized EXAFS spectra [98] in which oscillations originating at a particular scattered atom can appear or disappear depending on whether the electric field is directed towards that atom or not. This has been widely used in surface EXAFS studies to determine separately the interatomic distances parallel and normal to the surface plane [99, 100].

Using the polarization dependence of the absorption cross section in the XANES spectra of anisotropic clusters the multiple scattering contributions due to a set of atoms in a particular direction or on a plane can be selected. Therefore, the orientation and angular distribution of neighboring atoms can be easily determined by changing the relative position of the incident beam and the sample.

In Fig. 18 we report the vanadium K-edge XANES spectra of a spontaneously dehydrated sol-gel sample $V_2O_5 \cdot 1.6\, H_2O$. This sample is formed of oriented sheets separated by about 10 Å of water molecules. Each sheet consists of square plane bipyramids with the base on the xy plane and a short double bond of 1.58 Å on one

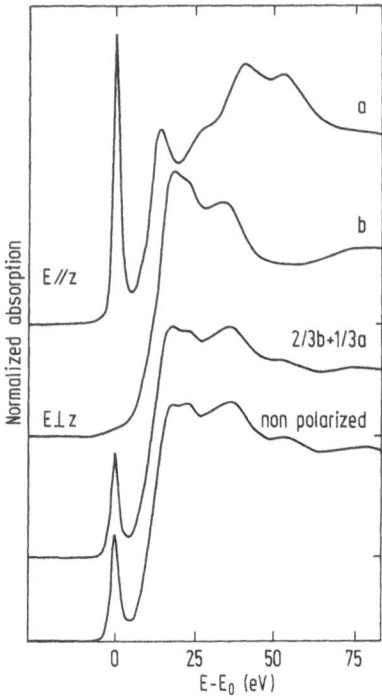

Fig. 18. Polarized experimental XANES spectra $V_2O_5 \cdot 1.6\ H_2O$ dehydrated sol-gel with the electric field **E** parallel to the normal **z** of the V_2O_5 layers, i.e., along the axis of the square pyramids (curve a); curve b represents the polarized spectrum with the elctric field **E** parallel to the V_2O_5 layers; the lowest curve is the unpolarized spectrum and above it the weighted spectrum from the two polarized spectra

side and a long bond with the oxygen of a water molecule on the other side, in the **z** direction. The experimental spectra for the electric field **E** oriented along the **z** direction (**E** || **z**) and parallel to sheet planes (**E** ⊥ **z**) are shown. These spectra are a good example of angular resolved XANES because the dichroism is very large and it appears both in the threshold region for the molecular like bound excitations, giving the while line, and in the continuum. The white line is completely suppressed for the polarization vector oriented on the plane, showing that it is produced by the short double V—O bond. The nonpolarized spectrum of the sol-gel in solution is reported and the features of the solution spectrum are reproduced by the weighted sum of the two polarized spectra.

The interpretation of polarized XANES spectra in terms of multiple scattering allows the determination of the role of a selected group of atoms. This is shown in Fig. 19 where the angular resolved XANES calculations are reported for a VO_6 cluster shown in the figure. The calculation clearly shows that the white line appears only with the electric field along the V—O double bond. The main features of the experimental spectrum (**E** || **z**) are reproduced by theoretical calculations for a simple cluster with the vanadium atom on the xy plane. The maximum at about 40 eV in the continuum is caused by the closest oxygen and the feature at about 60 eV originates from the longest bond in the z direction. The calculated XANES for multiple scattering with the four oxygen atoms on the xy plane are in qualitative agreement with the experimental spectrum. Agreement with the experimental spectra are improved when the calculation is done for a cluster in which the vanadium atom is out of the plane, as shown in the lower part of Fig. 19.

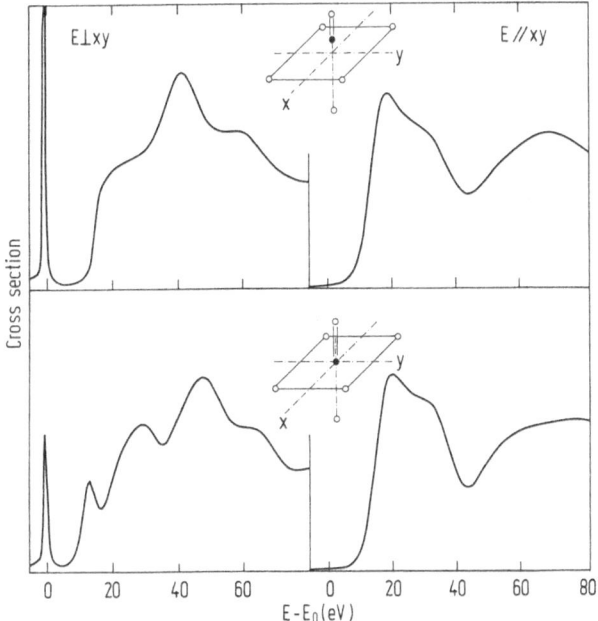

Fig. 19. The theoretical angular resolved multiple scattering calculation for a square bipyramid with a short double V—O bond (d = 1.58 Å) and with a long V—O bond (d = 2.6 Å) along the z axis due to the oxygen of a water molecule; and four oxygens on the xy plane at 2.0 Å. The upper curves show calculated polarized curves with the V atom out of plane by 0.5 Å and the lower curves are the same for a similar cluster with the same distances but with the V atom in the plane

The measurement and multiple scattering interpretation of angular resolved XANES is very useful for understanding and interpreting the spectra of unpolarized spectra in solution. In fact, some features in unpolarized spectra can be to assigned to multiple scattering with a selected number of atoms. For example, the features at about 25 eV and that at 60 eV in the unpolarized spectrum in Fig. 18 can be associated with multiple scattering in the plane and along the normal respectively.

Angular resolved XANES has been used to determine the orientation of chemisorbed molecules on surfaces. The absorption spectra at the carbon K-edge of CO on Ni(100) have been used to show that the CO molecule stands upright on the surface [101]. The sulfur $L_{2,3}$ edge and the carbon K-edge of thiophene molecules on the platinum Pt(111) surface has been used to investigate thermal decomposition of the molecules on the surface [102].

The angular resolved iron K-edge XANES spectra of carbonmonoxy-myoglobin single crystal and their interpretation have allowed the determination of the CO bonding angle on the porphyrin plane [93].

4 Band Structure Approach to XANES of Metals

4.1 Multiple Scattering Approach

XANES spectra in metals can be solved in real space with the multiple scattering approach. For f.c.c transition metals, multiple scattering calculations in real space have been carried out with the Durham et al. program [68,103], using a finite cluster approximation [104], and recent calculations by Kitamura et al. [105] confirm the first results. The size of the cluster has been found by adding successive shells to obtain good agreement with experimental data. A large cluster with four shells was necessary to reproduce the experiment. In this calculation an energy independent broadening parameter has been used to take into account the effects of inelastic electron-electron scattering and the core hole lifetime.

In Fig. 20 the experimental Cu K-edge XANES is reported in the upper part of the figure. The multiple scattering calculations for copper clusters of different size are reported, in the lower part [104]. The calculations began with a simple cluster including only first neighbors (curve a), and then two shells (curve b), three shells (curve c) and four shells (lower curve) were included.

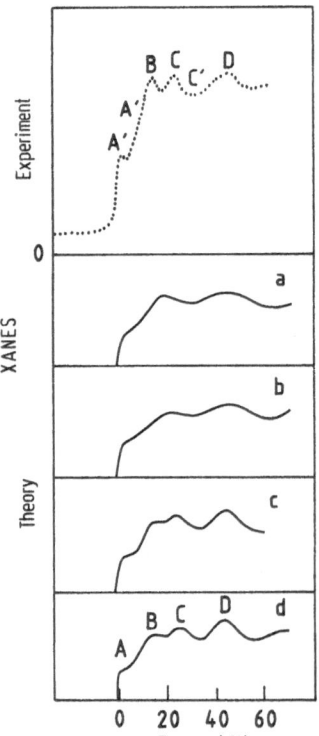

Fig. 20. K edge XANES spectrum of Cu. The experimental spectrum (upper curve) is compared with multiple scattering calculation for different cluster sizes; a) one shell; b) two shells; c) three shells and d) four shells

The great similarity [106] between the K-edge XANES spectra of transition metals with the same f.c.c. crystal structure shows that the experimental features of the spectra depend mainly on the atomic arrangements of the atoms, and the multiple scattering resonances are only shifted in energy depending on the interatomic distance.

The fact that the XANES spectra are determined by a small cluster is experimentally determined by the fact that the main features of the Cu metal K-XANES are present for a small metallic microcluster of 10 Å diameter in solid argon [107].

4.2 Band Structure Approach

The spectra of metallic crystals can be solved in k-space. Theoretical calculations of the partial and projected density of states of the crystal band structure were reported by several groups to interpret the XANES of metals [108-114]. We discuss here the band structure approach developed by Müller et al. [115-117] by which it is possible to calculate the absorption spectra over a large energy range. In this theory the selected angular moment and the local character of the final state are considered. The matrix element and the inelastic scattering are included in the theory, giving good agreement with experiments.

The central approximation of this theory is the one-electron formalism, and the x-ray absorption is described by single-particle processes. Deviations of the experiments from accurate single-particle calculations may point to many-body effects. One such effect involves the filling of the core hole and the decay of the electron excited in the absorption. This effect is incorporated in this theory by convoluting the single particle result with an energy-dependent function which takes account of the lifetime of the core hole and of the excited electron.

In the case of metals, the core hole in the static final state potential is fully screened by valence electrons close to the Fermi level. The attractive interaction between the electron and the core hole can be considered to be zero in this approximation and translational symmetry is conserved. Most of the XANES spectra can be interpreted in the frame of the Barth-Grossman final state rule [118,119], which states that the wave function of the excited photoelectron is determined by the final state potential of the core hole and the relaxed N-1 electrons. This final state rule has also been tested in metals in a detailed analysis of the $M_{2,3}$ edges of transition metals [120].

Also, in the band structure formalism, the spectra can be understood as the product of an atomic like term and a solid state term as in multiple scattering formalism. This factorization results from the localized nature of the core state involved in the x-ray transition. Further, since the dipole transitions dominate the process, excitation of a core state having angular momentum l probes the l $\pm$ 1 components of the conduction band.

The experimental spectra are described in the following way:

1) The overall magnitude and shape of a particular spectrum is determined by the corresponding atomic transition rate.
2) The fine structure of the spectrum is determined by a solid state factor which is proportional to the density of band states with l $\pm$ 1 orbital character.

The contribution of the core level c to the absorption coefficient $\alpha_c(E)$ can be expressed as

$$\alpha_c(E) = \{4\pi^2 \alpha n/\Omega\}\, F_c(E)$$

where α is the fine structure constant, Ω is the volume of the primitive shell, n is the number of contributing atoms in the primitive shell and $F_c(E)$ is the oscillator strength for unpolarized x-rays given by [116, 117, 120]

$$F_c(E) = (2m/3h^2)\,(\hbar\omega)\, \sum_{kj} \sum_M |<\varphi_{cM}|\, r\, |\Psi_{kj}>|^2\, \delta(E - E_{kj})$$

Neglecting the spin-orbit coupling for the band states $F_c(E)$ for a core state $c = (n, l, J)$ can be written as

$$F_c(E) = \{\omega(2J + 1)\}/\{6\,(2l + 1)\}\, \{[l/(2l - 1)]\, f_{c, l-1}(E)$$
$$+\, [(l + 1)/(2l + 1)]\, f_{c, l+1}(E)\}$$

where the partial strength $f_{c, l}(E)$ can be factorized into

$$f_{c, l}(E) = r_{cll}^2\, N_l(E)$$

Here $N_l(E)$ is the angular projected density of states, defined by

$$N_l(E) = 2 \sum_{k, j} \sum_m |<\gamma_{lm}|\, \Psi_{kj}>|^2\, d(E - E_{kj})$$

where the energy band states are labeled by the reduced wave vector **k** and the band index j. The effective matrix element $r_{c, l}(E)$ is given by

$$r_{c, l}(E) = <\varphi_c|\, r\, |\varphi_l(E)> <\varphi_l^2(E)>^{-1/2}$$

where the partial wave $\varphi_l(E)$ is a solution of the radial Schrödinger equation inside the muffin-tin sphere and γ_c is the core wave function.

For the purpose of comparison with the measured absorption coefficient, the theoretical spectra are convoluted with a Lorentzian broadening function $\Gamma(E)$. This function is the sum of two terms. The first takes account of the core hole width and the second term is the width of the excited band energy, which is a function dependent on the mean free path of the excited electrons, and takes account of the photoelectron inelastic scattering which is energy dependent and varies for each material as shown in Fig. 1. Note that in this theory any broadening effect due to the experimental resolution and many-body effects, such as the influence of the core hole on the band states, are not included.

With this formalism, the interpretation of the spectra in terms of the local partial density of states around a particular atom in periodic systems is evident. The connection between the solid state effect and the atomic effect is obtained by decomposition of the partial oscillator strength

$$f_{c, l}(E) = r_{c, l}^2\, N_l(E) = M_{c, l}(E)\, X_l(E)$$

with

$$M^2_{c,l}(E) = (2l + 1) \, N^{fe}(E) <\varphi_c| \, \mathbf{r} \, |\varphi_l(E)>^2$$

and

$$X_l(E) = N_l(E)/(2l + 1) \, N^{fe}(E) <\varphi_l^2(E)>$$

where N^{fe} is the free-electron density of states.

In band structure formalism the spectrum is expressed in terms of the partial density of states, $N_l(E)$ which includes both atomic and solid state effects. Thenormalization, reported above, permits the factorization of the spectrum into a solid state term X_l and an atomic matrix element which characterizes the effect of the central potential. This last term corresponds to the atomic factor α_0 in the multiple scattering approach. The solid state factor X_l can be described using either a scattering or a band structure approach. In this formalism X_l is proportional to the density of band states N_l with angular momentum l, determined by the orbital symmetry of the core state and the dipole selection rules. For the unoccupied states successive l-bands arise from additional nodes in the wave function in the region between the core region and the cell boundary, which produce features in N_l roughly periodic in $\sqrt{E}$. The amplitude of features due to specific bands are proportional to the strength with which they hybridize.

The spectra exhibits deviations from the single-particle picture of the potential for the ejected electron, which can be divided into two categories: i) broadening of the spectra due to lifetime effects, and ii) energy shifts due to exchange correlation effects]

4.3 Probing the Local Density of States

XANES spectra of different systems have been interpreted with the band structure aproximation [106-117]. As an example for a transition metal we discuss here palladium absorption edges. The comparison between the K [115] and L_1 [121] edge of Pd metal with the theoretical band approach is shown in Fig. 21. We can observe that the K and L_1 edges present the same spectral features and therefore contain identical information. In fact, the selection rule for electronic transitions selects the same $l = 1$ projected density of states. Because the L_1 edge occurs at lower energy a better instrumental energy resolution is obtained and the structures are better resolved.

In the figure the theoretical absorption coefficient before ($\mu(E)$) and after ($\overline{\mu(E)}$) it has been smeared out to take account of inelastic terms is plotted. The convoluted spectrum reproduces all the experimental features [115].

Because of the selection rules $\Delta l = \pm 1$, $\Delta S = 0$ and $\Delta J = \pm 1$ the edge spectra select the angular momentum l' and the total angular J of the local density of states. In Fig. 22 the L_3 edge spectrum of Pd metal [17,122] is compared with the one-electron band calculation of Muller [117]. The good agreement between theory and experiment is evident. The total density of states is plotted in Fig. 22 to show that the total DOS does not explain the experimental spectra because only the local $l' = 2$ and $l' = 0$ projected density of states contribute to the spectrum.

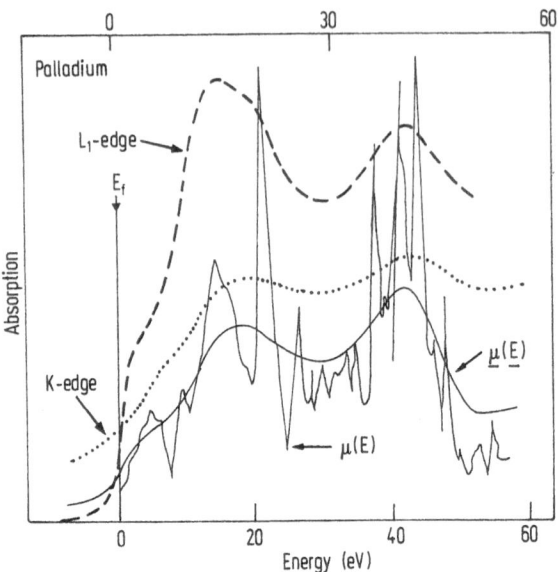

Fig. 21. Palladium L_1-XANES [21] and K-XANES [115] are compared with the calculated theoretical $\mu(E)$ from the p-like partial density of states taking account of the matrix element of the Pd crystal before and after it has been smeared to account for lifetime effects. The difference between the L_1 and the K-edge is due to the better instrumental energy resolution in the energy range of the L_1 edge and shorter core hole lifetime at the K-edge

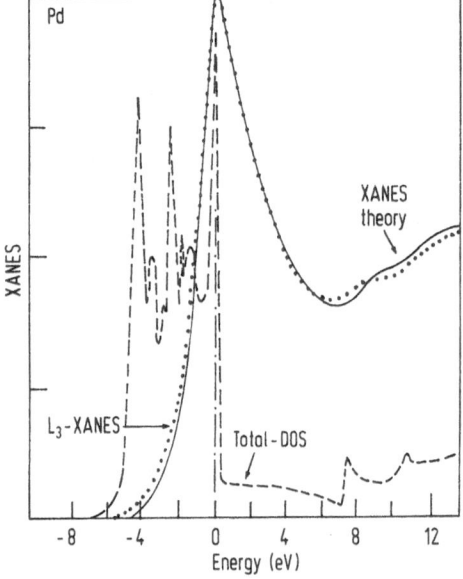

Fig. 22. L_3 palladium edge of Pd metal (dotted line) compared with one-electron band theory (solid line) taking account of the partial ($l' = 2$) local density of states, of the inelastic mean free path and of the core-hole lifetime. The dashed line shows the total density of states of palladium metal, which is quite different from the absorption spectrum. The zero of the energy scale is fixed at the Fermi energy

Comparison between the K (or L_1) and L_3 edges shows that there is a high density of d-like unoccupied electronic states at the Fermi level giving the maximum of the L_3 white line. The absorption at the K (L_1) threshold is due to the hybridization between the sp conduction band and the more localized d states. In the high energy XANES region the difference between the spectra is due to the difference in the scattering phase shift excited photoelectron between the $l' = 2$ and $l' = 0$ channel (at L_3 or L_2 edge) and the $l' = 1$ channel (at L_1 or K edge).

The large differences at the threshold region are due to electronic structure near the Fermi level. For a low p-like density of states only weak structures are observed in the L_1 or K edges, that is the case for the K-edge spectra of Ti and Fe in the alloy TiFe measured by Balzarotti et al. [123,124].

If the density of unoccupied states above the Fermi level can be described by a step-like function and the core-hole as a state with width because of its lifetime, the absorption edge can be fitted with an arctan function

$$G(h\nu) = A \arctan \{2(h\nu - E_0)/\Gamma\}$$

The fitting of experimental curves with this function is expected to be good in the low energy part of the spectrum and the inflection point determines the Fermi level position. This is not the case for the L_3 edges of transition metals in which a strong peaks appear at the edge. These strong peaks are called "white lines" and their large intensity is due to the atomic-like character of the d resonance. When the unoccupied nd states have a narrow bandwidth (the case of Ni, Pd and Pt) the white line maximum corresponds to the Fermi level E_f [115]. For elements with a wide unfilled nd band it is more difficult to locate the Fermi level on the rising edge of the experimental spectrum. This fact is clearly demonstrated by the case of the L_3 edge of Pd where the Fermi level is at the white line maximum as obtained by comparison with the calculated spectrum. The changes in the shape of the white line with increasing atomic number are mainly determined by the localization and hybridization of the d unoccupied states and by the progressive filling of the d band.

Since $\Delta J = \pm 1$ and $\Delta S = 0$, edge spectroscopy provides detailed information on the total angular momentum J of the empty electronic states above the Fermi level. The $J = 3/2$ and $J = 5/2$ density of unoccupied states are probed by the L_3 edge and the $J = 3/2$ density by the L_2 with the same $l' = 2$. The L_3/L_2 intensity ratio is given by the statistical ratio $2:1$, due to the $2j + 1$ degeneracy of the inital core states. But if the number of holes in the d band is such that the ratio of holes $H_{5/2}/H_{3/2}$ changes a deviation from the statistical ratio 2 to 1 is expected. In Fig. 23 the L_3 and L_2 edge spectra of Pd are shown where the L_2 spectrum was multiplied by the statistical weight. The anomalous L_3/L_2 ratio 2.75 at the white line maximum of Pd [17] shows the predominance of the $J = 5/2$ total angular momentum of the empty 4d hole over the $J = 3/2$ one, not only in the spike at the Fermi level, but also in the p-like conduction band up to 8 eV. Above this region the L_3/L_2 intensity ratio is 2.1, very close to the statistical ratio.

In Ta [125], in rare earth compounds [126] where the 5d band is completely empty, in Ni, [109] in Fe, [127] and in other metals [116] the L_3/L_2 intensity ratio is close to 2.

In the early 3d elements [44,128−131], a large deviation from the statistical L_3/L_2 ratio has been found. In these systems the small spin-orbit splitting is comparable

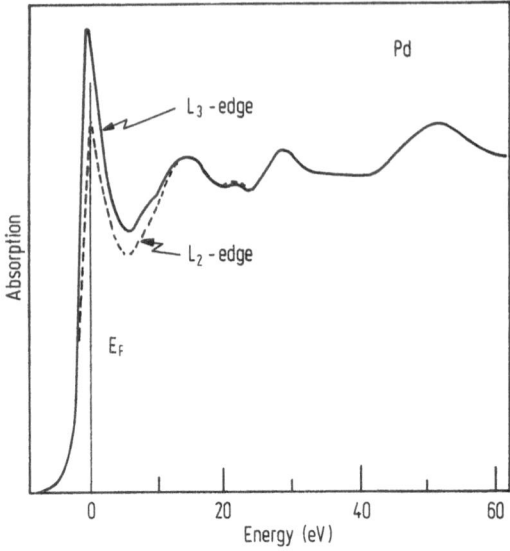

Absorption

Pd

L₃ -edge

L₂ -edge

E_F

0   20   40   60
Energy (eV)

Fig. 23. Comparison of the L_3 and L_2 absorption spectra. The L_2 spectra has been multiplied by a factor of 2.1 to overlap the high energy part of the spectra. A difference can be observed at the white line over a range of about 10 eV. At the white line maximum the intensity ratio L_3/L_2 is 2.75, indicating a larger density of unoccupied states with total angular momentum number $J = 5/2$

to the exchange energy therefore the two final states are mixed and the anomalous L_3/L_2 intensity ratio can not be assigned to a different $J = 3/2$ and $J = 5/2$ density of states but to a breakdown of the one electron picture [132-133]. For the $L_{2,3}$ absorption of the early 3d transition metals (Ca, Ti and V) a strong interplay of the atomic exchange, spin orbit and solid state band structure causes the observed preculiar structures which are intermediate between those expected for the free atom and those expected for the d band density of states.

Multiplet splitting of the white line of Mn impurities in metals due to localization of 3d orbitals has been observed [134]. Recently a theory which explains the rather anomalous line shape observed is emerged taking into account on an equal footing the atomic Coulomb and exchange interactions and the band structure [133]. Evidence of large deviation of the L_3/L_2 intensity going from the Cr metal to chromium in alloys has been correlated with the degree of spin pairing in the 3d band [131].

In metallic compounds formed of different atoms the XANES spectrum probes the local density of unoccupied states as has been observed in different materials [123]. In Fig. 24 the K-edge spectra of the Mn site and of the Ga or Cu site in the metallic perovskites Mn_3GaC and Mn_3CuN are shown [135]. The absorption K-edge spectra probe the states with p-symmetry in these two different sites. In the case of the Mn K-edge in the threshold region more spectral features have been resolved than in the Cu or Ga edges showing a complicated band structure at the magnetic Mn atom originating from hybridization of the sp band with the split d band. In the case of Cu or Ga atoms only one structure (A′) is observed, motivated by hybridization of a single d empty band with the large sp band at the metal site.

An example in which information on the unoccupied states have been deduced from XANES spectra is the system PdH_x, x about 0.6 [136]. The L_3 XANES spectrum of hydrogenated Pd is reported in Fig. 25. Comparing this spectrum with the Pd L_3 spectrum the presence of a new peak at 5.9 eV from the white line is induced by hydrogen. The peak at 5.9 eV can be correlated with the presence of a high density of the

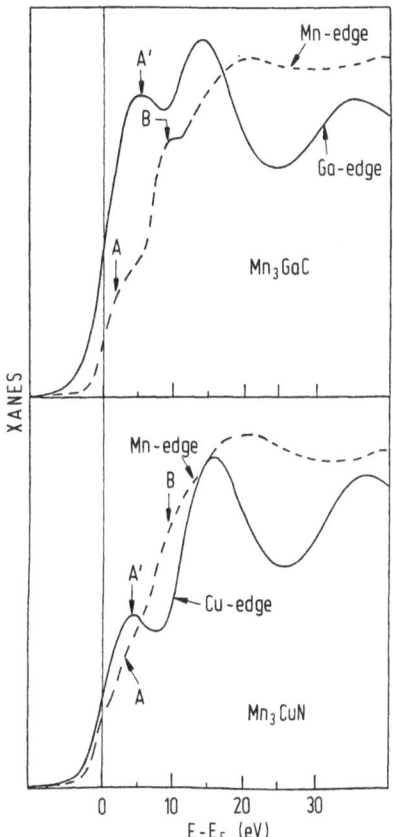

Fig. 24. K-edge absorption spectra probe the local density of states at selected atomic sites. The gallium K-edge and manganese K-edge of Mn_3GaC shown give the respective local densities of states. In the lower part the local density of states at the Mn and Cu sites of Mn_3CuN are measured by manganese and copper K-edges

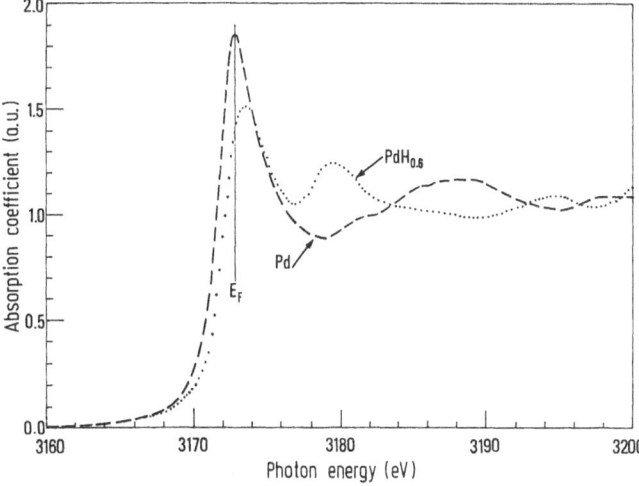

Fig. 25. Pd L_3-edge absorption spectrum (dotted line) of $PdH_{0.6}$ compared with the Pd metal (dashed line). The peak at 5.9 eV above the white line in the L_3 XANES spectrum of $PdH_{0.6}$ gives experimental evidence for the unoccupied states formed by hydrogen

unoccupied states at 5.9 eV above the Fermi level. This interpretation is in agreement with band structure calculations [137,138] which predict for the PdH system a peak in the toal DOS at 5.2 eV above E_f.

The shift of the white line toward high energy, observed in PdH_x, has been observed also going from Pd to Pd_2Si [139,140]. Recent calculations of the XANES spectra of Pd_2Si using a band structure approach [141] have shown that the L_3-edge spectra can be interpreted in terms of the DOS calculated from the ground state potential without any inclusion of many body final state effects.

Because of the small role of unoccupied s states the white line intensity is due to empty d states. Comparing the origin of the white line in Pd and in Pd_2Si the nature of the empty states involved in the transition is different. In Pd the empty states are the tail of the main d band while in Pd_2Si they are antibonding d states well above the Fermi level.

5 Conclusion

In this short review of recent advances in XANES we have discussed the use of this spectroscopy to probe the geometry of local structure via multiple scattering effects in the real space and and the local partial unoccupied density of states in metals via the band structure approach. Several important aspects of x-ray absorption spectroscopy have not be considered such as the excitonic effects, the many body infrared singularities in free electron metals and the many body final state configurations in solids with high electronic correlation such as the valence fluctuating metals and correlated oxides as CeO_2 and the new high T_c oxide superconsuctors. These effects require to go beyond the one-electron aproximation discussed here but give important information on the local electronic structure of materials.

6 References

1. Fano U, Cooper JW (1968) Rev. Mod. Phys. *40*: 441
2. Samson JAR (1982) Atomic Photoionization. In: Melhorn W (ed) Corpuscles and Radiation in Matter I Springer, Berlin, Handbuch der Physik vol. 31, p. 123
3. Starace AF (1982), Theory of Atomic Photoionization. In: Mehlhorn W (ed) Corpuscles and Radiation in Matter I Springer, Berlin, Handbuch der Physik, vol. 31, p. 1
4. Robin MB (1975) Chem. Phys. Lett. *31*: 140
5. Schwarz WHE (1974) Angew. Chem. Int. Edit. Engl *13*: 454
6. Koch EE, Sonntag BF (1982) In: Kunz (ed) Synchrotron Radiation Technique and Applications, Springer, Berlin p. 1
7. Bianconi A (1987) In: Prinz R, Konigsberger D (eds) X-ray Absorption: Principles and Techniques of EXAFS, SEXAFS and XANES, J. Wiley, New York
8. Dehmer JL; Dill D, Parr AC (1983) In: McGlynn S, Findley G, Hueber R (eds) Photo-physics and Photochemistry in the Vacuum Ultraviolet, Reidel Dordrecht, Holland
9. Kronig R de L (1920) Z. Physics *1*: 119
10. Parratt LG (1959) Rev. Mod. Phys. *31*: 616
11. Azaroff LV (ed) (1974) X-ray Spectroscopy, McGraw-Hill, New York
12. Kunz C (ed) (1979) Synchrotron Radiation, Springer, Berlin (Topics in Current Physics vol. *10*)
13. Winick H, Doniach S (eds) (1980) Synchrotron Radiation Research, Plenum Press, New York

14. Bianconi A, Incoccia L, Stipcich S (eds) (1983) EXAFS and Near Edge Structure, Springer, Berlin, Springer Series in Chemical Physics vol. *27*
15. Hodgson KO, Hedman B, Penner-Hahn JE (eds) (1984) EXAFS and Near Edge Structure III Springer Berlin (Springer Proc. Phys. vol. 2)
16. Bruhn R, Sonntag B, Wolff HW (1978) Phys. Lett. *69A*: 9
17. Benfatto M, Bianconi A, Davoli I, Incoccia L, Mobilio S, Stizza S (1983) Solid State Commun. *46*: 367
18. Kaindl G, Brewer WD, Kalkowski G, Holtzberg F (1983) Phys. Rev. Lett. *51*: 2056
19. Esteva JM, Karnatak RC, Fuggle JC, Sawatzky GA (1983) ibid. Lett. *50*: 910
20. Thole BT, van der Laan G, Fuggle JC, Sawatsky GA, Karnatak RC, Esteva JM (1985) Phys. Rev. B *32*: 5107
21. Powell CJ (1974) Surf. Sci. *29*: 44
22. Bauer RS, McMenamin JC, Bachrach RZ, Bianconi A, Johansson L, Petersen H (1979) Inst. Phys. Conf. Ser. *43*: 797
23. Norman D, Woodruff DP (1977) Solid State Commun. *22*: 711
24. Bianconi A, Doniach S, Lublin D (1978) Chem. Phys. Lett. *59*: 121
25. Bianconi A (1980) Appl. Surf. Sci. *6*: 77
26. Belli M, Scafati A, Bianconi A, Mobilio S, Palladino S, Reale A, Burattini E (1980) Solid State Commun. *35*: 355
27. Kutzler FW, Natoli CR, Misemer DK, Doniach S, Hodgson KO (1980) J. Chem. Phys. *73*: 3274
28. Durham PJ, Pendry JB, Hodges CH (1981) Solid State Commun. *38*: 159
29. Bianconi A, Dell'Ariccia M, Durham PJ, Pendry JB (1982) Phys. Rev. *B26*: 6502
30. Natoli CR (1983) In: Ref. 14, p. 43
31. Bianconi A (1984) In: Ref. 15, p. 167
32. Benfatto M, Natoli CR, Bianconi A, Garcia J, Marcelli A, Fanfoni M, Davoli I (1986) Phys. Rev. B *34*: 5774
33. Bianconi A, Garcia J, Marcelli A, Benfatto M, Natoli CR, Davoli I (1985) J. Phys. (Paris) *46*: C9-101
34. Sayers DE, Stern EA, Lytle FW (1971) Phys. Rev. Lett. *27*: 1204
35. Lee PA, Pendry JB (1975) Phys. Rev. *B11*: 2795
36. Prinz R, Konigsberger D (eds) (1986) X-ray Absorption: Principles and Techniques of EXAFS, SEXAFS and XANES, Wiley, New York
37. Barchewitz R, Cremonese-Visicato M, Onori G (1978) J. Phys. C. *11*: 4439
38. Dartyge E, Fontaine A, Tourillon G, Cortes R, Jucha A (1986) Phys. Lett. *113A*: 384
39. Heald SM, Keller E, Stern EA (1984) ibid. *103A*: 155
40. Bianconi A, Bachrach RZ (1979) Phys. Rev. Lett. *42*: 104
41. Phys. Today (1984) September 1984, p. 24
42. Stohr J, Noguera C, Kendelewicz T (1984) Phys. Rev. B *30*: 5571
43. De Crescenzi M (1983) In: Ref. 14, p. 43
44. Fink J, Muller-Heizerling Th, Scheerer B, Speier W, Hillebrecht FU, Fuggle JC Zaanen J, Sawatzky GA (1985) Phys. Rev. *B32*: 4899
45. Kohra K, Ando A, Matsushita T, Hashizume H (1978) Nucl. Instrum. Methods *152*: 161
46. Calas G, Petiau J (1983) Solid. State Commun. *48*: 625
47. Greaves GN, Diakun GP, Quinn PD, Hart M, Siddons DP (1983) Nuclear Instrum. Methods *208*: 335
48. Dartyge E, Depautex C, Dubuisson JM, Fontaine A, Jucha A, Leboucher, Tourillon G (1986) ibid. *A246*: 452
49. Tourillon G, Dartyge E, Fontaine A, Jucha A (1986) Phys. Rev. Lett. *58*: 737
50. Johnson KH (1966) J. Chem. Phys. *45*: 3085 and Advances Quantum Chem. (1973) *7*: 143
51. Dill D, Dehmer JL (1974) Chem. J. Phys. *61*: 692
52. Natoli CR, Misemer Dk, Doniach S, Kutzler FW (1980) Phys. Rev. *A22*: 104
53. Clementi E, Roetti C (1974) At. Data Nucl. Data Tables *14*: 1
54. Schwartz K (1972) Phys. Rev. *B 5*: 2466
55. Hedin L, Lundqvist BI (1971) J. Phys. *C4*: 2466

56. Natoli CR, Benfatto M, Doniach S (1986) Phys. Rev. A *34*: 4692
57. Breit G, Bethe HA (1954) Phys. Rev. *93*: 888
58. Schaich WL (1984) ibid. *B29*: 6513
59. Natoli CR, Benfatto M (1986) J. Phys. (Paris) *47*: C8–11
60. Best P (1966) J. Chem. Phys. *44*: 3248
61. Lytle FW (1967) Acta Crystallogr. *22*: 321
62. Garcia J, Benfatto M, Natoli CR, Bianconi A, Davoli I, Marcelli A (1986) Solid State Commun. *58*: 595
63. Rabe P, Tolkien G, Werner A (1979) J. Phys. *C12*: 1173
64. Teo BK (1981) J. Am. Chem. Soc. *103*: 3990
65. Teo BK (1983) In: Ref. 14, p. 1
66. Biebesheimer VA, Marques EC, Sandstrom DR, Lytle FW, Greegor RB (1984) J. Chem. Phys. *81*: 2599
67. Bianconi A, Dell'Ariccia M, Durham PJ, Pendry JB (1982) Phys. Rev. B *26*: 6502
68. Durham PJ, Pendry JP, Hodges CH (1982) Comput Phys. Commun. *25*: 193
69. Durham PJ (1987) In: Ref. 36
70. Kohn W, Sham LJ (1965) Phys. Rev. *140*: A1133
71. Kiriyama R, Kiriyama H, Wada T, Niizeki N, Hirabashi H (1964) J. Phys. Soc. Japan *19*: 540
72. Figgis BN, Gerloch M, Manson R (1969) Proc. R. Soc. London Ser. A *309*: 91
73. Davoli I, Marcelli A, Bianconi A, Tomellini M, Fanfoni M (1986) Phys. Rev. B *33*: 2979
74. Grunes LA, Lepman RD, Wilker CN, Hottmann R, Kunz AB (1982) ibid. B *25*: 7157
75. Colliex C, (1984) Adv. Opt. Electron Microsc. *9*: 65
76. Norman D, Stohr J, Jeager R, Durham PJ, Pendry JB (1983) Phys. Rev. Lett. *51*: 2052
77. Vvedensky DD, Pendry JB (1985) ibid. *54*: 2725
78. Knapp GS, Veal BW, Pan HK, Klippert T (1982) Solid. State Commun. *44*: 1343
79. Norman D, Garg KB, Durham PJ (1985) Solid State Commun *56*: 895
80. Bianconi A (1981) in Proc. of the meeting on EXAFS for Inorganic Systems, Daresbury, March 28–29, 1981, Daresbury Laboratory Report No. Serc DL/SCI/R17 1981 Gardner CD, Hasnain SS (eds) p. 13
81. Bianconi A, Dell'Ariccia M, Gargano A, Natoli CR (1983) In: Ref. 14, p. 57
82. Natoli CR (1984) In: Ref. 15, p. 38
83. Stohr J, Gland JL, Eberhardt W, Outka D, Madix RJ, Sette F, Koester RJ, Doebler W (1983) Phys. Rev. Lett. *51*: 2414
84. Sette F, Stohr J, Hitchcock AP (1984) Chem. Phys. Lett. *110*: 517
85. Hitchcock AP, Beaulieu S, Steel T, Stohr J, Sette F (1984) J. Chem. Phys. *80*: 3927
86. Bianconi A, Fritsch E, Calas G, Petiau J (1985) Phys. Rev. B *32*: 4292
87. Piancentini M (1983) In: Ref. 14, p. 193
88. Templeton DH, Templeton LK (1982) Acta Crystallogr. *A38*: 62
89. Penner-Hahn JE, Scott RA, Hodgson KO, Doniach S, Desjardins SR, Solomon EI (1982) Chem. Phys. Lett. *88*: 595
90. Smith TA, Penner-Hahn JE, Berding MA, Doniach S, Hodgson KO (1985) J. Am. Chem. Soc. *107*: 5945
91. Kosugi N, Yokoyama T. Asakura K, Kuroda H (1984) Chem. Phys. *91*: 249
92. Kutzler FW, Scott R.\. Berg JM, Hodgson KO, Doniach S, Cramer SP, Chang CH (1981) J. Am. Chem. Soc. *103*: 6083
93. Bianconi A, Congiu Castellano A, Durham PJ, Hasnain SS, Phillips S (1985) Nature *318*: 685
94. Scott RA, Penner-Hahn J, Doniach S, Freeman HC, Hodgson KO (1982) J. Am. Chem. Soc. *104*: 5364
95. Stizza S, Benfatto M, Garcia J, Bianconi A (1986) J. de Phys (Paris) *47*: C8–691
96. Stizza S, Benfatto M, Davoli I, Mancini G, Marcelli A, Bianconi A, Tomellini M, Garcia J (1985) J. de Physique *46*: C8–255
97. Penner-Hahn JE, Benfatto M, Hedman B, Takahashi T, Doniach S, Groves JT, Hodgson KO (1986) Inorg. Chem.
98. Lee PA (1976) Phys. Rev. *B13*: 5261
99. Heald SM, Stern EA (1978) ibid. *B17*: 4069

100. Stohr J (1986) In: Ref. 36
101. Stohr J, Jaeger R (1982) Phys. Rev. *B26*: 4111
102. Stohr J, Gland JL, Kollin EB, Koestner RJ, Johnson AL, Muetterties, Sette F (1984) Phys. Rev. Lett. *53*: 2161
103. Durham PJ (1984) In: Ref. 14, p. 37
104. Greaves GN, Durham PJ, Diakun G, Quinn P (1981) Nature *294*: 139
105. Kitamura M, Muramatsu S, Sugiura C (1986) Phys. Rev. B *33*: 5294
106. Grunes LA (1983) ibid. B *27*: 2111
107. Montano PA, Shenoy GK. Alp FF. Schulze W, Urban J (1986) Phys. Rev. Lett. *56*: 2076
108. Szmulowicz F, Segall B (1980) Phys. Rev. *B21*: 5628
109. Szmulowicz F, Pease DM (1978) ibid. B *17*: 3341
110. Gupta RP, Freeman AJ (1976) Phys. Lett. *59A*: 226
111. Gupta RP, Freeman AJ (1976) ibid. *59A*: 223
112. Gupta RP, Freeman AJ (1976) Phys. Rev. Lett. *36*: 1194
113. McCaffrey JW, Papaconstantopoulos DA (1974) Solid State Commun. *14*: 1055
114. Papaconstantopoulos DA (1973) Phys. Rev. Lett. *31*: 1050
115. Muller JE, Jepsen O, Andersen OK, Wilkins JW (1978) ibid. *40*: 720
116. Muller JE, Jepsen O, Wilkins JW (1982) Solid State Commun. *42*: 365
117. Muller JE, Wilkins JW (1984) Phys. Rev. *B29*: 4331
118. von Barth U, Grossmann G (1979) Solid State Commun. *32*: 645
119. von Barth U, Grossmann G (1982) Phys. Rev. *B25*: 5150
120. Dietz RE, McRae EG, Weaver JH (1980) ibid. *B21*: 2229
121. Davoli I, Stizza S, Bianconi A, Benfatto M, Furlani C, Sessa V (1983) Solid State Commun. *48*: 475
122. Sham TK (1985) Phys. Rev. B *31*: 1888
123. Balzarotti A, De Crescenzi M, Incoccia L (1982) ibid. *B25*: 6349
124. Motta N, De Crescenzi M, Balzarotti A (1983) ibid. B *27*: 4712
125. Wei PSP, Lytle FW (1979) ibid. *B19*: 679
126. Bianconi A, Modesti S, Campagna M, Fisher K, Stizza S (1981) J. Phys. *C14*: 4737
127. Wakoh S, Kubo Y (1978) Japan Journ. Appl. Phys. 17, Suppl. 17-2 193
128. Leapman RD, Grunes LA, Fejes PL (1982) Phys. Rev. *B26*: 614
129. Leapman RD, Grunes LA (1980) Phys. Rev. Lett. *45*: 397
130. Morrison TI, Brodsky MB, Zaluzek NJ, Sill LR Phys. Rev. B in press
131. Pease DM, Bader SD, Brodky MB, Budnick JI, Freeman AJ preprint
132. Wendin G (1984) Phys. Rev. Lett. *53*: 724
133. Zaanen J, Sawatzky GA, Fink J, Speier W, Fugle JC (1985) Phys. Rev. B *32*: 4906
134. Thole BT, Cowan RD, Sawatzky GA, Fink J, Fuggle JC (1986) Phys. Rev. B *31*: 6586
135. Garcia J, Bianconi A, Marcelli A, Davoli I, Bartolome J (1986) Nuovo Cimento D *7*: 493
136. Davoli I, Marcelli A, Fortunato G, D'Amico A, Coluzza C, Bianconi A to be published
137. Papaconstantopoulos DA, Klein BM, Faulkner JS, Boyer LL (1978) Phys. Rev. B *17*: 141
138. Gelatt CD Jr., Ehrenreich H, Weiss JA (1978) ibid. B *17*: 1940
139. Rossi G, Jaeger R, Stohr J, Kendelewicz T, Lindau I (1983) ibid. B *27*: 5154
140. De Crescenzi M, Colavita E, Del Pennino U, Sassaroli P, Valeri S, Rinaldi C, Sorba L, Nannarone S (1985) ibid. B *32*: 612
141. Bisi O, Jepsen O, Andersen OK (1987) ibid. B to be published

Characterization of Heterogeneous Catalysts: The EXAFS Tool

Dominique Bazin, Herve Dexpert and Piere Lagarde

L.U.R.E. Laboratoire CNRS, CEA, MEN Bat. 209D, Université de Paris-Sud
91405 ORSAY Cedex, FRANCE

Table of Contents

The first observations of an increase of the reactivity between gases in the presence of a metal were made, almost one hundred years ago, by chemists like Davy or Thenard. This pioneer work led Berzelius and Mitscherlich to define the concept of "decomposition of species by contact under a catalytic force". Since then much work has been done in order to understand the behavior of these small transition metal clusters which are able to promote some chemical reactions. In fact, catalysis has become a field of enormous economical interest. It spans a wide variety of areas from oil reforming to the preparation of synthetic fibers or fertilizers. Theoretical research as well as chemical engineering are therefore deeply involved and many current studies deal with the knowledge of such materials.

It is not always easy to characterize the electronic and crystallographic structures of very small aggregates. Their size (a few nanometers) is due to the fact that as many atoms as possible must be active and therefore must be at the surface. Moreover, the analysis has to be done in situ, under the true reaction conditions, in order to build a physical model for the role of the catalyst. Then, many experimental techniques have been used, including most recently electron microscopy and X-ray absorption. We focus our attention here on the EXAFS (Extended X-ray Absorption Fine Structure) technique and its possibilities for the study of supported metal catalysts. Most of the examples come from a collaboration between LURE and some public CNRS laboratories (Strasbourg, Meudon) and a private one (IFP — Rueil Malmaison). We begin with some generalities about the technique and the type of catalysts studied, then move to several examples of application.

Topics in Current Chemistry, Vol. 145
© Springer-Verlag, Berlin Heidelberg 1988

1 The EXAFS Technique in Catalysis

X-ray absorption is not really a new technique since the first absorption spectra were obtained by Fricke, Hertz or Lindh at the beginning of the century, but the EXAFS technique appeared as a definite tool for experimentalists in the 1970s when, at the same time, synchrotron radiation and a new physical model became available.

The basic model is the following: the absorption of an X-ray photon by an atom ejects a photoelectron, which is scattered by neighbours. An interference process builds up between the wave function of the outgoing electron and its scattered parts, leading to a modulation of the absorption coefficient. Far beyond the absorption edge, typically a few hundred eV, the electron is assumed to be free and thus described by a plane wave; the scattering is weak and involves only one neighbour atom (single scattering approximation).

Within these restrictions, the modulations of the absorption coefficient above the edge can be expressed by the following simple expression [1, 3]:

$$X(k) = -\frac{1}{k} \sum_j \frac{n_j}{R_j^2} \sin\left(2kR_j + \varphi(k)\right) A(k)\, e^{-2\sigma^2 k^2}\, e^{-R/\lambda}$$

where:

k is the wave vector of a free electron with an energy E_c. In our model, E_c is the kinetic energy of the photoelectron, i.e. the difference between the photon energy and the energy of the absorption edge;

A(k) and $\Phi(k)$ are the modulus and the phase of the scattering function of the electron when scattered by the neighbouring atom;

n is the number of these scattering atoms, located at a distance R_j from the atom which has absorbed the photon. These distances R can be affected by a spread σ;

λ is the mean free path of the photoelectron in the medium. It has a small value (a few angstroms), so the scattering process affects only the first few neighbour shells around the central atom.

Between X(k) and the radial distribution function around the central atom, there is just a simple Fourier transform relationship. Figure 1 gives the different steps of an EXAFS analysis.

Figure 1a is the raw experimental data Log (I_0/I) = f(E) with the edge step and the EXAFS modulations beyond.

In Fig. 1b, the EXAFS signal X(k) has been extracted by subtraction of the non-oscillatory part and by normalisation.

Figure 1c is the pseudo-radial distribution function obtained by Fourier transforming X(k). Exact positions of the peaks are shifted because of $\Phi(k)$; amplitudes are affected by the leading terms in the above equation.

Figure 1d is the inverse Fourier transform of Fig. 1c after a filtering of the signal coming from one give shell of atoms. This function is then simulated using the theoretical expression.

We have therefore shown that EXAFS is a structural probe, sensitive only to the order around one given type of atom in the medium (well defined by its X-ray absorption edge), and sensitive only to the local order around this atom (because of the mean free path term λ). Like other structural techniques, it allows a determination of the

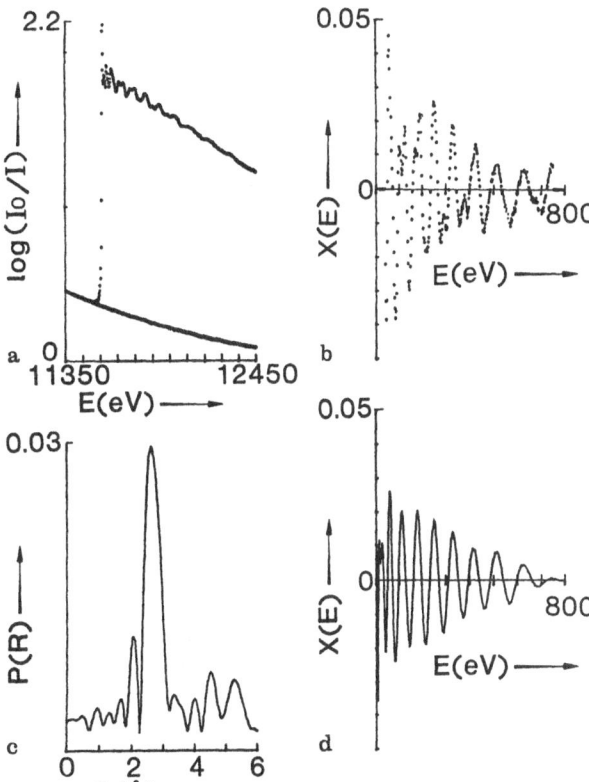

Fig. 1. The different steps of an EXAFS analysis

kind, number, and distances of the neighbours of one given type of atom, but without any need for long-range order as in X-ray or electron diffraction. Moreover, since the basic process in a photon absorption, an EXAFS experiment can be done in situ, without any special sample preparation as in electron microscopy.

All these features explain why the applications of EXAFS to the study of very dispersed materials have increased so much in the last few years. Figure 2 shows the tremendous growth of the number of papers on catalysits characterization by EXAFS coupled to the use of synchrotron light. This figure comes from the very detailed and exhaustive review Bart and Vlaic recently published on the subject [4].

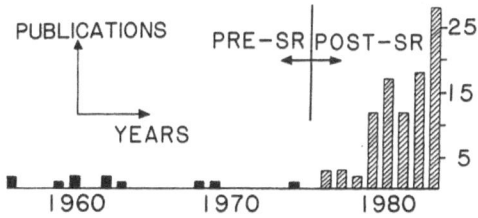

Fig. 2. Relation between EXAFS studies of catalysts with the use of synchrotron radiation (from Ref. (4))

71

Dominique Bazin, Herve Dexpert and Piere Lagarde

At this point, we must recall what a supported heterogeneous catalyst is. It consists of a porous oxide like alumina or silica, with a very large specific area (200 m²/g typically), at the surface of which are dispersed the clusters of the active metal. In oil reforming, the metals used are essentially transition metals of group VIII, the main one being platinum, with a loading of the order of 1 wt.-%.

Three main steps characterize the life of a catalyst: the preparation, the steady state and aging, and finally the regeneration. Up to now, it is mainly the first and the last operations which have been studied by EXAFS. The first one is divided into three steps:

a) Impregnation of the active element, i.e. fixation of isolated complex ions at the open surface of the oxide platelets.

b) Drying and calcination. The goal is to eliminate the water as well as many different anions coming from the precursor (Cl^-, NO_2^{2-}, etc.)

c) Reduction under hydrogen, which ends up with small metal clusters at the oxide surface.

These three steps have been carefully studied now [5], in particular for multimetallic catalysts which are now the most widely used (see for example [6-8]). The following examples will show that EXAFS experiments can help to understand the structure and

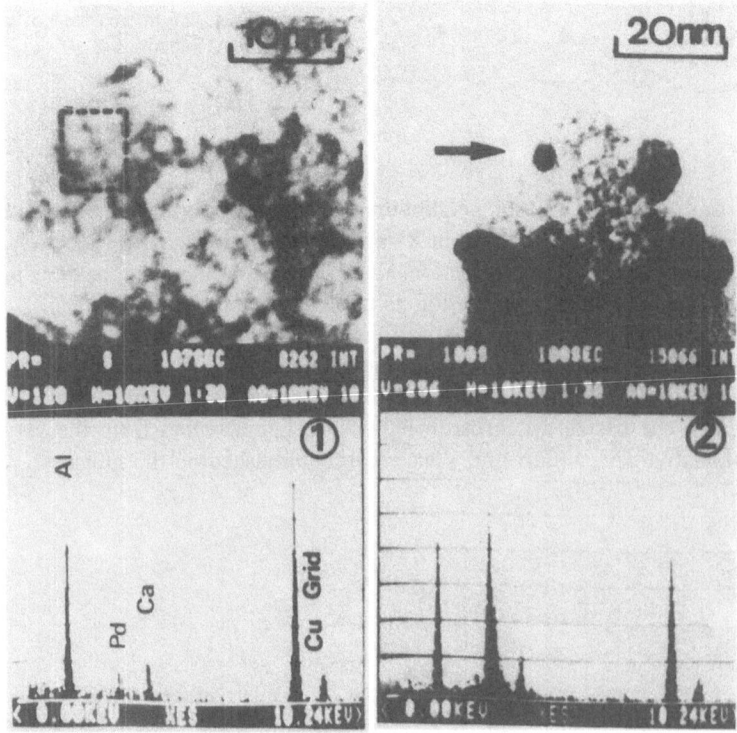

Fig. 3. STEM micrographs and EDX analysis of a dried Pd/Al_2O_3 catalyst

behaviour of these catalysts. Although the interesting element (active metal, e.g. Pt, Pd, Re . . .) is present at a very low concentration (less than 1 %), the contrast between the atomic numbers of the heavy metal and of the oxide support is large enough to allow absorption experiments with a good signal-to-noise ratio in situ within the reacting cell.

Figure 3 shows two scanning electron microscopy (STEM) pictures of a 0.3% of palladium ex-acetylacetonate (ex-acac) on alumina. On the left, the dashed rectangular zone has been analyzed by X-ray emission, without any evidence (lower part) of a palladium signal: the copper signal detected is due to the sample holder. After about 100 s the same region is analyzed again; the impinging electron beam has heated the sample enough to produce the clustering of a palladium-containing particle, indicated by the arrow in the upper right part and quantitatively analyzed in the lower one: a microdiffraction study of this particle confirms that palladium oxide is formed [9]. This example shows the strengths but also the limitations of STEM analysis for the study of very reactive systems like the catalysts. While accurate information can be extracted from a very localized area, any reactive experiments are practically impossible due to the ultra high vacuum constraints electron microscopy imposes for its use.

Figure 4 shows then the role EXAFS can play in the in situ description of this calcination process [10]. It is composed of various pseudo-radial distribution functions, uncorrected from phase shifts, obtained for various loadings, various temperatures and different precursors. First of all, two successive steps are clearly evidenced:

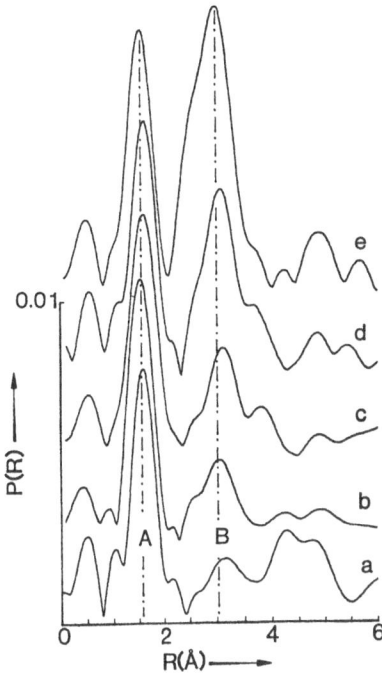

Fig. 4. Radial distribution functions of various Pd/Al$_2$O$_3$ catalysts

a) Drying at about 100 °C. The most significant result is obtained with ex-acac catalysts, for which different processes had previously been imagined. The impregnation breaks the palladium acetylacetonate into two parts with formation of acetylacetone. EXAFS analysis leads to an increase of the coordination number of palladium to oxygen, rising from 4 oxygens at 1.96 Å in the acetylacetonate solution to 5 at 1.98 Å in the dried sample. The only structural model is the insertion of the metal into octahedral aluminium vacant sites at the alumina surface.

b) Calcination. This step further decomposes the precursor, which loses the second acetylacetonate molecule. We point out here:

(i) The growth of the oxide phase, as seen by the increase of the second Pd—Pd shell of the PdO oxide: the peak denoted B grows continuously from curve 4b to curve 4d, which show the same 0.7% Pd catalyst calcined respectively at 300°, 500° and 700 °C. This is confirmed by a quantitative analysis of the first shell, which gives between 4 and 4.5 atoms at 2.00 Å, i.e. a tendency to a situation close to the oxide one (4 neighbours at 2.02 Å).

(ii) At the same calcination temperature, the catalyst ex-nitrate (4e) looks like the oxide while the catalyst ex-acac (4c) has only 2/3 of the overall Pd—O amplitude. In this last case, a very dispersed oxide phase is present.

(iii) The higher the metal loading, the more complete is the oxide formation. For instance, curves 4a and 4b are both ex-acac catalysts, but curve 4a relates to a 0.3% while curve 4b relates to a 0.7 wt.-% palladium sample.

We therefore see how EXAFS can precisely describe the build-up of a chemical phase. Seven years ago, Greegor and Lytle showed that, in some cases, the morphology of such clusters can even be determined [11]. In our samples, looking at the reduced stage of an ex-H_2PtCl_6 catalyst on different aluminas (gamma and eta), we have seen that the carrier can influence the final structure of the clusters [6]. In the case of gamma alumina, the hydrogen reduction builds tridimensional cuboctahedra,

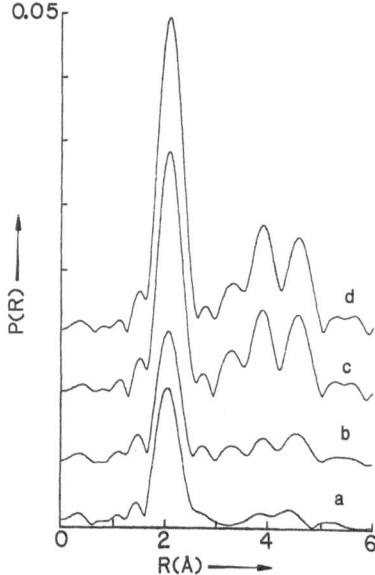

Fig. 5. Radial distribution functions (uncorrected from phase shifts) at the nickel edge of $CeNi_5$ for different $CO + H_2$ reduction times

while the lack of a third shell in the radial distribution function of clusters deposited on eta alumina is thought to lead to the formation of (110) platelets.

The quantitative aspect of the EXAFS technique is also well known and the literature gives several studies where chemisorption and EXAFS measurements are compared (see for example [12]). We can illustrate this particular contribution of the spectroscopy by a study of rare earth transition metal catalysts prepared from intermetallic $LaNi_5$-type compounds. The three classical preparation steps are here skipped with a carbon monoxide hydrogenation reaction. The intermetallic phase is transformed into a rare earth oxide upon which the transition metal is left as metallic clusters which form the active species. This transformation has been followed as a function of the time reaction [13]. In Fig. 5 we plot the Fourier transforms of $CeNi_5$ at the nickel edge before the reaction (a), after 10 hours (b) and after 27 hours (c) under the $CO + H_2$ mixture. These are all compared to elemental nickel (d). The increase of the amplitude of the first peak and the growth of three new ones at greater distances are the consequence of the formation of nickel particles. A careful analysis of these four shells has allowed us quantitatively to estimate the fraction of extracted nickel during the reaction as 30% after 10 hours and 80% after 27 hours on a $CO + H_2$ flux at 350 °C.

2 The Case of Bimetallic Aggregates: Some Data on the Intermetallic Competition

Bi- or multimetallic catalysts are of growing importance for the oil industry. We give here two cases where we have been able to build models for the role played by the metals versus the carrier or the reactant. These results rest on a comparison between a monometallic ex-H_2PtCl_6 catalyst and two bimetallic Pt-Re and Pt-Rh systems. At the dried stage we have shown the existence of two fixation sites for the metal while the study of the reduced catalyst concludes an inhomogeneous repartition of the two metals inside the aggregate.

2.1 Metal to Site Affinity

In the case of the monometallic samples, the drying basically keeps the integrity of the complex $PtCl_6^{2-}$ ion which sticks onto the support (some oxygen atoms may substitute the chlorine), and the calcination process in air progressively induces the formation of PtO_2 clusters with less and less chlorine around [5]. However, we found for a bimetallic Pt—Rh with two chlorinated precursors that the metal environment looks, at the dried stage, like the platinum one at the calcined step in the monometallic case. Therefore, the second metal brings about a situation in which the chlorine is removed from the close environment of the platinum at a temperature several hundred degrees lower. The platinum is then enchored to the support by metal oxygen bonds [14]. As a function of different Pt, Re, Rh and Cl loadings, the process can be described as if the support were offering two types of fixation sites:

a) Alumina platelets enriched with so-called strong sites, where the Al—O bond is strong. These sites fix the $PtCl_6^{2-}$ ion electrostatically without affecting it. These sites are a minority.

b) Weak sites where the Al—O bond is less stable, e.g. in the vicinity of a kink, a hole, or an impurity.

At low metal loading, it is the strong sites which intervene: the close environment of the platinum in either the mono- or the bimetallic systems, stays mainly chlorinated. When the metal loading increases, weak sites are used and one sees a change of the platinum chlorine environment to a platinum oxygen one. This is clear in Fig. 6 where the number of first chlorine (crosses) and oxygen (circles) neighbours around one platinum atom has been plotted as a function of the NRh/NPt atomic ratio. We see the chlorine being replaced by oxygen when the amount of rhodium increases since rhodium uses the strong sites first. One understands then why the addition of a second metal favours the platinum dispersion onto the support and, consequently, its sintering properties.

2.2 Repartition of Two Metals in the Cluster

The reduction step is a fundamental one in the catalyst preparation, and the main problem is to define how the catalytic function is related to the chemical relationship between the two metals, so that investigations on the relationship between the two metals inside the cluster are of great importance.

Let us consider here the Pt—Re case. Fig. 7 gives the radial distribution functions around the platinum and the rhenium atoms at two different reduction temperatures. One sees, first, that these treatments do not induce a complete reduction of the rhenium, since the corresponding Fourier transform does not look like the metal one. This is corroborated by examination of the amplitude of the white line at the Re L_3 edge. The widths of these Fourier transforms come from the limited energy domains available between the Pt L_3 and Re L_2 edges. The rhenium atom is still partially oxidized, while a quantitative analysis shows that the amount of oxygen around platinum is less than in the corresponding monometallic case. These results suggest that rhenium is partially involved in an oxidized phase linked to the carrier; this phase acts as a support for the platinum atoms. Therefore, the stability of the rhenium oxidized phase

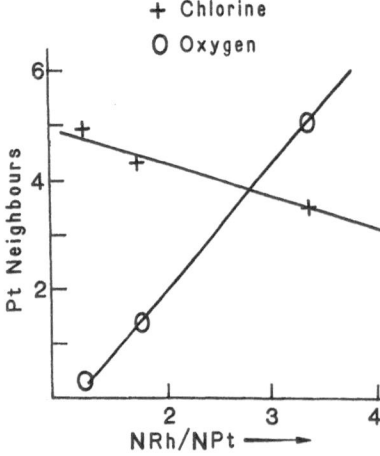

Fig. 6. Change of the platinum environment as a function of the NRh/NPt ratio for different Pt—Rh on aluminium catalysts

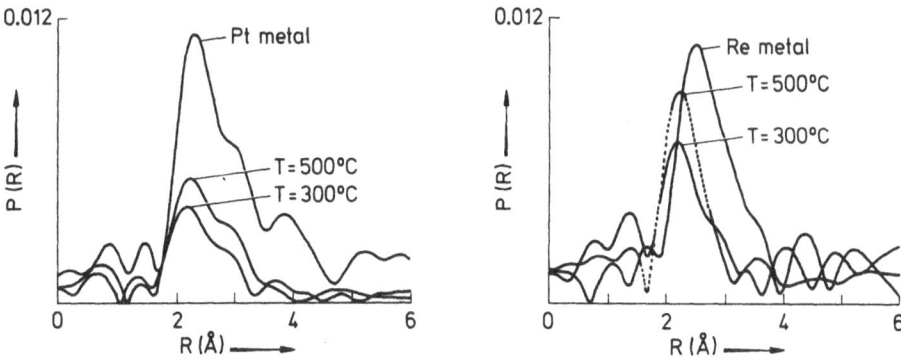

Fig. 7. Pt—Re on alumina catalysts at the reduced step compared to the metal. Radial distribution functions (uncorrected from phase shifts) at the L_3 edges of Pt and Re for two reduction temperatures

induces that of the platinum clusters at the top, and the increased stability to the sintering of the bimetallic catalyst compared to the monometallic one is thus explained [15].

3 Kinetic Studies

Because of the strong dependence of the intensity of the white line at the L_3 edge of metals like platinum or rhenium upon the chemical situation of this element, it is worth comparing the electronic state of the metal (edge region) with its structural situation (EXAFS) during a reaction process. This has to be done under

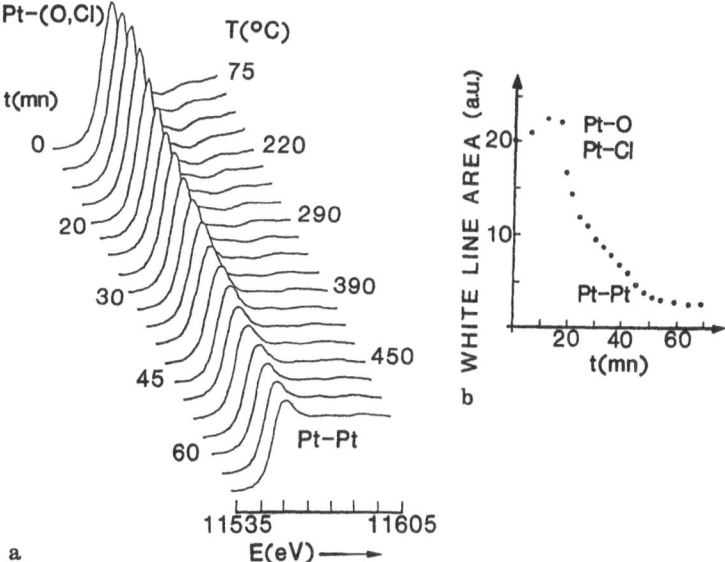

Fig. 8. Direct reduction from the dried step of a monometallic 1.5% by weight Pt on alumina catalyst. Edge absorption change with time and temperature

real physico-chemical conditions, and we have used at the same time either the conventional step-by-step method or the new EXAFS collection in dispersive mode.

3.1 Direct Reduction Studies

This new technique has been developed at Lure by A. Fontaine and E. Dartyge [16]. Figure 8a shows the evolution of the Pt L_3 edge of a dried catalyst which has been reduced, the rate of increase of temperature under a hydrogen flux being 6°/min: the decrease of the amplitude of the white line is complete after one hour. Figure 8b shows the variation of the area of this white line, normalised to that of the elemental metal. Two points are worth mentioning here:

a) The white line area increases during the first 20 min: this corresponds to the removal of chlorine by hydrogen at low temperature, and the bonding of the metal to the oxygen from the support.

b) The reduction of this last species occurs with two speeds: this may either originate from two different chemical species, or it may correspond to two different sizes of particles.

3.2 Dispersive EXAFS Studies

Schematically, the studies proceed as follows: the white synchroton radiation beam is at the same time focused and dispersed in energy by a curved crystal. The sample is located at the focus point, and the detection is done by a positon-sensitive detector, such as a photodiode array, that is able to withstand a very high photon flux. Therefore, the relationship Bragg angle–energy of the diffracted beams is translated into a correlation energy–position in the detector. In the best cases, like an elemental copper foil, only a few milliseconds are needed to acquire an EXAFS spectrum.

We recently went through a set of experiments to follow the activation, poisoning and regeneration of platinum [17] and iridium-copper catalysts [18]. In the latter case catalytic tests were done at the same time using ^{13}C isotopic tracing in order to understand the correlation between reaction mechanisms and electronic properties. These experiments were undertaken either on 2% and 10% Pt on alumina, or on 5% Ir and (Ir 0.2, Cu 0.8) on alumina with a 5% total loading. Hydrocracking, hydrolysis, isomerisation, aromatisation and hydrogenation reactions of cyclopentane, methyl-cyclopentane, hexane, heptane and benzene have been followed at a constant hydro-carbon pressure. Reaction products were analyzed by gas phase chromatography, and isotopic ^{13}C products determined by mass spectroscopy [13]. Some preliminary results are reported in Fig. 9, where the direct reduction of a monometallic 2% ex-H_2 $PtCl_6$ is given. With a first experiment, we have determined the temperature of the starting point of the reduction process, i.e. 210 °C for monometallic, or 85 °C for the Pt—Rh bimetallic system. Then, in a second run, keeping this temperature constant, we collected 5 s shots every 30 s without the hydrogen flux and then with the hydrogen flux. The white line strongly decreases from a Pt^{4+} state to the Pt^0 state, and we have compared this evolution with the structural result coming from the corresponding EXAFS analysis. The characteristic peak of the Pt—Cl bonds stays

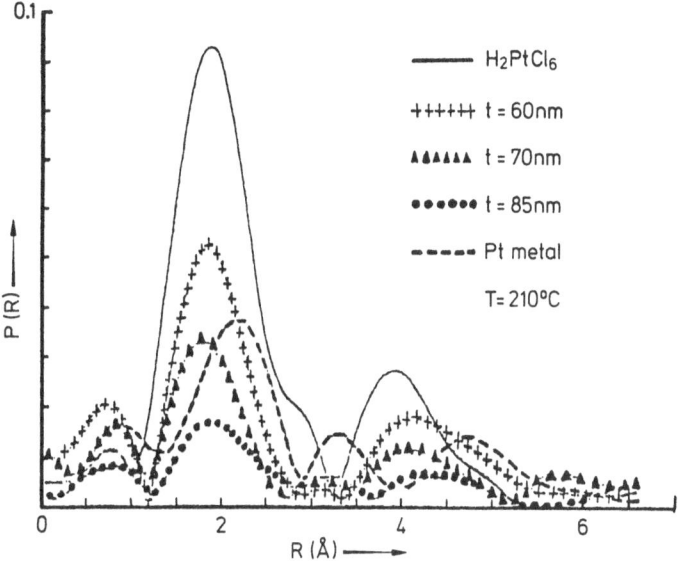

Fig. 9. Kinetic study of the direct reduction of a 2 % Pt on alumina catalyst. Only three steps of the hydrogen consumption are reported. The platinum-platinum bonds begin to occur on the right side of the Pt—Cl amplitude only after 85 min of hydrogen flow

constant in the absence of hydrogen, and suddenly decreases when hydrogen is admitted on the sample. Nevertheless, it has been observed that the white line area starts to decrease before hydrogen gets into the catalyst, indicating that EXAFS and edge changes do not have the same kinetics. At the end of the experiment, the white line is characteristic of metallic platinum while the Fourier transform still has some features of the Pt—Cl bonding. In the Pt—Rh case, we observed conversely a more synchronous evolution of both the white line and the Fourier transform.

4 Conclusion

The few examples discussed here, added to the large number anyone can find in the literature, are definite proof of the usefulness X-ray spectroscopy has today in the field of catalyst characterization. We have restricted our illustrations to the EXAFS domain but it becomes clearer every day that the edge region is also of great interest: XANES (X-ray Absorption Near Edge Structure) provides information on the electronic states taken by the active species during the reaction. As we very briefly reported, the number of empty d states can be followed accurately and relations with the different chemical pathways of the reaction may be established (see for example [19]). The whole will undoubtedly contribute to the development of time-resolved studies done under real reaction conditions, so that kinetic measurements will be one of the major uses of the technique in the near future.

5 References

1. Ashley CA, Doniach S (1975) Phys. Rev. B *11*: 1279
2. Lee PA, Pendry JB (1975) ibid. *11*: 2795
3. Lytle FW, Sayers DE, Stern E (1975) ibid. *11*: 4836
4. Bart JCJ, Vlaic G (1985) Adv. Catalysis 34
5. Lagarde P, Murata T, Vlaic G, Freund E, Dexpert H, Bournonville JP (1983) J. Catalysis *84*: 333
6. Lagarde P, Dexpert H (1984) Adv. Physics *33(6)*: 567
7. Sinfelt J (1985) Pour la Science 68
8. Prins R, Koningsberger DC (1986) X-ray Absorption: Principles, Applications, Techniques of EXAFS, SEXAFS and XANES, Wiley, New. York
9. Dexpert H, Freund E, Lynch J (1985) Quant. Mic. Met. Soc. 101
10. Lesage-Rosenberg E, Vlaic G, Dexpert H, Lagarde P, Freund E (1986) Appl. Catalysis *22*: 211
11. Greegor RB, Lytle FW (1980) J. Chem. Phys. *71*: 690
12. Mansour AN (1983) Thesis North Carolina State Univ.
13. Paul Boncour V, Percheron Guegan A, Achard JC, Barrault J, Guilleminot A, Dexpert H, Lagarde P (1984) EXAFS and Near Edge Structure III; 199 Springer Berlin
14. Bazin D, Dexpert H, Lagarde P, Bournonville JP (accepted 1986) J. Catalysis
15. Bazin D (1985) Thesis, Orsay Univ.
16. Dartyge E, Fontaine A, Jucha A, Sayers D (1984) EXAFS and Near Edge Structure III, 472 Springer Berlin
17. Sayers DE, Bazin D, Dexpert H, Jucha A, Dartyge E, Fontaine A, Lagarde P (1984) EXAFS and Near Edge Structure III, 209 Springer Berlin
18. Maire G, Garin F, Bernhardt P, Girard P, Schmitt JL, Dartyge E, Dexpert H, Fontaine A, Jucha A, Lagarde P (1986) Appl. Catalysis *26*: 305
19. Lytle FW, Greegor RB, Marques EC, Sandstrom DR, Via GH, Sinfelt JH (1985) J. Catalysis *95*: 546

X-Ray Absorption Studies of Liquids:
Structure and Reactivity of Metal Complexes in Solution and X-Ray Photoconductivity of Hydrocarbon Solutions of Organometallics*

Tsun-Kong Sham**

Chemistry Department, Brookhaven National Laboratory Upton, New York 11973, USA

Table of Contents

Applications of X-ray absorption spectroscopy in two areas of liquid state studies using synchrotron radiation are reviewed. One area concerns with the determination of the local structure, bonding and dynamics at the metal site of transition metal complexes using Extended X-Ray Absorption Fine Structures (EXAFS) and X-Ray Absorption Near Edge Structures (XANES). Emphasis are placed on the implications of the derived parameters to the reactivity of the complexes in exchange reactions (ligand and electron) in solution and to the subtle electronic and structural change of the complexe upon dissolution. The other area deals with the principles and practices of X-ray induced photo-conductivity of hydrocarbon solutions. The conductivity measurments of EXAFS of organo-metallics in hydrocarbon solution is described. The special feature of the technique. EXAFS measure-ment under the condition of total X-ray absorption, is discussed in some details. Current developments is related areas are also noted.

* Research Supported by U.S. DOE at Brookhaven National Laboratory under Contract DE-AC02-76CH00016.

** Present address: Canadian Synchrotron Radiation Facility Synchrotron Radiation Center University of Wisconsin Stoughton, Wisconsin, 53589 U.S.A.

1 Introduction

Recent advancement in synchrotron radiation studies [1, 2] has opened up many new avenues for research in Chemistry and Biology as evident from the various contributions in this two volumes. The recognition of the usefulness [3] of X-ray absorption spectroscopy (XAS) with synchrotron radiation is one of the important elements that lead to this development. It is timely to discuss here some unique applications of XAS in liquid state studies.

X-Ray absorption spectroscopy (XAS) concerns the measurement and interpretation of the X-ray absorption coefficient of elements in various chemical situations. The absorption coefficient often exhibits extended X-ray absorption fine structures (EXAFS) [4−6]. For many pratical and theoretical reasons [3], it has been a common practice to divide an XAS spectrum into two regions. The first 50 ∼ 80 eV above the edge is the near edge region or often referred to as X-Ray Absorption Near Edge Structure (XANES) while the oscillations beyond the XANES are the EXAFS. EXAFS arises from the backscattering of the outgoing photoelectrons and can be adequately described by a single-particle, single scattering short-range order theory [6−9]. If we assume harmonic motion, the EXAFS function $\chi(k)$ for K edge absorption of an atom in a spherically average (random) sample is given by

$$\chi(k) = A \cdot \sin \Phi = -\sum_i \frac{N_i}{kr_i^2} |f(k, \pi)| \, e^{-2\sigma_i^2 k^2} \, e^{-2r_i/\lambda_i(k)} \cdot \sin(2kr_i + \varphi(k))$$

(1)

In Eq. (1), k is the photoelectron wave vector relative to E_0 (k = 0) N_i is the the number of neighboring atoms of the same kind at a distance r_i, σ_i^2 is the mean-square relative displacement (MSRD) of the absorber-scatterer atom pair from their equilibrium inter-atomic distance or in molecular spectroscopy terminology, the mean-square amplitude of vibration [9], other terms have their usual meaning [10]. Using standard Fourier transform and curve fitting procedures, we can derive the coordination number, bond length and local dynamics (MSRD) from EXAFS.

XANES study deals with the absorption coefficient in the vicinity of the absorption edge. These structures can be generally regarded as measurements of the distribution of the density of unoccupied states. Although many XANES discussions [3] usually do not distinguish between states below the vacuum level and those just above the vacuum level, we find it convenient to treat the energy position of the edge jump, the width and intensity of the peaks (some times called whitelines) just above the edge as chemically sensitive parameters. Information such as the effective charge at the metal site and the nature of the density of states (p, d characters for example) can be derived from the analysis of these parameters [11, 12]. XANES features above the vacuum level have a more complex nature and can be regarded as resonance states (virtual molecular orbital) [3] or in the view of electron scattering theory, multiple scattering states (MS) [3, 13, 14]. The entire absorption spectrum in fact can be alternatively viewed as the result of photoelectron scattering processes of which MS is enhenced at low photoelectron kinetic energy (XANES region), single scattering (standard EXAFS formalism) predominates at higher energies (EXAFS region) and there is a region in between. This MS teatment of the XANES region has been dealt with by Bianconi et al. [13, 14]. We only concentrate on the absorption features within 20 eV above the edge jump.

Two specific applications are discussed here. Both concern XAS studies of liquids. In one application, emphasis is placed on the determination of the structure and dynamic behavior of transition metal complexes in aqueous solution and their implication to the reactivity of metal complexes. The other application aims at the development of a new technique, photoconductivity measurements of XAS in hydrocarbon solutions.

This article is arranged as follows: In Sect. 2, the chemical implication of EXAFS parameters of metal complexes in solution is discussed, this includes applications to ligand exchange and electron exchange reactions, correlation of XANES with bonding, and secondary interactions in solids vs in solution. In Sect. 3, the photoconductivity technique for X-ray absorption measurments is presented. Specific examples involving organometallic solutions are discussed. Finally, summary and recent developments in related areas are given in Sect. 4.

2 Structure, Bonding and Reactivity of Metal Complexes in Solution from XAS Studies

2.1 General Considerations

The application of synchrotron XAS to solve structural problems in solution was demonstrated by several groups during the early development of synchrotron spectrometry [15-18]. In this section, the application of XAS to the studies of a number of metal complexes in solution is described in connection with the structural and electronic aspects of chemical reactions such as substitution reactions [19] and electron exchange reactions [20-22]. The fundamental understanding of the reaction rate is greatly facilitated by the quantitative information about the inner-shell configurations of the metal complexes derived from EXAFS. This information, which is usually obtained from crystal structure of X-ray [23] and neutron [24] diffraction techniques, is not readily available for most complexes in solution. In addition, EXAFS experiment takes shorter time (typically, several minutes) to perform than what is required by the diffraction techniques and has a better sensitivity at dilute concentrations.

XAS spectra of solution speciemens are usually recorded in a transmission mode in which the incoming (I_0) and the transmitted (I) photon flux are monitored with gas ionization chambers and the absorption is expressed as

$$\mu t(h\nu) = \ln (I_0/I) \tag{2}$$

For dilute solutions (<0.01 M), fluorescence detection can be routinely used [25]. For liquid samples, we find it very convenient to use sealable Mylar bags (made of mixture of Mylar and Polyethylene) as sample containers. The speciemen is sandwiched between two plastic holders so that the thickness and orientation of the cell can be easily adjusted for both transmission and fluorescence experiments.

Standard Fourier Transform (FT) and Curve Fitting (CF) techniques are used concurrently in the EXAFS analysis with either theoretical or model-compound phase shift and amplitude. The procedure is well established [10] and is not discussed here any further. We instead describe a special procedure used in the derivation of the bond length difference (Δr) between two closely related systems. This Δr is a particularly desired parameter in the studies of the role of inner-shell configuration changes in the electron exchange reactions of metal complexes $ML_6^{n+}/ML_6^{(n+1)+}$ (L = ligand) in solution. Δr can be accurately determined from the phase functions ($\Phi(k) = (2kr + \varphi(k))$, Eq. 1) of the M—L pairs derived from FT. It happens that $\Phi(k)$ vs. k is nearly a straight line for low Z (atomic number) scatterers. This is because 2kr is much greater than $\varphi(k)$. This situation holds for all the transition metal complexes with low Z nearest neighbors. Since $\varphi(k)$ is linear in k ($\varphi(k) = \alpha_1 + \alpha_2 k$, α_1, and α_2 being constant), one empirical rule can be derived for the separation of the maxima in $\chi(k)$.

$$\Delta k_{max} = \frac{\pi}{(2r + \alpha_2)} \tag{3}$$

Based on Eq. (3) where k_{max} is the position of the $\chi(k)$ oscillation maxima, it can be generally stated that for identical absorber-scatterer pairs, the longer the bond, the closer the EXAFS oscillations in k space (hence energy space) and vice versa. Examples for the use of this rule are illustrated in Fig. 1 for absorber-scatterer pairs with typical (2.11–2.14 Å in $Ru(NH_3)_6^{3+/2+}$) and extremely short (1.62–1.66 Å) in $MnO_4^{2-/1-}$) bonds. With good data, this technique is sensitive to $\Delta r = 0.01$ Å.

Similarly, comparison of first shell coordination numbers and mean-square displacements along the metal-ligand bond can be obtained by comparing the amplitude functions A_1 and A_2 of the two systems (Eq. (1))

$$\ln \left(\frac{A_1}{A_2} \right) = \ln \left[\frac{N_1 f_1(\pi, k)}{N_2 f_2(\pi, k)} \cdot \frac{r_2^2}{r_1^2} \right] + 2(\sigma_2^2 - \sigma_1^2) k^2 \tag{4}$$

By plotting $\ln (A_1/A_2)$ vs. k^2, we should get a straight line with a slope $(\sigma_2^2 - \sigma_1^2)$ and an intercept, $\ln (N_1 r_2^2/N_2 r_1^2)$. Such a comparison for the $Fe(H_2O)_6^{2+/2+}$ system, both in the solid ($Fe(NO_3)_3 \cdot 9\, H_2O$ and $Fe(NH_4)_2(SO_4)_2 \cdot 6\, H_2O$ where N = 6) and in solution indicates that within 5% accuracy these complexes remain six coordinate in solution. This comparision, however has a greater uncertainty than that in the determination of Δr due to uncertainty in the experimental amplitude function [10].

2.2 EXAFS Parameters of Metal Ions in Solution and Water Substitution Reactions

The chemical implications of the EXAFS bond length and the mean-square relative displacement of the absorber-scatterer pair are best illustrated with a system of metal ions in aqueous solution and its correlation with the rate of water substitution reaction [19]

$$M(H_2O)_6^{n+} + H_2O^* \xrightarrow{k_1} M(H_2O)_5(H_2O^*)^{n+} + H_2O \tag{5}$$

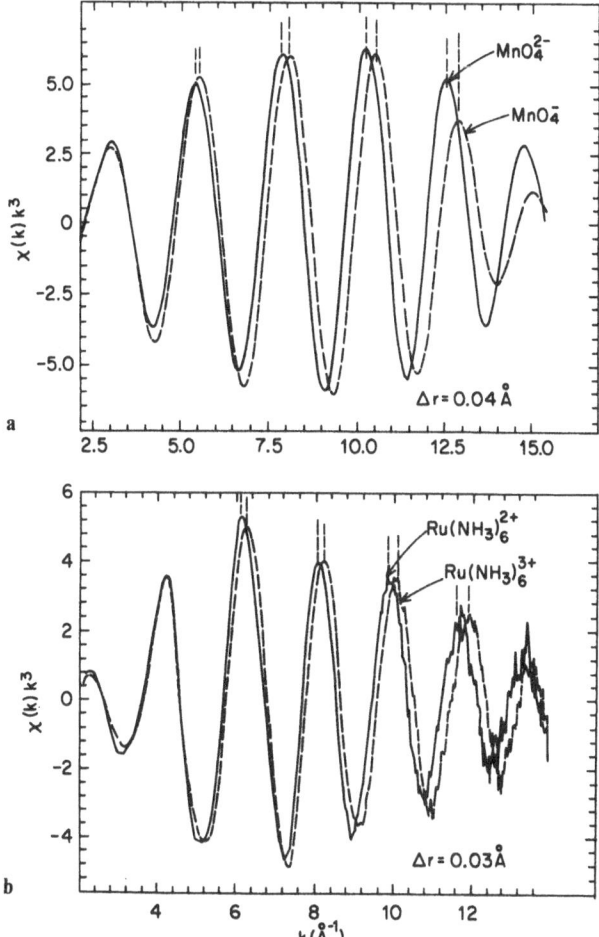

Fig. 1. Comparison of EXAFS oscillations for identical absorber-scatterer pairs having a) extremely short bonds $(MnO_4^{-/2-}$, 1.62–1.66 Å Ref. 21) and b) typical $(Ru(NH_3)_6^{3+/2+}$, 2.11–2.14 Å, Ref. 22) bond lengths. It can be seen that the progressive mismatch in k_{max} (Eq. 3) is quite sensitive to bond length difference Δr

where H_2O^* denotes the water molecule from the bulk and k_1 is the exchange rate constant. This reaction is known to proceed via a S_N1-Type mechanism for most octahedral aquo transition metal ions. The difference in the rate constants among the aquo ions is largely dependent upon their dissociation activation energy E_a which is closely related to the strength of the $M—H_2O$ bond and hence the bond length and the root-mean-square relative displacement (RMSRD), σ_i.

EXAFS metal-ligand bond length and RMSRD for several $M(H_2O)_6^{n+}$ (n = 2, 3) ions are given in Table 1. There are two classes of ions: the octahedral (site symmetry O_h) and the tetragonally-distorted (Jahn-Teller, site symmetry D_{4h}) $Cr(H_2O)_6^{2+}$ and $Cu(H_2O)_6^{2+}$ complexes. From Table 1, two interesting features are noted. First, among O_h complexes, $Mn(H_2O)_6^{2+}$ (d^5, high spin) has the longest $M—H_2O$ bond indicating a small E_a for bond dissociation (S_N1 mechanism). Second, the M^{2+} ions have longer bonds and larger RMSRD than those of the M^{3+} ions. σ_i^2 in solution can be expressed [20, 21] as

$$\sigma_i^2 = \sigma_{vib}^2 + \sigma_{stat}^2 + \sigma_{dyn}^2 \qquad (6)$$

Table 1. EXAFS parameters of metal ions in aqueous solution

Aquo Ion[a]	Ionic Radius[b] $r(M^{n+})$, Å	Bond Length[c] $r(M-H_2O)$, Å	RMSRD[d] $\sigma_i \pm 0.02$ Å	Rate Constant[b] k_1 (sec^{-1})
$Cr(H_2O)_6^{2+}$	0.80	eq. 2.07(1)	0.036	7×10^9
		ax. 2.30(5)	0.15	
$Cr(H_2O)_6^{3+}$	0.65	1.98	0.068	5×10^{-7}
$Mn(H_2O)_6^{2+}$	0.91	2.18	0.078	3×10^7
$Fe(H_2O)_6^{2+}$	0.83	2.10	0.081	3×10^6
$Fe(H_2O)_6^{3+}$	0.67	1.98	0.055	3×10^3
$Co(H_2O)_6^{2+}$	0.82	2.05	0.082	1×10^6
$Ni(H_2O)_6^{2+}$	0.78	2.05	0.070	3×10^4
$Cu(H_2O)_6^{2+}$	0.72	eq. 1.96(1)	0.036	8×10^9
		ax. 2.60(3)	0.12	

[a] Room temperature measurements from Ref. 19, 20 and unpublished results, the shorter axial distance in $Cr(H_2O)_6^{2+}$ reported in Ref. 20 was due to partial oxidation.

[b] From Ref. 26, the total rate is the coordination number times k_1, S_N2 mechanism is responsible for the slow k_1 in $Cr(H_2O)_6^{3+}$. The bond length difference Δr for the $Fe(H_2O)_6^{2+/3+}$ couple is 0.13 Å from the phase-difference analysis; this value is used in the electron-exchange reaction analysis (Fig. 3).

[c] Uncertainty is ± 0.01 Å unless indicated in the parenthesis.

[d] RMSRD = Root-Mean-Square-Relative-Displacement

where σ_{vib}^2, σ_{stat}^2 and σ_{dyn}^2 are the mean-square displacement of the equilibrium $M-H_2O$ distance resulting from all molecular vibration modes, static disorder and inner-shell/outer-shell exchange between coordinated H_2O and solvent H_2O molecules respectively. In the limit of spherical screening of the octahedral complex by the solvent molecules, σ_{stat} (this may exceed 0.01 Å in anisotropic solid) vanishes; further, σ_{dyn}^2 is small for strong $M-H_2O$ bonds and slow ligand exchange rate constants. Therefore, the most significant contribution to the σ_i values of the stable octahedral ions in Table 1 should arise from vibrational modes affecting the bond length. In the context of the harmonic model, σ_i^2 is a measure of the mean-square average of the relative displacement of the absorber and scatterer atom pair from their equilibrium position in a particular normal mode and is inversely proportional to the squre root of the force constant. From Table 1, one can infer that, for highly symmetrical complexes the bigger the σ_i the smaller the force constant and the weaker the bond. The fact that octahedral M^{2+} ions have larger r, bigger σ_i, and faster water exchange rate constants than M^{3+} ions is in qualitative accord with the S_N1 mechanism in which the rate determining step requires a coordinated H_2O as a leaving group. The extremely slow rate for $Cr(H_2O)_6^{3+}$ indicates a S_N2 rather than a S_N1 mechanism. [26]

We now compare the vibrational amplitude σ_{vib} with EXAFS σ_i. Using the totally symmetric stretching mode of $Fe(H_2O)_6^{2+}$ ($\bar{v}_s = 390$ cm^{-1}) and $Fe(H_2O)^{3+}$ ($\bar{v}_s = 523$ cm^{-1}) [27] we can calculate the mean square vibrational amplitude of the breathing motion of these complexes with [28]

$$\sigma_s^2 = [h/8\pi^2 \mu_M v_s] \coth (hv_s/2kT) \tag{7}$$

where μ_M is the reduced mass, $h\nu_s$ is the photon energy ($\nu_s = \bar{\nu}_s c$, c = velocity of light). The σ_s calculated for $Fe(H_2O)_6^{2+}$ and $Fe(H_2O)_6^{3+}$ at 25 °C from Eq. (7) are 0.0233 and 0.0187 Å respectively. These values are small compared with the experimental σ's of 0.081 and 0.055 Å, respectively for these two complexes as expected from Eq. (6). Quantitatively however, the vibrational contribution of the normal modes does not seem large enough to account for σ_i, although inclusion of the anti-symmetric stretching motion in some cases improves the agreement somewhat [21]. Since the EXAFS parameter in Table 1 are derived on the basis of Eq. (1) using theoretical phases and amplitudes and has large uncertainty in σ_i this discrepancy is most likely due to the inadequacy of the theoretical parameters (many electron effects are not accounted for) and the harmonic model, the neglect of other vibrational modes and the σ_{dyn} term in Eq. (6), and experimental error. More quantiative information must be obtained from temperature dependent measurements [9]. For symmetric one-distance systems however it is safe to infer systematic bond strengths from EXAFS bond length and RMSRD values.

EXAFS studies of the Jahn-Teller distorted ions $Cr(H_2O)_6^{2+}$ and $Cu(H_2O)_6^{2+}$ are worth of mentioning. This situation is particularly interesting because it not only involves one short (strong) and one long (weak) bond within the same complex but it also shows the limitations of EXAFS. It has been reported that the presence of the "weak bond" is not apparent in the Fourier transform of $\chi(k) k^2$ due to the N_i/r_i^2 dependence and the large σ_i^2 term, and more detailed modeling is needed for the extraction of the EXAFS parameters [19]. In fact, both 4-ligand (equatorial) and 6-ligand (2 axial, 4 equatorial) fits have been attempted in the analysis of the EXAFS of these ions. In turns out that the tetragonally-distorted octahedron fits the data best. It should be noted that while the EXAFS bond lengths of $Cr(H_2O)_6^{2+}$ and $Cu(H_2O)_6^{2+}$ listed in Table 1 are in accord with literature values, the σ_{ax} and σ_{eq} values are strongly correlated with the amplitude function and the coordination number, their usage is at best qualitative. The axial RMSRD, σ_{ax}, in particular cannot be literally interpreted as the harmonic mean displacement of a weak axial bond; σ_{ax} is better described in terms of contributions from σ_{dyn} (Eq. (6)). The position of the loosely bonded axial ligands can in principle be treated as a non-gaussian distribution in distance space ($e^{-\sigma^2 k^2}$ in Eq. (1) is replaced by an integral) [9, 10]. Qualitatively, however, the long axial bond, the large σ_{ax}, the short equatorial bond, and the small σ_{eq} obtained in the analysis together with the fast exchange rate support the view that only axial ligands can be directly involved in the fast Cu^{2+} and Cr^{2+} water exchange [29].

2.3 Application to Electron Exchange Reactions

The perhaps most useful application of EXAFS to the solution chemistry of transition metal complexes is the determination of the bond length difference Δr between transition metal oxidation-reduction couples. Despite the various theories proposed for the interpretation of the electron exchange rates there is a general agreement that the equilibrium nuclear configurations of the reactants change when electron transfer takes place [30–33]. Thus the energetics involved for the reactants to achieve a favorable configuration are intimately related to equilibrium bond lengths

$r(M-L)$ (L = ligand) of the reactants and particularly to the bond length difference Δr between them. A typical electron exchange reaction for octahedral complexes can be expressed as

$$ML_6^{2+} + ML_6^{3+} \xrightarrow{k_{obs}} ML_6^{3+} + ML_6^{2+} \tag{8}$$

The rate of electron exchange can be described by a semiclassical model [30-33] in which the observed rate k_{ob} is expressed in terms of a preequilibrium constant K_A and a first-order rate constant for electron transfer k_{el} ($k_{el} = v_n \varkappa_{el} \varkappa_n$)

$$k_{obs} = K_A v_n \varkappa_{el} \varkappa_n \tag{9}$$

where K_A is the equilibrium constant for the formation of reactant pairs separated by a distance between d and δd (Eq. 10)

$$ML_6^{2+} + ML_6^{3+} \underset{}{\overset{K_A}{\rightleftharpoons}} ML_6^{2+} \text{----} ML_6^{3+} \tag{10}$$

v_n is an effective nuclear frequency, $\varkappa_n$ is the nuclear factor, and $\varkappa_{el}$ is the electronic factor. Electron transfer reaction rates in general have been recently reviewed [32]. Here, we only focus on the effects of the inner-shell configuration changes ($\varkappa_n$) that greatly influence k_{el} (by nine orders of magnitude). A pictorial representation is shown in Fig. 2 [34].

The nuclear factor $\varkappa_n$ contains both solvent and inner-shell contributions.

$$\varkappa_n = \Gamma_\lambda \exp\left[-(\Delta G_{out}^* + \Delta G_{in}^*)/RT\right] \tag{11}$$

$$\Delta G_{out}^* = \left(\frac{\Delta q^2}{4}\right)\left(\frac{1}{2a_2} + \frac{1}{2a_3} - \frac{1}{d}\right)\left(\frac{1}{D_{op}} - \frac{1}{D_s}\right) \tag{12}$$

$$\Delta G_{in}^* = \frac{1}{2}\sum f_i[(\Delta r)_i/2]^2 \tag{13}$$

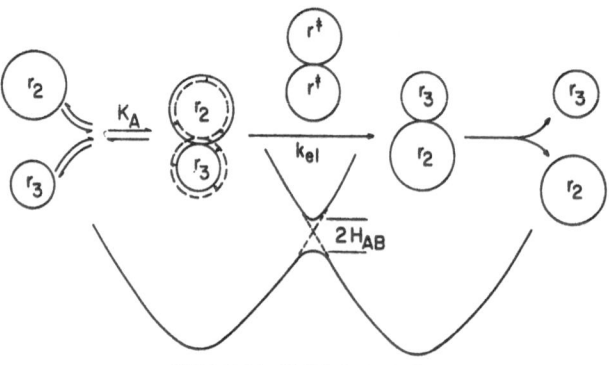

NUCLEAR CONFIGURATION

Fig. 2. Pictorial representation of electron exchange reaction involving identical reactants and products. r^* is the bond length of the activated complex. K_A is the preequilibrium constant (Eq. 10) and H_{AB} is the electronic coupling matrix element (Ref. 32)

where Γ_λ is the nuclear tunnelling factor. ΔG^*_{out} and ΔG^*_{in} are the free energy of reorganization for the solvent and inner-shell contributions respectively, Δq is the difference in charge between the two oxidation states, a_2 and a_3 are the radii of the two reactants, $d = (a_2 + a_3)$, D_s and D_{op} are the static and optical dielectric constants of the medium respectively, $f_i = 2f_2f_3/(f_2 + f_3)$ is a reduced force constant for the ith inner-sphere vibration and $r_2 - r_3 = \Delta r$ is the corresponding bond difference in the two oxidation states.

The nuclear frequency v_n is related to the solvent and inner-shell reorganization energies as well as the corresponsing vibration frequencies. The electronic factor χ_{el} can be described on the basis of the Landau-Zener framework and is related to the electronic coupling matrix element H_{AB}.[32]

We now discuss the implication of Δr to the exchange rate in terms of the reorganization of the inner shells (shortening and lengthening of the bonds) of the reactants in two redox couples $Fe(H_2O)_6^{2+}/Fe(H_2O)_6^{3+}$ and MnO_4^-/MnO_4^{2-}. Prior to electron transfer in these systems, an activated complex in which both reactants have rearranged their inner-coordination shells to an identical bond length r^* must be formed (Fig. 2). If we only consider the symmetrical breathing motion of the first (inner) hydration shell of the reactants, the common bond length r^* in the activated complex for the $Fe(H_2O)_6^{2+}/Fe(H_2O)_6^{3+}$ couple is given by

$$r^* = \frac{f_2r_2 + f_3r_3}{f_2 + f_3} \tag{14}$$

where f_2 and f_3 are the symmetric $M-H_2O$ stretching force constants in the 2+ and 3+ complexes respectively[31]. The results are listed in Table 2 together with EXAFS σ_i and vibrational amplitudes σ_s calculated using Eq. (7). From Table 2 it is evident that the reorganization $(r - r^*)$ for the $Fe(H_2O)_6^{2+}/Fe(H_2O)_6^{3+}$ (0.078 Å/0.043 Å) couple is much greater than $(r - r^*)$ for the MnO_4^-/MnO_4^{2-}

Table 2. EXAFS Bond Length[a], Root-Mean-Square Relative Displacement, Reorganization and Symmetric Vibrational Amplitude of $Fe(H_2O)_6^{2+/3+}$ and $MnO_4^{-/2-}$

System	Bond Length		Reorganization	Root-Mean-Square Relative Displacement	Symmetric[c] Vibrational Amplitude	
	Reactant	Activated[b] Complex				
	r, Å	r*, Å	r* − r, Å	σ_i	σ_s, Å	
					(25 °C)	(100 °C)
$Fe(H_2O)_6^{3+}$	1.98[d]	2.023	0.043	0,055	0.0187	0.0197
$Fe(H_2O)_6^{2+}$	2.10[d]	2.023	0.078	0.081	0.0233	0.0251
MnO_4^-	1.624	1.644	0.020	0.054	0.0180	0.0184
MnO_4^{2-}	1.666	1.644	0.022	0.056	0.0184	0.0188

[a] Uncertainty for r and σ_i is ≤ 0.01 Å for $MnO_4^{-/2-}$.
[b] Values calculated on the basis of Eq. 14 and known frequencies (Ref. 20, 21).
[c] Values calculated from Eq. 7 and known frequencies (Ref. 20, 21).
[d] The bond length difference Δr of 0.13 Å was obtained from the phase-difference analysis. This value was used in the electron-exchange reaction analysis (Fig. 3).

(0.020/0.022 Å) couple. It is interesting to compare these values with the amplitude (σ_s) of the total symmetric breathing of the complex, because this vibrational mode would most likely lead to the formation of the activated complex (Fig. 2). It is apparent that while σ_s for the $Fe(H_2O)_6^{2+}/Fe(H_2O)_6^{3+}$ couple is considerably smaller than the reorganization, σ_s for the MnO_4^-/MnO_4^{2-} couple is comparable to the reorganization, indicating that there is sufficient energy in the lower vibrationally excited state in the latter system to supply the reorganization energy even at room temperature while the former system requires higher vibrationally excited states. If we consider $\varkappa_n$ as the dominant factor in the determinatiion of the rate, and Δr dominates $\varkappa_n$ (Eq. 9, and Eqs. 11–13) we have $\varkappa_n(Fe(H_2O)_6^{3+/2+})/\varkappa_n(MnO_4^{1-/2-}) \ll 1$. Thus the MnO_4^-/MnO_4^{2-} exchange should proceed faster than the $Fe(H_2O)_6^{2+}/Fe(H_2O)_6^{3+}$ exchange. [21] This is consistant with the observed rate ($k = 710 \, M^{-1} \, sec^{-1}$ for the former and $k = 1.1 \, M^{-1} \, sec^{-1}$ for the latter).

Quantitative systematics of the effect of Δr on the reaction rate have been studied [22]. From Eq. (9) we can rearrange the logarithm to give

$$-\left[\ln\left(\frac{k_{obs}}{K_A}\right) + \frac{\Delta G^*_{out}}{RT}\right] = \frac{\Delta G^*_{in}}{RT} - \ln(\varkappa_{el}\nu_n) \qquad (15)$$

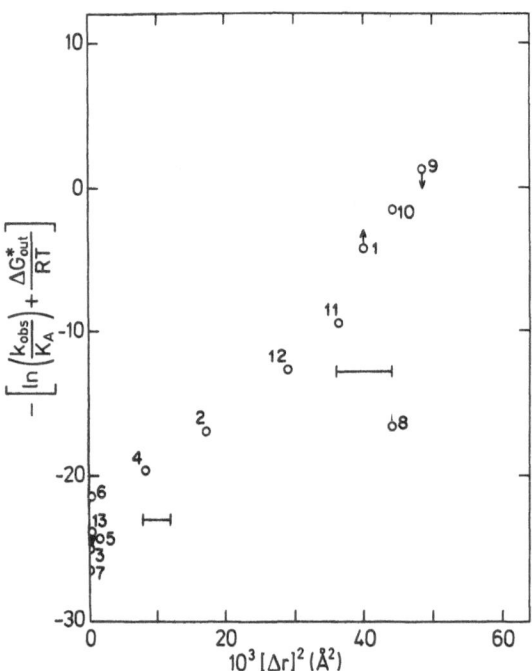

Fig. 3. Correlation of the observed exchange rate constant (corrected for the stability of the precursor complex plus the outer-shell barrier divided by RT) with $(\Delta r)^2$, the square of the difference of the metal-ligand bond distances in the two oxidation states (Ref. 22).
(1) $Cr(H_2O)_6^{2+/3+}$; (2) $Fe(H_2O)_6^{2+/3+}$; (3) $Fe(phen)_3^{2+/3+}$; (4) $Ru(H_2O)_6^{2+/3+}$; (5) $Ru(NH_3)_6^{2+/3+}$; (6) $Ru(en)_3^{2+/3+}$; (7) $Ru(bpy)_3^{2+/3+}$; (8) $Co(H_2O)_6^{2+/3+}$; (9) $Co(NH_3)_6^{2+/3+}$; (10) $Co(en)_3^{2+/3+}$; (11) $Co(bpy)_3^{2+/3+}$; (12) $Co(sep)^{2+/3+}$; (13) $Co(bpy)_3^{1+/2+}$, en = ethylenediamine, bpy = 2,2'-bipyridine and sep = "sepulchrate", phen = 1,10-phenanthroline

where ΔG_{in}^* is a function of Δr^2 (Eq. 13). EXAFS study of a number of redox pairs shows that a good correlation exists between the exchange rate constant (corrected for the stability of the precursor complex and outer-shell barrier) and Δr^2 as expressed in Eq. (15) [22, 32]. The results are plotted in Fig. 3. It can be seen from Fig. 3 that, with the exception of $Co(H_2O)_6^{2+/3+}$, the correlation is remarkably good for the rate constants that span a range of 15 orders of magnitude. The behavior of the cobalt couple has been known to be anomalous and other mechanisms have been proposed for this reaction [32].

2.4 Other Applications

Structures of lanthanide ions in solution are also of particular interest because these ions are highly coordinated (coordination number n = 8 or 9) [35]. From the few existing crystal structures, it is found that hydrated lanthanide ions have two sets of inequivalent bond lengths (differ by 0.04 Å). No distinctly different bond lengths are observed from an EXAFS study of Ce(III) aquo ions [36]. Smooth single oscillations are observed in the X-ray absorption spectrum of $Ce(H_2O)_n^{3+}$. This result suggests that the cerous ion is more spherically symmertrical in solution than in the solid where lattice distortion is an important factor. A bond length of 2.50 ± 0.02 Å has been obtained for $Ce^{3+}-H_2O$ and a bond length difference Δr of 0.07 ± 0.02 Å for the $Ce(H_2O)_n^{3+}/Ce(H_2O)_n^{4+}$ couple has also been derived [36].

The XANES study of the mixed oxidation $Ce(H_2O)_n^{3+}/Ce(H_2O)_n^{4+}$ solution is also of great interest. It reveals the atomic nature of the 5d orbitals in lathanide compounds and the chemical sensitivity of the $2p \rightarrow 5d$ transition. It is found that the L_2 and L_3 XANES of the mixture exhibit large chemical shifts (7.3–9.0 eV) between the $2p \rightarrow 5d$ resonance ("white line") of $Ce(H_2O)_n^{3+}$ and $Ce(H_2O)_n^{4+}$ in different acidic media (Fig. 4). These values are small compared with atomic calculation but are slightly larger than those of mixed-valent solid systems. This study has two interesting implications. First the chemical shift between Ce^{3+} and Ce^{4+} in solution is given rise by a initial state effect in contrast to metallic mixed-valent systems where final state screening by conduction electrons may be responsible for additional white-line features. Second, this system is an excellent example for the illustration of the equivalent core approximation as applied to X-ray photoelectron spectroscopy (XPS) and X-ray Absorption studies. L_3 edge chemical shifts are summarized in Table 3 [36].

For first row transition metal complexes, XAS results are currently limited to the K edge absorption measurements, and the K edge XANES structure does not probe the d states directly. Although the XANES of these complexes can be carefully analyzed in terms of multiple scattering, through which structural information can be revealed, this approach has already been discussed by Bianconi in this volume and is not discussed here. Rather, we want to present an interesting example concerning the chemical sensitivity of the energy position of the edge jump threshold E_{th} to the local charge distribution at the metal site of complexes with and without π bonding ligands. Figure 5 shows the Fe K edge XANES of two Fe redox couples [34], $Fe(H_2O)_6^{2+/3+}$ and $Fe(phen)_3^{2+/3+}$ (phen = 1,10-phenanthroline) in solution. E_{th} shifts to the higher binding energy side from Fe(II) to Fe(III) in the high-spin

Table 3. L_3 edge chemical shifts of lanthanide ions in solutions and in solids

Sample	ΔE (eV)[a]		I_{ratio} [a]
	observed	calculated[b]	
Liquid mixed oxidation solution			
Ce^{3+}/Ce^{4+}, 0.5 N $H_2Ce(ClO_4)_6$ in 6 N $HClO_4$	9.0	11.0	0.4
Ce^{3+}/Ce^{4+}, 0.1 N $H_4Ce(SO_4)_4$ in 1 N H_2SO_4	7.3	11.0	1.1
Solid mixed valence compound			
Sm^{2+}/Sm^{3+} in SmB_6[c]	7	9.3	1.9 ± 0.2
Sm^{2+}/Sm^{3+} in $Sm_{0.75}Y_{0.25}$[d]	7	9.3	1.3 ± 0.1
Tm^{2+}/Tm^{3+} in $TmSe$[e]	7	9.3	1.4 ± 0.1

[a] The difference and the ratio are between high and low oxidation states. I_{ratio} is the relativity intensity of the white line. I_{ratio} for solution samples is dependent upon the relative concentration of Ce^{3+} and Ce^{4+} (Ref. 36).
[b] The calculation was done with the Herman Skillman program, for example, configurations $2p^{5.5}4f^16d^{0.5}$ and $2p^{5.5}$, $4f^04d^{0.5}$ for Ce^{3+} and Ce^{4+}, respectively were used; only the difference in the transition energy was considered.
[c] Vainshtein EE et al (1964) Sov. Phys. Solid State *6*, 2318
[d] Martin R et al. (1980) Phys. Rev. Lett. *44*, 1275
[e] Launois H et al. (1980) Phys. Rev. Lett. *44*, 1271

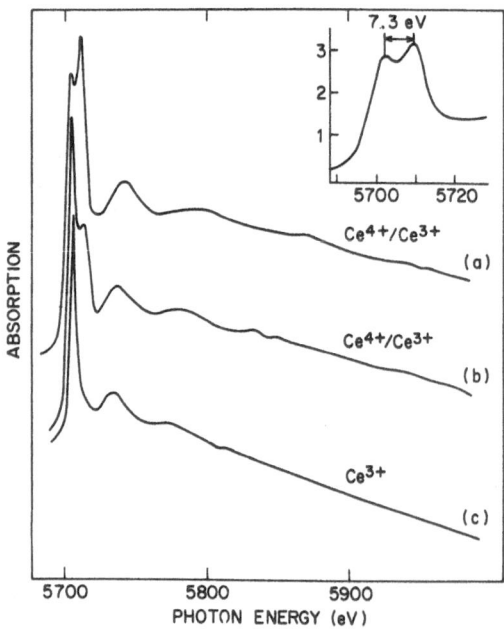

Fig. 4. XAS spectra of mixed oxidation Ce^{4+}/Ce^{3+} solutions (a) in H_2SO_4 and (b) in $HClO_4$ (see table 3), and (c) aqueous $Ce(NO_3)_3$, ~1 M. (Ref. 36)

aquo complexes and the corresponding EXAFS bond length difference is 0.13 Å, whereas in the case of the low-spin pair, only a small shift and no EXAFS bond length difference is observed. This observation indicates that significant charge redistribution involving the ligand π orbitals and the metal d orbitals takes place so that the effective potential experienced by the Fe atoms in the two oxidation states are very similar. The double peak feature at $50 \sim 80$ eV above E_{th} is due to multiple scattering (see Bianconi, this series).

XANES has also been applied to study secondary interactions (such as the changes in bonding resulting from changes beyond the first coordination shell). In a recent experiment [37], in which the PtL_{III} XANES of the square planar "$PtCl_4$" moiety in solid K_2PtCl_4 and aqueous solution of K_2PtCl_4 were measured, it was found that there were two visibly different XANES features between the solid and solution samples of K_2PtCl_4 (Fig. 6). First, the intensity of the white line that probes the unoccupied d character at the Pt site increases by 20% going from solid to solution. This observation immediately indicates a depletion in d character (occupied) at the Pt site upon dissolution. Second, sharp features appearing at the XANES of the solid sample disappear in the XANES of the solution sample despite the fact that the local square planar structure of "$PtCl_4$" remains unchaged as shown by the EXAFS. This result can be attributed to final state effects and secondary interaction in-

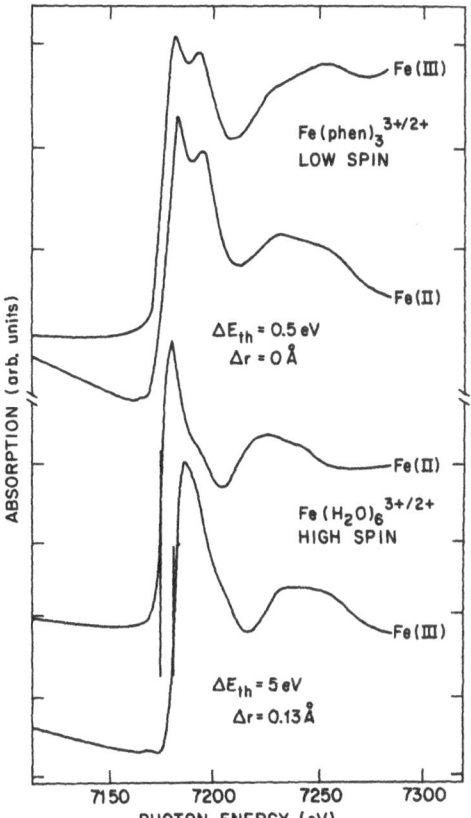

Fig. 5. XANES of $Fe(H_2O)_6^{2+/3+}$ (weak π bonding) and $Fe(phen)_3^{2+/3+}$ (extensive π bonding). The E_{th} (marked with vertical bar) energy shift ΔE_{th} and corresponding EXAFS bond length difference Δr are also given

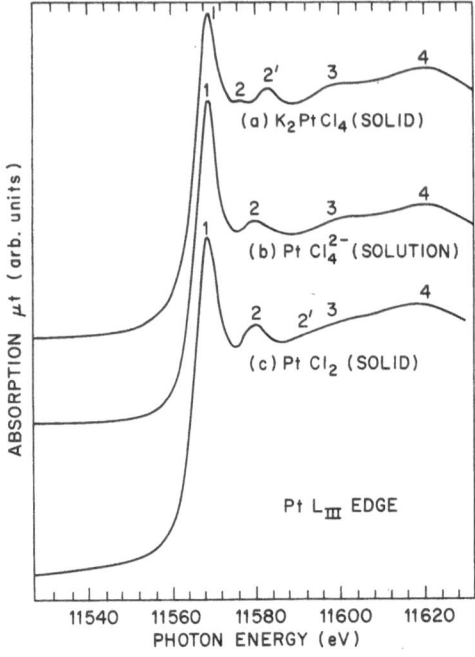

Fig. 6. Pt L_{III} edge XANES of K_2PtCl_4: (a) in the solid state and (b) in solution. The Pt atoms in the solid are nearly colinear. (c) The Pt L_{III} edge XANES of $PtCl_2$ of which the Pt atoms form a slightly distorted octahedron

volving Pt—Pt along the one dimensional Pt chain and Pt—K interaction in the solid [37]. Similar phenomenon has been previously observed in solid $KMnO_4$ and its solution [21].

3 Photoconductivity Measurements of X-Ray Absorption Spectra

3.1 General Considerations and Experimental Techniques

In connection with applications described in Section 2, it is natural to proceed to apply the technique to systems involving different solvents. One recent development is the investigation of the X-ray induced ionization behavior of organometallics in hydrocarbon solution. Since organometallics are often insoluble in H_2O, but are soluble in organic solvents, we have devised an alternative scheme to obtain XAS spectrum in organic solution [38-41]. The objectives of this research are (a) to study the implication of different decay channels to the yield spectrum, particularly in the X-ray Absorption Near Edge Structures (XANES) region and (b) to investigate the ion yield of the pure hydrocarbon liquids and their solutions with the expectation that this study may lead to the development of liquid ionization chambers and related techniques for X-ray studies.

Photoconductivity measurement of X-ray absorption of a liquid involves the utilization of liquid cells equipped with parallel plate electrodes. Two configurations have been used for these measurements [39] (with electrodes parallel or perpendicular

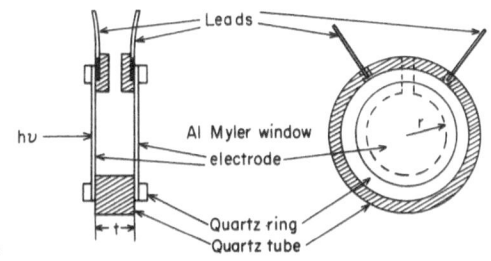

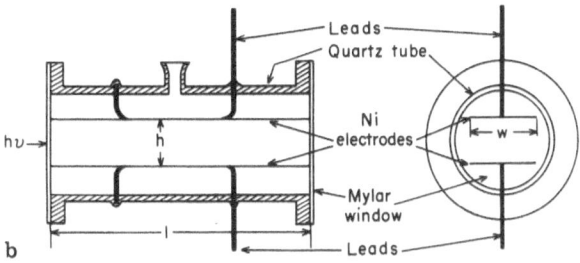

Fig. 7a and b. Schematics of the solution cells with **a)** electrode surface perpendicular to the direction of the beam and (b) electrode surface parallel to the beam[39]; 0.1 to 0.25 mil Mylar windows were used. Electrode separations of 1 mm to 20 mm in case (a) and 1 mm to 5 mm in case **b**) have been studied. $l = 2$ to 3 cm is commonly used in case (b)

to the incoming beam). Schematics for the typical solution cells are shown in Fig. 7. Two cell parameters are important to the measurements. One is the thickness of the cell, the other is the separation between the electrodes. The photocurrent of the cell is extracted by applying a voltage to one of the electrodes. The field strength V/L (V = bias voltage, L = electrode separation) is an important experimental parameter. Since these are DC measurements and we are dealing with a small signal, the liquids employed in these measurements are hydrocarbons with very low background conductivity (picoamperes). The photocurrent signal I_c measures the X-ray induced ionization in the solution (including primary and secondary ionization) and is related to the absorption μt of the liquid. I_c can be expressed as

$$I_c = Y \cdot f(P, h\nu, I_0) \cdot I_0[1 - e^{-\mu t}] \qquad (16)$$

where Y is the ion yield per eV of photon absorbed, f, the collection efficiency of the cell, is a function of the cell parameters, P_c (P_c includes the electrode separation, thickness, bias voltage and the properties of the liquid), the photon energy $h\nu$ and the flux I_0. The collection efficiency of the cell for uniform ionization can be expressed as

$$f = \frac{2}{[1 + (1 + 2/3\xi^2)^{1/2}]} \qquad (17)$$

95

where

$$\xi^2 = \frac{e}{\varepsilon_0 \varepsilon_r} \frac{L^4 q}{V^2} \frac{2}{\mu^\pm} \qquad (18)$$

and ε_0, ε_r are the permittivity of vacuum and the dielectric constant of the liquid respectively, L is the separation between the electrodes; q is the rate of charge production by the incoming X-ray in ion-pairs/cm^3 sec and is proportional to $I_0(1 - e^{-\mu t})$, $\mu^\pm$ is the ion mobility. Since the incident X-ray photon flux suffers a rapid exponential decay as a function of the depth of penetration, the overall efficiency f_s in a liquid cell has to be integrated over the entire cell [39],

$$f_s = \frac{\int_0^t \left[\frac{\partial q(t)}{\partial t}\right] f \, dt}{\int_0^t \left[\frac{\partial q(t)}{\partial t}\right] dt} \qquad (19)$$

where $\partial q / \partial t = \mu I_0 \cdot h\nu \cdot e^{-\mu t} \cdot G \cdot 10^{-2}$ is a measure of the amount of ion pairs created per unit volume per second between a distance t to $t + dt$ in the cell; μ is in cm^{-1}, I_0 is in number of photons/sec cm^2, G is the G value, a quantity commonly used in radition chemistry and is a measure of the number of ion pairs created for every 100 eV of photon energy absorbed by the system. In situations where the cell collection efficiency is low (this is common for liquid studies), Eq. (19) can be approximately expressed as

$$f_s = K(\mu I_0)^{-1/2} \frac{(1 - e^{-\mu t/2})}{(1 - e^{-\mu t})} \qquad (20)$$

Substituting Eq. (20) into Eq. (16) we have

$$I_c = K'(I_0/\mu)^{1/2} (1 - e^{-\mu t/2}) \qquad (21)$$

where I_0 is the incident flux and $K' = YK$ (K is a constant). It is apparent from Eqs. (16)–(21) that I_c depends on two factors given fixed cell configurations. Ons is the ionization yield Y, the other is the cell efficiency which determines how effective the ions can be accounted for by the cell. Although most of the measurements are made under inefficient conditions ($f < 1$), it is possible to adjust the experimental parameters such as the horizontal slit and the applied voltage so that f is close to unity. Let us consider in the following first the effect of f_s (when $f_s < 1$) on the photocurrent I_c, then the behavior of the yield Y as a function of the absorption coefficient below and above an absorption edge under efficient conditions ($f_s \sim 1$). From Eqs. (20) and (21), we can consider two limiting cases. When the cell is thin, that is μt is small, we have

$$I_c = \frac{K'}{2} (I_0 \mu)^{1/2} t \qquad (22)$$

and for a thick cell (optically black), which absorbs practically all the incident X-rays, we have

$$I_c = K'(I_0/\mu)^{1/2} \tag{23}$$

From these considerations, it is apparent that the current will drop at an absorption edge when the cell satisfies the total absorption conditions.

When a very thin slab ($\sim$0.1 mm$\times \sim$7 mm) of photons with a flux of 10^8 to 10^9 photons/sec is allowed to incident upon a parallel plate cell (Fig. 7b), f $\cong$ 1 can be achieved [40] and Eq. (16) becomes

$$I_c = Y \cdot I_0(1 - e^{-\mu t}) \tag{24}$$

From Eq. (24) Y can be determined from

$$Y = \frac{I_c}{I_0(1 - e^{-\mu t})} \tag{25}$$

Since Y is the total ionization yield (in no. of charges/photon), it takes into account yields resulting from all the components, (Y_M for the solute metal complex and Y_S for the solvent) in solution, and

$$Y = Y_M x + Y_S(1 - x) \tag{26}$$

where x is the fraction of photon intensity absorbed by the metal and $(1 - x)$ the fraction absorbed by the surrounding. If we label the parameters with superscripts a and b for absorption above and below the edge respectively, we can write

$$Y^b = Y_M^b x^b + Y_S^b(1 - x^b) = \frac{I_c^b}{I_0(1 - e^{-\mu^b t})} \tag{27}$$

$$Y^a = Y_M^a x^a + Y_S^a(1 - x^a) = \frac{I_c^a}{I_0(1 - e^{-\mu^a t})} \tag{28}$$

Again, we consider two limiting cases. If the cell is very thin (μt being very small) we have

$$\frac{I_c^b}{I_c^a} = \frac{Y_M^b x^b + Y_S^b(1 - x^b)}{Y_M^a x^a + Y_S^a(1 - x^a)} \frac{\mu^b}{\mu^a} \tag{29}$$

where $\mu^b < \mu^a$ determines the ratio. Therefore the I_c spectrum should resemble the absorption spectrum recorded in the transmission mode (positive edge jump). If the cell is thick (large μt) so that all the X-rays are absorbed, we have

$$\frac{I_c^b}{I_c^a} = \frac{Y_M^b x^b + Y_S^b(1 - x^b)}{Y_M^a x^a + Y_S^a(1 - x^a)} \tag{30}$$

Based on Eq. (30), it can be shown that $I_c^b > I_c^a$ despite the expected edge jump ($\mu^a > \mu^b$). This is because $Y_S > Y_M$, $(1 - x^b) > (1 - x^a)$, $Y_S^b \cong Y_S^a$ and $Y_M^b > Y_M^a$.

The reasons are (a) the primary photoelectrons produced following the X-ray absorption of solvent molecules have more energy than those of the metal; and (b) fluoresence X-ray from the metal can escape the solution. Thus, we have $Y_s > Y_M$ (we found this was true even in pure $(CH_3)_4Sn$ where Sn was the metal and $-CH_3$ was the "solvent"). Similar argument can be used to infer $Y_M^b > Y_M^a$ because, for absorption above the edge, more energy would escape without producing ionization through additional X-ray fluorescence channels. Returning to Eq. (24), it becomes clear that the photocurrent may increase or decrease above an absorption edge depending on the absorption thickness of the speciemen.

X-ray absorption measurement of a number of liquids can be made using cells similar to those described in Fig. 7. Hydrocarbon liquids such as hexane, 2,2,4-TMP (TMP = tetramethylpentane) and 2,2,4,4-TMP, or even toluene can be conveniently used as solvents. Current signals from the incident and the transmitted ionization chambers as well as the liquid cell should be recorded simultaneously. A typical set of cell parameters (Electrode separation: 1 mm to 30 mm, field: 0.1–1 KV/mm are the usual ranges) has a background current of several to tens of picoamperes (PA) and a signal of several hundred PA (with a flux of $\sim 10^9$ photons/sec). Since these systems have very low conductivity, it is important that the measurement is made in a dry environment with adequate electronic shielding. The cell assembly should therefore be kept in a metal sample chamber with two opposing windows (Be, Mylar, or Kapton) all the time. Singals can be conveniently collected with current amplifiers.

3.2 Applications

Several cases are presented here to illustrate the application of the technique in terms of a wide range of values of relevant variables. Shown in Fig. 8 is the

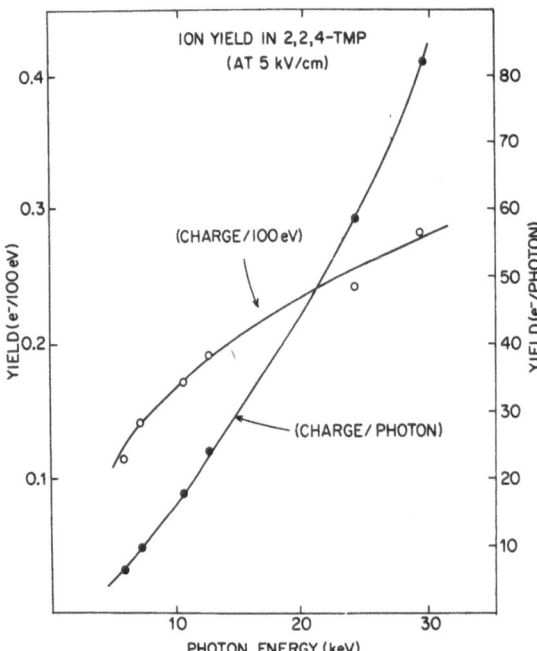

Fig. 8. Ion Yield in 2,2,4-TMP; the left and ritht ordinates are the G values and the yield per photon absorbed respectively[41].

ionization behavior of pure solvent 2,2,4-TMP [40]. The most important observation is that for a large range of X-ray energies the ionization yield per unit energy absorbed (G value) is larger at higher energies. For instance, the ionization yield of one 20 KeV photon is greater than that of two 10 KeV photons. The implication is that the yield is higher for photoelectrons with higher kinetic energy (in hydrocarbon liquids the absorption of two 10 KeV photon produces two $\sim$9.7 KeV photoelectrons and a 20 KeV photon produces one $\sim$ 19.7 KeV photoelectron because the binding energy of carbon is $\sim$0.3 KeV). Shown in Figs. 9, 10, 11 are the

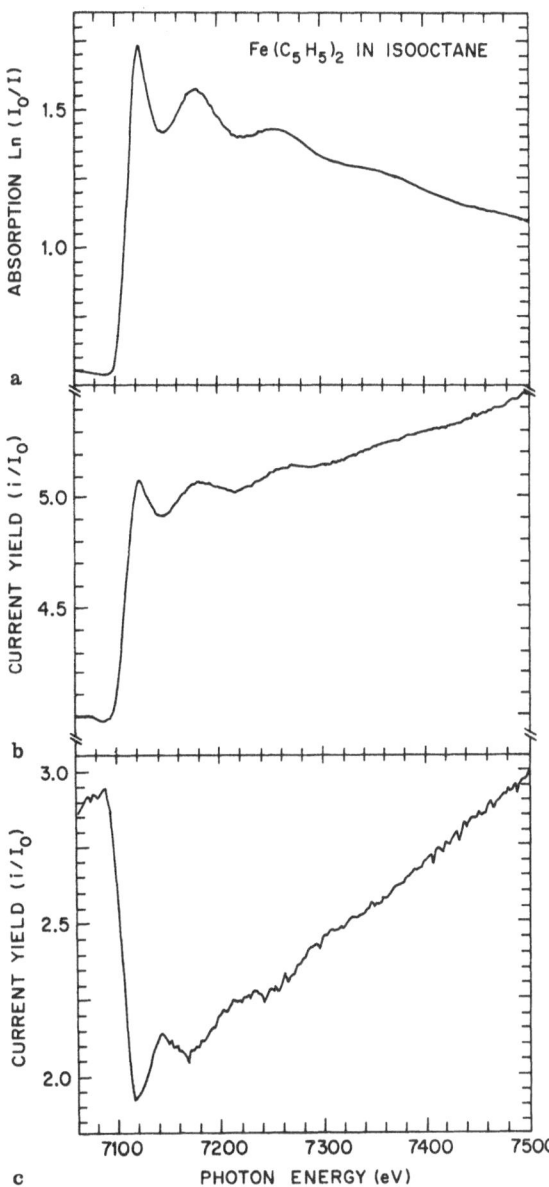

Fig. 9a–c. Fe K edge absorption spectrum of 0.136 M $Fe(C_5H_5)_2$ in 2,2,4-TMP: a) transmission, b) current yield with a t = 2 mm cell, and c) current yield with a t = 30 mm cell, see Table 4.

99

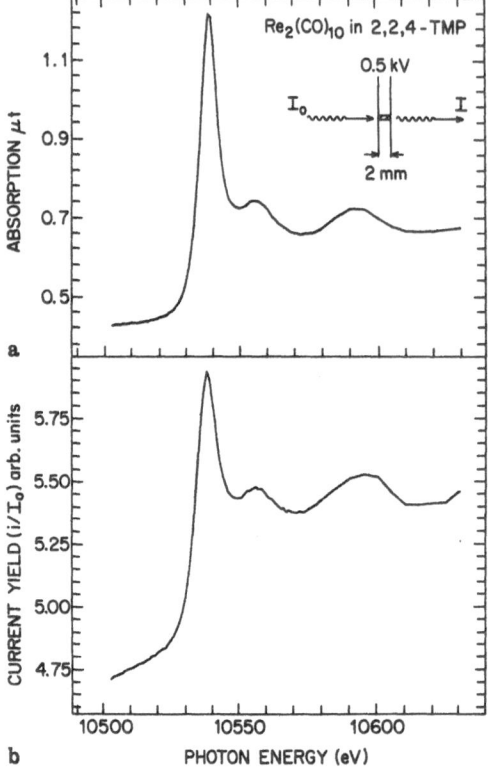

Fig. 10a and b. Re L_{III} edge spectra of $Re_2(CO)_{10}$ in 2,2,4 TMP (0.0364 M): a) transmission and b current yield

XAS spectra of ferrocene and rhenium carbonyl in 2,2,4-TMP recorded under low efficiency conditions with cell configurations illustrated in Fig. 7. The thickness of the cell was varied from a thin cell (not quite the very thin cell limit given by Eq. (22)) to a thick cell. The corresponding transmission spectra are shown for comparison. In Fig. 11c, the $1/(I_c/I_0)^2$ spectrum which should be related to μ (Eq. (23)) is also shown. In Fig. 12 is shown the measurement of the same $Re_2(CO)_{10}$ solution with high cell efficiency (f ~ 1) using a configuration in which the electrodes were parallel to the direction of the incoming beam. A very thin (0.08 mm) photon slab was used in recording the spectra shown in Figs. 12a and b. Relevant parameters are summarized in Table 4 for some typical measurements.

Several interesting features are immediately noted from Figs. 9–11 and Table 4. First, in a thin cell, the current yield spectrum reproduces a spectral profile similar to a normal transmission spectrum. Second, the current yield edge jump is only a fraction of that of the tramission spectrum. Third, in the case of the thick cell, the transmission spectrum (Figs. 10, 11) suffers from severe thickness effect as expected. Fourth, the signal to background ratio increases drastically in a dry environment and finally, the current yield in the thick cell drops sharply at the edge. This latest observation now appears to be a common phenomenon for optically black cells and its implication is discussed later.

Returning to Eqs. (21–33), it becomes apparent that the above mentioned results can be understood on the basis of these equations. Fig. 10 for example is very

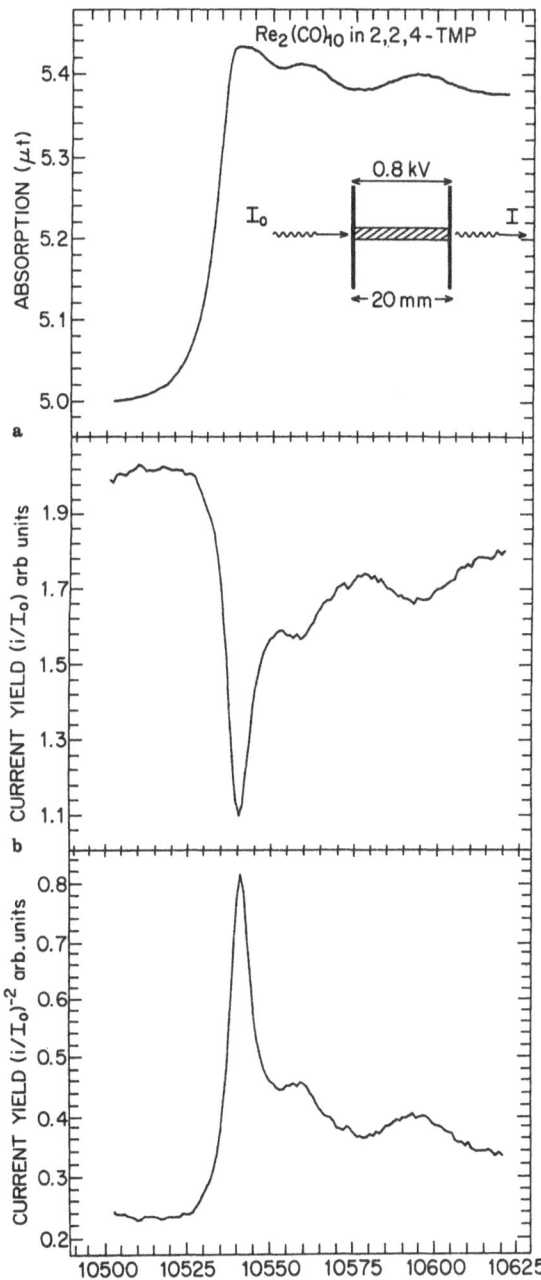

Fig. 11a–c. Re L_{III} edge spectra of $Re_2(CO)_{10}$ in 2,2,4-TMP in a 20 mm cell: **a)** transmission, **b)** current yield and **c)** $\mu \sim 1/(\text{current yield})^2$ (Ref. 39b)

close to the situation described by Eq. (22) because the series expansion of $e^{-\mu t/2}$ converges very rapidly. It is therefore the square of I_c/I_0 that is related to μt. In a thick cell (Fig. 11), the data can be conveniently expressed in terms of μ by taking the reciprocal of the square of the current yield. This is shown in Fig. 11c. It is clearly

101

seen that the spectral profile of Fig. 11c is nearly identical to that of the normal transmission spectrum (Fig. 10a), any discrepancy can be attributed to Y (Eq. 16).

When the measurements were made with an efficient cell, similar inverse appearance was again observed for a thick cell (Fig. 12)[39]. Since the thickness of the cell approaches the $1 - e^{-\mu t} = 1$ limit, no XANES is seen in the transmission spectrum except an edge jump. The current yield again shows sharp but inversed XANES structures characteristic of the spectrum obtained with low cell efficiency.

We now look into the origin of these current yields. There are three processes that produce charges in the cell. These are (a) direct ionization, (b) X-ray fluorescence and (c) Auger decay [(b) and (c) produced secondary charges]. Below the Re L_{III} edge for example, a combined contribution of these processes in both the solvent and the solute gives rise to the current. In fact the radition chemistry G value is a mesure of the overall effect. At the Re L_{III} edge, however, the fluorescence and the Auger decay are primarily responsible for the edge jump (no direct ionization occurs in a Re 2p → 5d bound to bound transition and the slow photoelectrons produced above the ionization threshold do relatively little in generating secondary ionization). Since the fluorescence X-ray has a probability of escaping reabsorption, particularly in the

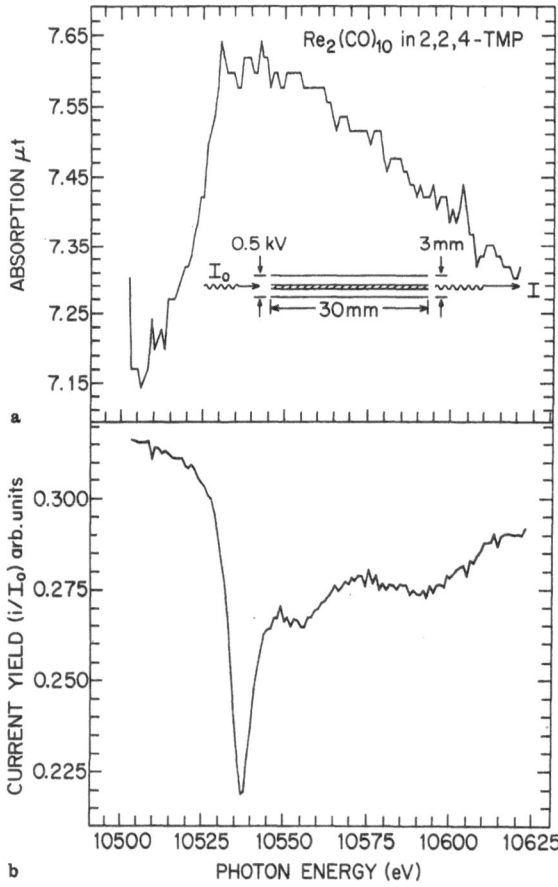

Fig. 12a and b. Re L_{III} edge spectra of $Re_2(CO)_{10}$ in 2,2,4-TMP with efficiency $f \cong 1$: a) transmission, b) current yield (Ref. 39b)

Table 4. X-ray photoconductivity parameters for $Fe(C_5H_5)_2$ and $Re_2(CO)_{10}$ in 2,2,4-TMP under various condition

Sample	Cell Configuration[a]	Field Strength V/L	Absorption (μt)[b]		Photon flux[c]	Current[d] (nanoamp)		
			Below edge	Above edge	photon	Dark	Below edge	Above edge
$Fe(C_5H_5)_2$ ambient, wide slit	(a) $t_\perp = 0.2$ cm (b) $t_{//} = 3$ cm	200 V/0.2 cm 750 V/0.75cm	1.4 17.2	1.7 25.2	5.9×10^9 6.3×10^9	0.23[e] 0.053[e]	0.66 0.64	0.77 0.47
$Re_2(CO)_{10}$ dry, narrow slit	(a) $t_\perp = 0.2$ cm (b) $t_{//} = 3$ cm	500 V/0.2 cm 500 V/0.3 cm	0.46 8.1	0.84 14.3	4.5×10^8 3.1×10^8	0.079[f] 0.007[f]	0.32 0.58	0.39 0.42

[a] See Fig. 7 and Ref. 39.
[b] Calculated with μ from McMaster WH, Kerr Del Grande N, Mallet JH, Hubbell JH Compilation of X-Ray Cross Sections Nat Techn. Inform Serv, Springfield, VA, and a density of 0.693 for 2,2,4-TMP.
[c] Calibrated with N_2 and Ar ion chambers in the $Fe(C_5H_5)_2$ and $Re_2(CO)_{10}$ experiments respectively (Ref. 39).
[d] Current measured with Keithley amplifiers, uncertainty is $\sim 5\%$.
[e] In ambient environment, the high dark current is due to leakage.
[f] Under dry N_2 atmosphere, dark current is significantly reduced.

vicinity of the first window where the absorption is the highest, the absolute current yield edge jump is expected to be less than that in the transmission spectrum in a thin cell configuration (Fig. 10). In the case of Fig. 12, the efficiency f is nearly unity. Returning to Eq. (25), it is immediately seen that $I_c/I_0 \propto Y$. Therefore the ion yield per eV photon absorbed in Fig. 12 drops across the edge. Bearing in mind that every photon is absorbed by the solution and both the solute and solvent are competing for the incident flux, this inversion can be understood in terms of fluorescence escape and relatively ineffective energy transfer through Auger cascade and photo-electrons above the edge where Re absorbs 75% of the photons in this solution. As already discussed above (Eq. (30)), these processes are less effective in the production of secondary ions than direct ionization of the solvent. The fact that Y itself carries the information of $1/\mu t$ of the solute is rather interesting. It may arise from cell inefficiency near the window where absorption is the strongest. Further study is now underway. It should be noted that the observation that the conductivity measurement yields sharp XANES and EXAFS structure for a thick cell (total absorption) indicates very interesting possibilities. For example, for elements of low atomic numbers in compounds such as organo-phosphorus, -sulfur and -chlorine compounds, conventional transmission spectra are not readily obtainable; this technique which makes use of a thick cell would be a very good alternative for obtaining K edge XANES of S, P and Cl. Another possibility is to develop liquid ionization chambers and to investigate related phenomena.

4 Summary and Recent Developments

Two types of applications of XAS in solution chemistry have been discussed here with one emphasis on the study of structure and bonding of transition metal complexes in solution and its correlation with reactivity. The unique application of EXAFS to derive Δr for closely related system can be extended to temperature-dependent studies of local dynamics and the study of oxidation reduction couples in general. Another emphasis involves the development of a technique in measuring XAS of organometallics in solvents of low background conductivity. This technique should be applicable to a large number of situations where thin sample cannot be easily prepared for transmission measurements. A detail experiment has recently been carried out in which we showed that both Eqs. (29) and (30) are valid and the behavior of the edge jump can be accurately described by these equations [41].

Finally, a recent development in connection with the photoconductivity technique should be noted. This new development involves the measurement of the X-ray induced optical photons (scintillation) from the sample solution of organometallics in toluene which contains a small amount of organic scientillators. In a recent experiment, the optical yield and the ionization yield of $(CH_3)_4Sn$ in toluene (containing a small amount of anthracene) were simultaneously monitored with a photomultiplier and direct DC photocurrent amplifier respectively [42]. The Sn K-edge XANES spectra recorded simultaneously in the scintillation and the photoconductivity mode show that the scintillation yield XANES spectra are practically voltage

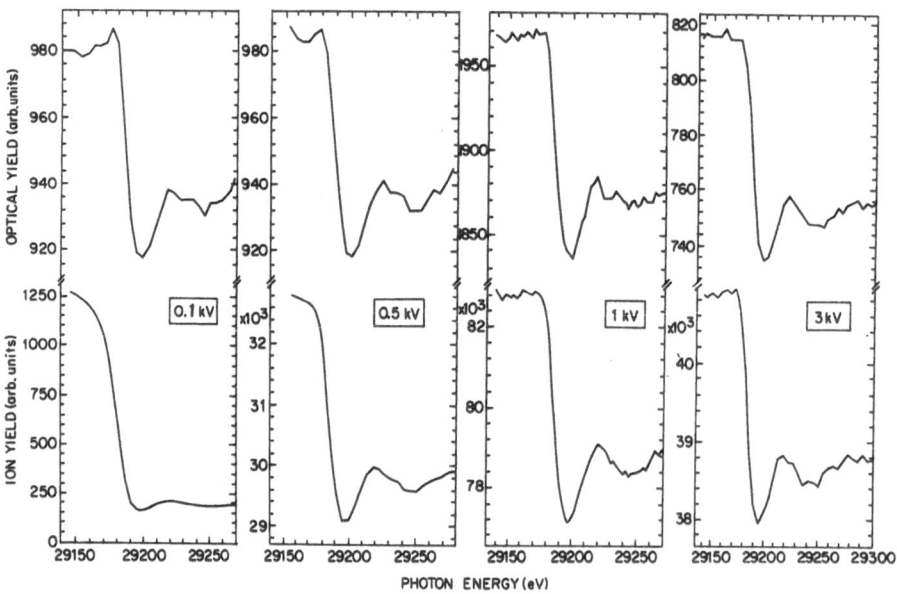

Fig. 13. Sn K edge XANES Spectra from simultaneous optical (luminescence) and ion yield measurements of an anthracene/toluene solution (1 g/l) of tetramethytin (0.5 ml $(CH_3)_4Sn$ in 20 ml solution) in a parallel plate liquid ionization chamber at various voltages across the electrodes. The separation of the electrodes and the length of the cell are 3 mm and 2 cm respectively

independent and resemble those of high efficient (high voltage f $\simeq$ 1) ionization yield measurements (Fig. 13). These results [42] strongly suggest that both the ionization and the scintillation yield will have future applications in various chemical situations.

5 Acknowledgement

I am indebted to my colleagues Drs. B. Brunschwig, C. Creutz, J. Hastings, S. Heald, R. Holroyd, D. Macartney, M. Newton, M. Perlman, and N. Sutin for their collaboration, suggestions and comments in various stages of the work presented here. I am particularly grateful to the late Dr. Morris Perlman for his constant inspiration in the early stages of this work. EXAFS experiments were carried out at the Stanford Synchrotron Radiation Laboratory (SSRL) which is supported by DOE and at the Cornell High Energy Synchrotron source (CHESS) which is supported by NSF. Assistance of their staff is acknowledged. This research was carried out at Brookhaven National Laboratory under contract DE-AC02-76CH00016 with the U.S. Department of Energy and supported by its Division of Chemical Sciences, Office and Basic Energy Science. Support from the Department of Chemistry. The Chinese University of Hong Kong where part of the manuscript was prepared is also acknowledged.

6 References

1. Synchrotron Radiation Research (1980) (Ed) Winick H, Doniach S, Plenum Press, New York
2. Handbook on Synchrotron Radiation Vol. 1 (1983) (Ed) Koch EE, North-Holland, Amsterdam
3. (a) EXAFS and Near Edge Structure (1983) (Ed) Bianconi A, Incoccia L Stipcich S, Springer New York;
   (b) EXAFS and Near Edge Structure III (1984) (Ed) Hodgson KO, Hedman B, Penner-Hahn JE, Springer New York. These are the proceeding of two recent international conferences on the subject.
4. Kronig R de L, (1931) Z Physik 70: 317, (1932) 75: 191, (1932) 75: 468
5. Stern EA (1974) Phys. Rev. B 10: 3027
6. (a) Sayers DE, Stern EA, Lytle FW (1971) Phys. Rev. Lett. 27: 1204; (b) Lytle FW, Sayers DE, Stern EA (1975) Phys. Rev. B 11: 4825; (c) Stern EA, Sayers DE, Lytle FW (1975) ibid. 11: 4836
7. Ashley CA, Doniach S (1975) Phys. Rev. B 11: 1279
8. Lee PA, Pendry JB (1975) ibid. 11: 2795
9. Beni G, Platzman PM (1976) ibid. 14: 1514
10. The technique is now well established for general use; for a review see Lee PA, Citrin PH, Eisenberger P, Kincaid BM (1981) Rev. Mod. Phys. 53: 769
11. Sham TK (1983) J. Am. Chem. Soc. 105: 2269
12. Sham TK (1985) Phys. Rev. B 31: 1888; (1985) ibid. 31: 1903
13. Bianconi A, Dell' Ariccia M, Durham PJ, Pendry JB (1982) ibid. 26, 6502
14. Benfatto M, Natoli CR, Bianconi A, Garcia J, Marcelli A, Fanfoni M Davoli I, ibid. to be published; Bianconi A this volume
15. Eisenberger P, Kincaid BM (1975) Chem. Phys. Lett. 36: 134
16. Sandstrom DR, Dogen HW, Lytle FW (1977) J. Chem. Phys. 67: 473
17. Fontaine A, Legarde P, Raoux D, Fontana MP, Maisano G, Milgiliardo P, Wanderlingh F (1978) Phys. Rev. Lett. 41: 504

18. Morrison TI, Reis, Jr. AH, Knapp GS, Fradin FY, Chen H, Klippert TE (1978) J. Am. Chem. Soc. *100*: 2362
19. Sham TK, Hastings JB, Perlman ML (1081) Chem. Phys. Lett. *83*: 391
20. Sham TK, Hastings JB, Perlman ML (1980) J. Am. Chem. Soc. *102*: 5904
21. Sham TK, Brunschwig BS (1981) ibid. *103*: 1590
22. Brunschwig BS, Creutz C, Macartney DH, Sham TK, Sutin N (1982) Faraday Discuss. Chem. Soc. *74*: 113
23. For example, see Ohtaki H, Yamaguchi T, Maeda M (1976) Bull. Chem. Soc. Jap. *49*; 701
24. Enderby JE Neilson GW (1980) Adv. Phys. *29*: 323
25. Jaklevic J, Kirby TA, Klein, MP, Robertson AS, Brown GS Eisenberger P (1977) Solid State Commun. *23*: 679
26. Basolo F, Pearson RG (1967) Mechanism of Inorganic Reactions, Wiley, New York
27. Nakamoto K (1978) Infrared and Raman Spectra of Inorganic and Coordination Compounds 3rd Ed., Wiley, New York; the Fe^{3+} value was recently reported by Best SP, Beattie JK Armstrong RS (1984) J. Chem. Soc. Dalton Trans. 2611
28. Cyvin SJ (1968) Molecular Vibrations and Mean Square Amplitudes, Elsevier, Amsterdam
29. Eigen MV (1963) Ber. Bunsenges, Physik. Chem. *67*: 753; Poupko R Luz Z 1972) J. Chem. Phys. *37*: 307
30. Marcus RA (1966) Ann. Rev. Nucl. Sci. *15*:
31. Sutin N (1982) Acc. Chem. Res. *9*: 275
32. Newton MD Sutin N (1984) Ann. Rev. Phys. Chem. *35*: 437
33. Jortner J (1979) Philos. Mag. *40*: 319
34. Sham TK (1986) Acc. Chem. Res. *19*: 99
35. Habenschuss A, Spedding FH (1980) J. Chem. Phys. *73*: 442
36. Sham TK (1983) ibid. *79*: 1116
37. Sham TK (1986) ibid. *84*: 7054
38. Sham TK, Heald SM (1983) J. Am. Chem. Soc. *105*: 5142
39. Sham TK Holroyd RA (1984) J. Chem. Phys. *80*: 1026; (b) in Ref. 3b) p. 504
40. Holroyd RA Sham TK (1985) J. Phys. Chem. *89*: 2909
41. Sham TK, Holroyd RA, to be submitted for publication
42. Sham TK, Holroyd RA Munoz RC (1986) Nucl. Inst. Meth. A *249*: 530; (1986) J. de Phys. Coll. C8-153

Order and Disorder in Low Dimensional Materials: Beyond the First Coordination Sphere with EXAFS

Alain Michalowicz[1, 2], Michel Verdaguer[3, 4], Yves Mathey[3] and René Clement[3]

1 L.U.R.E., Universite de Paris-Sud, 91405 Orsay, France
2 Laboratoire de Physicochimie Structurale, Universite de Paris Val de Marne, 94000 Creteil, France
3 Laboratoire de Spectrochimie des Elements de Transition, U.A. C.N.R.S. 420, Universite de Paris-Sud, 91405 Orsay, France
4 Ecole Normale Superieure, Le Parc, 92211 Saint-Cloud, France

Topics in Current Chemistry, Vol. 145
© Springer-Verlag, Berlin Heidelberg 1988

The strengths and weaknesses of EXAFS in determining the radial distribution of the first nearest neighbours in coordination compounds are well known.

We point out here how local information given by EXAFS upon the first coordination sphere can be used to characterize short or long range order or disorder in low dimensional materials.

Low temperature measurements, use of several absorbing species in the same compound, comparison between seemingly contradictory X-Ray diffraction and EXAFS data allow to solve three dimensional structural problems related to physical properties.

The materials studied are insulating uni- or bimetallic magnetic chains and pure or intercalated MPS$_3$ layers.

1 Introduction

Low-dimensional (LD) materials are of renewed interest for chemists and phycisists. On the one hand, one-dimensional (1D) linear chains [1] or two-dimensional (2D) planar arrays [2] (Fig. 1) allow to study fundamental properties of matter more simply than three-dimensional (3D) lattices: much synthetic work has been stimulated by the search for materials with structures and properties close to the 1D and 2D theoretical models, simpler than the 3D ones. On the other hand, such systems present by themselves new, exciting, and sometimes predictable and tunable physical and chemical properties.

We are ourselves engaged in such a synthetic endeavour [3-12] and we report here on some typical structural problems encountered in coordination and solid state chemistry of these low-dimensional materials. Indeed, the properties of these compounds are very sensitive to order and disorder, purity and defects and chemical modifications such as doping, insertion or intercalation; their understanding needs careful structural characterization. Extended X-Ray Absorption Fine Structures (EXAFS) spectroscopy possesses some unique features which can contribute to the solution of well-defined cases to set up correlations between structural and physical properties.

(1) The most frequently encountered problem is the obtention of structural local information on badly organized solid materials, when single crystals are not available.

When various possible coordination sites can accomodate different absorbing atoms, EXAFS is able to collect independent as well as overlapping informations by a multiedges study of the same material. Sect. 2.1 shows why addition and cross-checking of independent data is of the utmost interest, precisely to solve structural order and disorder problems beyond the first coordination sphere.

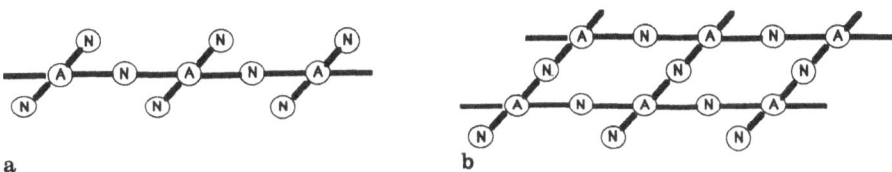

a b

Fig. 1. Low-dimensional materials: **a)** 1D chain; **b)** 2D array

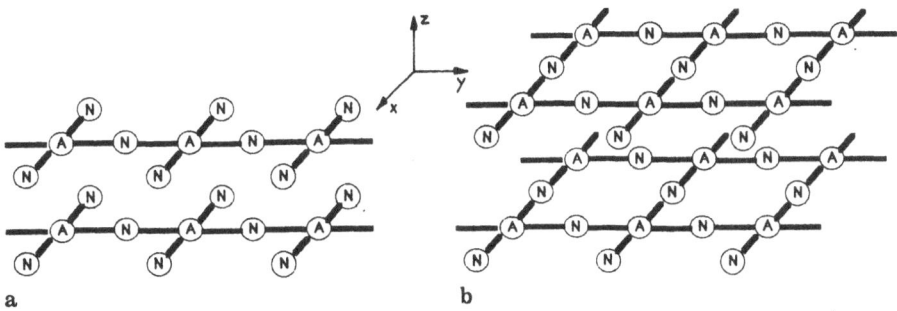

a b

Fig. 2. Anisotropy in low-dimensional materials: **a)** axial anisotropy in 1D; **b)** planar anisotropy in 2D

In microcrystalline or amorphous materials, EXAFS has proved its usefulness since a long time. Limits and weaknesses of the technique have also been put forward: variation of the EXAFS signal with temperature, occurrence of multiple scattering or destructive interference effects which complicate the extraction of the information. Sect. 2.2 presents how it is possible to overcome these apparent drawbacks and to obtain from them more information about the coordination site, its possible distortion and even about 3D organization of the material.

(2) Another common question is the resolution of the disorder present in 1D or 2D single crystals made up of different coordination sites with several transition ions. In this case, the trouble rises from seemingly contradictory results between X-Ray Diffraction (XRD) and EXAFS data. Sect. 2.3 discusses how short range order (SRO) and long range order (LRO) pictures can merge to give a more accurate description of the disordered structure.

We apply these considerations to solve structural problems in real systems in Sect. 3 (uni or bimetallic linear chains) and in Sect. 4 (bidimensional layers).

(3) The anisotropy of 1D and 2D materials provides advantage to solve the above mentioned structural problems: the surroundings of the absorbing atom is very different in the three crystallographic directions (Fig. 2). Generally, this does not occur at the first coordination sphere level but beyond this first shell. This consequence of anisotropy is crucial to collect information not only about local arrangement but also about 3D order or disorder.

Most of the experimental work and analyses were performed at the french synchrotron radiation facility, the Laboratoire pour l'Utilisation du Rayonnement Electromagnetique issu des anneaux de collision d'Orsay (LURE). The appendix gives a short account of the recording and processing of the experimental data.

2 EXAFS, a Short Range Order Method Used to Solve 3D Problems: Some Useful Features

2.1 Introduction

EXAFS was transformed from a laboratory curiosity to an useful structural technique when Sayers, Stern and Little proposed to Fourier transform the high energy oscillatory part of the absorption spectrum $\chi(k)$, function of the wave vector of the ejected photoelectron, k [13-17]:

$$\chi(k) = \sum_{j=1}^{N} \frac{1}{kR_j^2} \cdot N_j \cdot S_i(k) \cdot F_j(k) \cdot \exp - (2\sigma_j^2 k^2) \cdot \exp(-2Rj)/\lambda(k)$$

$$\times \sin(2kRj + \Phi_{i,j}(k)) \tag{1}$$

Formula (1) is the basis of the short-range-order (SRO), single-electron, single-scattering formulation of EXAFS. It gives the number of neighbours N_j, their distances from the absorbing atom i, R_j and the corresponding Debye-Waller factor σ_j, when the amplitude factor $F_j(k)$, the phase shift $\varphi_{i,j}(k)$, the electron mean free path $\lambda(k)$, the many-body effects amplitude reduction factor $S_i(k)$ are known.

EXAFS gives therefore the radial distribution function (RDF) centered at the absorber and involving only pairs of atoms formed by the absorber itself and the other atoms. It is highly specific in giving local structural information in the immediate vicinity of each absorbing centre with an accuracy of 0.01–0.03 Å up to 3–4 Å and semiquantitative results up to around 5–6 Å, whatever the organization of the material may be (crystalline, amorphous, solution . . .).

This key advantage of EXAFS for investigating the structure of condensed systems where local informations are sought allows to deal with periodic systems where bond characteristics or mean square displacements modifications upon pressure, temperature, irradiation, etc. are needed [18]. This applies also when investigating and modeling highly disordered or aperiodic systems. As a matter of fact, despite severe limitations (either inherent to the technique — loss of low momentum transfer information — or related to the relative uncorrectness of the data treatment) EXAFS is unrivalled for the study of glasses [19], molten salts [20], electrolytes [21] or biological materials [22].

In coordination chemistry also, this local probe has been used to study badly crystallized compounds.

The first problem is of course to obtain indications, routinely given by room temperature EXAFS data to characterize coordination spheres and their possible local distortion by chemical modifications. As a matter of fact, in an isolated mononuclear complex, EXAFS at the K edge of the metallic atom would give the RDF corresponding to the successive coordination spheres of light atoms, while other remote mononuclear complexes would be ignored (Fig. 3).

It is possible to go beyond this point and to gather more information by using other ressources of the technique. For example, in polynuclear complexes, either binuclear, linear or planar (Fig. 2), the picture can be enlarged since in some conditions it is possible to "see" also the next neighbour metal (Fig. 3).

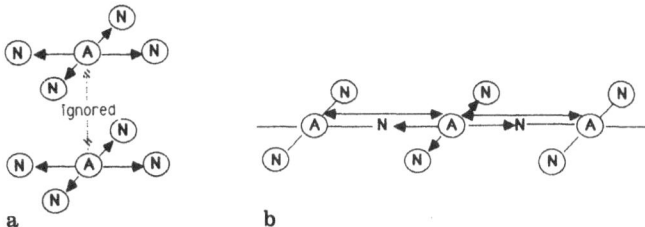

Fig. 3. EXAFS in mononuclear and polynuclear complexes: a) remote complexes are ignored; b) next nearest heavy neighbours can be seen

2.2 Absorption at Several Edges

The possibility for EXAFS to collect independent data using different edges is well-documented [23a] and is not developed at length here.

We recall that it allows independent study of the local structure around each absorber and we mention a few illuminating examples of such a multiedge approach:
(i) in a bimetallic compound where the two absorbing centres are far from each other (Fig. 4a), the surroundings of the two metals can be explored independently; a special application is the study of impurities and doping elements in a matrix;

(ii) in a binuclear complex with two absorbers close from each other an interesting cross-checking of the data allows to find out the AB distance and sometimes to find the ANB angles (Fig. 4b et c);

(iii) in a bimetallic (A, B) material where the spectrum displays an heavy atom peak around absorber A: the absence of an heavy atom peak at the AB distance in the B spectrum can be used to identify A as the first neighbour of A [5];

(iv) in a mononuclear complex where the ligand bears an absorbing atom N, metal-centered and ligand-centered views of the complex become possible (Fig. 4d et e) [24, 25];

(v) in a 1D chain with rigid ligands, the possible distortion of the metallic site upon chemical modification can be put in evidence (Fig. 4f–i); this opportunity is taken in the study of copper(II) bromanilate chains [4].

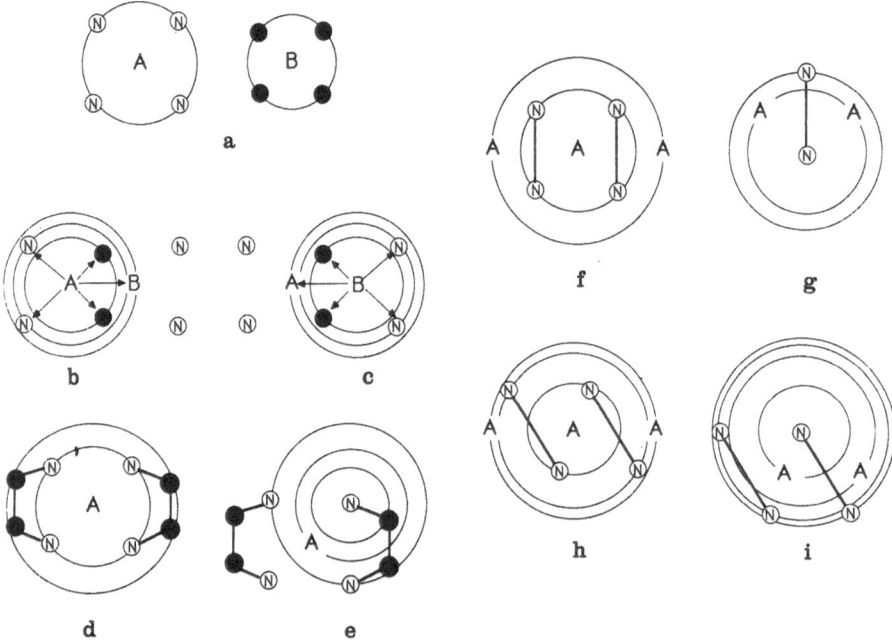

Fig. 4. Use of several edges spectra: **a)** bimetallic compounds with two remote absorbers A and B; only nearest neighbours N are seen; **b)** A-centered view and **c)** B-centered view in a bimetallic complex with two close absorbers A and B; **d)** Metal-centered view and **e)** ligand-centered view in a complex; **f–h)** Metal-centered and **g–i)** N ligand-centered views of the local distortion in a 1D compound: both spectra gain new peaks; if the rigid ligand bears two N absorbers, the N—N distance remains unchanged

2.3 Undesirable Effects at Work to Gain Information

We turn now to effects which are commonly considered as a complication of EXAFS spectra and make the structural interpretation difficult. Among these effects which disturb the information, three are of the utmost importance:

a) the temperature dependence of the Debye-Waller factor [26];
b) the multiple scattering in colinear systems [27];
c) the destructive interference effect [28].

We show hereunder that they can be used to bring valuable new informations beyond the classical first distances and coordination numbers.

2.3.1 Temperature Dependence of the Debye-Waller Factor

The Debye-Waller factor can be considered to have two components $\sigma(stat)$ and $\sigma(vib)$ arising from static disorder and thermal vibrations respectively. In first approximation (symmetric pair distribution and harmonic vibration):

$$\sigma^2 = \sigma^2(stat) + \sigma^2(vib) \tag{2}$$

An increase of σ — due to static disorder or to thermal motion — results in a loss of amplitude of the corresponding RDF peak rather than in a broadening of this signal. This well-known behaviour [23b] is due to the finite k wave vector range available, particularly in the absence of any signal in the 0–$3\,\text{Å}^{-1}$ range [26]. Moreover, when the Debye-Waller term is too large, the EXAFS signal completely vanishes. This has been considered as a serious limitation of EXAFS. In the case of LD coordination compounds, the thermal variation of the Debye-Waller factor can be used as a significant advantage

In such compounds, two kinds of EXAFS signals are observable: those coming from scatterers belonging to the molecule of the absorber (INTRAmolecular signals) and those belonging to other molecules (INTERmolecular signals).

As the INTRAmolecular movements are largely correlated, their corresponding Debye-Waller terms are only slightly temperature dependent. On the other hand, INTERmolecular signals (metal-metal signal from two adjacent metallic sites in a solid for example) are expected to be very sensitive to thermal motion. For example, Fig. 5 illustrates some aspects of the effect: at high temperature (Fig. 5a) only the N neighbours are seen from A, with static and thermal disorder. Other neighbours (X, A) can be ignored if their Debye-Waller factor is too large. At low temperature (Fig. 5b), static disorder can be resolved on N; the X neighbours, close to A and bound to it, can appear as well as remote heavy atoms A. Fig. 6 presents the simulation of an hypothetical molecule with a central copper atom, 4 oxygen atoms at 2 Å and 4 car-

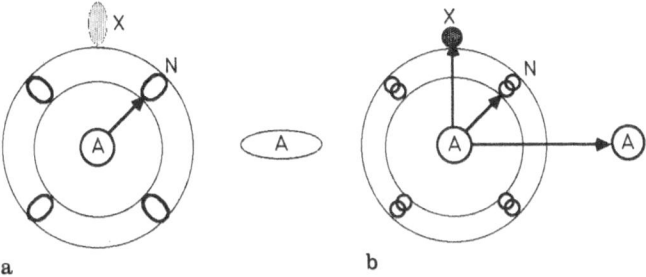

a   b

Fig. 5a and b. Temperature dependence of the SRO picture in ordered materials: a) isolated mononuclear complex; b) appearance of new heavy atoms peaks at low temperature

113

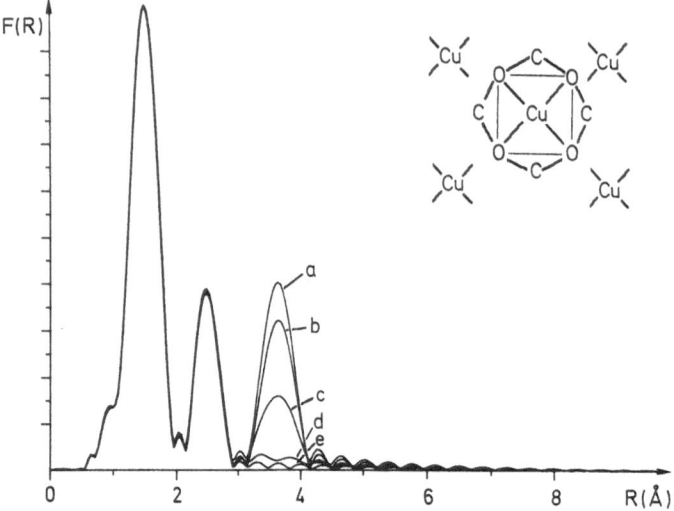

Fig. 6. Simulation of the thermal variation of the Debye-Waller factor in the case of a mononuclear complex surrounded by 4 identical entities (see text)

bon atoms at 3 Å, each molecule is surrounded by 4 equidistant identical molecules with a Cu—Cu distance of 4 Å. The two first Debye-Waller factors are considered as temperature independent ($\sigma = 0.02$ Å) and σ(Cu) varies from 0.02 at "low temperature" (Curve 6.a) to 0.2 Å at "high temperature" (Curve 6.d). When σ(Cu) $= 0.2$ Å, the Cu—Cu signal is almost absent and the spectrum looks like the simulated spectrum free from the copper signal (Curve 6.e).

Most of the coordination compounds EXAFS spectra reported in the literature were recorded at room temperature and most of the structural informations, up to the very recent period, is thus concentrated on the two or three INTRAmolecular coordination shells. Thus, reported metal-metal signals are essentially those of metals directly bound or interacting via a monoatomic bridging ligand. By cooling the sample at nitrogen or helium temperature one may expect to observe INTERmolecular metal-metal contributions characterizing the molecular packing (Fig. 3). Comparison between room and low temperature data is thus a good way to discriminate between INTRA and INTERmolecular EXAFS contributions, and therefore to get 3D information with a local probe.

2.3.2 Multiple Scattering Effect

The classical EXAFS formula (1) assumes that only single backscattered photoelectrons contribute to the signal. This is generally true in an energy range far from the edge — where a multiple scattering mode occurs —, except in systems where two scattering atoms are aligned, or nearly aligned with the absorber. In such linear A—B—C systems, the EXAFS signal of C is perturbed both in phase, with an apparent decrease of the A—C distance, and in amplitude, with an enhanced intensity of the C peak. The effect is very sensitive to the A—B—C angle [23c]. Once again, multiple scattering is considered as an undesirable complication. A complete quantitative treatment of this effect needs the use of a more complicated formalism than formula (1). However, it has been shown that an empirical treatment with the classical formula and with parameters extracted from model compounds can account for this effect [27]. Monodentate ligands such as CO, CN—, pyridine, histidine, are known to produce this effect on the second or on the third coordination shell. Since the multiple scattering effect is very sensitive to angles, it can be used as a signature of such ligands and their coordination mode to the metal. Chelating ligands can also present such an effect but in very accute conditions: we shall see that it can be disclosed in oxalato or chloranilato complexes with precise configurations of the metallic ion and of the ligand.

Furthermore, in 1D or 2D complexes the alignement of heavy atoms can be displayed by such an effect, as shown schematically in Fig. 7 or in Fig. 23 in the experimental case of tetramethylammonium manganese chloride (TMMC), a well-known 1D antiferromagnetic chain.

 **Fig. 7.** Multiple scattering effect in a 1D linear chain

2.3.3 Destructive Interference Effect

When two EXAFS contributions lie at close frequencies with a phase difference close to π, a partial cancellation of both signals occurs and there is loss of information [28]. In coordination chemistry, this effect was observed by Goulon et al. on porphyrins [28b]. A simulation of destructive interference is shown in Fig. 8: the structure is the same as in Fig. 6 but we add now an extra signal from a light atom shell called E, lying at 4.05 Å with a constant Debye-Waller factor. When $\sigma(Cu) = 0.02$ Å, the Cu—Cu and Cu—E signals have similar amplitudes and are partially cancelled (Fig. 8a). When $\sigma(Cu)$ increases, the Cu—Cu signal disappears progressively as in Fig. 6, and the Cu—E signal apparently raises (Fig. 8d).

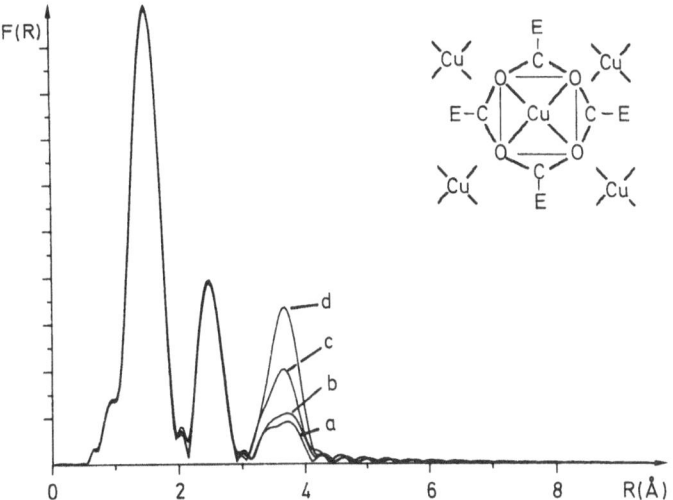

Fig. 8. Simulation of the destructive interference effect (see text)

Identification and analysis of this interference destructive effect are not obvious. Since the signal is partially lost in the experimental spectrum, it cannot be recovered without complementary information, coming for example from other structural data or from chemical evidences. In the simulation presented above, any attempt to find by fitting procedures both the number of neighbours and the Debye-Waller factors for the Cu and E shells will fail. However, if these values are known for one of the neighbours, they can be used to extract those of the other by differential fit [28].

Once more, an unwanted effect, properly handled, results in supplementary information about the first coordination sphere and beyond.

2.4 Discrepancies in Short Range and Long Range Order Pictures

We turn now to another very useful tool to solve disorder problems, namely the apparent discrepancy appearing sometimes between EXAFS and X-Ray diffraction data:
— on the one hand, as shown in Sect. 2.1, EXAFS is particularly suitable to investigate the various coexisting local arrangements (around dilute species in solid state solution for example); if we ignore the near edge structures, where higher correlation effects are present, the EXAFS signal is a pair correlation function between the absorber and its nearest neighbours:

$$\text{EXAFS SIGNAL} = \text{RDF} \sum_{\text{nearest } j} A_i N_j \tag{2}$$

where A_i stands for the absorber and N_j for the nearest scatterers;
— on the other hand, XRD reflects the translational symmetry of the investigated system. It provides interatomic distances but the measured scattering amplitude leads to a Distribution Function (DF) weight-averaged over the various pairs of scattered atoms:

$$\text{XRD SIGNAL} = \text{DF} \sum_{\text{all } i} \sum_{\text{all } j} A_i \cdot A_j \tag{3}$$

where A_i and A_j are the various scatterers.

Consequently, XRD will not be able to appreciate or even, in some cases, to suspect the existence of substitutional disorder or nonhomogeneous thermal disorder.

In other words, EXAFS allows a detailed insight into the short range order (SRO) of the system centered at the absorbing atom while XRD assumes the existence of a long range order (LRO) when such order exists but gives information averaged over all the unit cells of the investigated system (Fig. 9).

A wide open series of solid solution systems, such as ionic alkali halides $KCl_{(1-x)}Br_x$ [29], binary and pseudobinary metallic or semiconducting alloys $Ag-Cu$ [30], $Al-Cu$ [31], $Cu-Ti$ [32], $Ga_x In_{(1-x)}As$ [33], $GaAsySb_{(1-y)}$ [34], carbides TiC_x [35], intermediate valence rare earth compounds $Sm_{(1-x)}Y_xS$ [36], rare gases $Xe-Ar$ [37], etc. has already been investigated by EXAFS spectroscopy.

In some cases, the picture of the local arrangement extracted from EXAFS data (SRO description) does not coincide with the picture provided by XRD experiment (LRO description). This puzzling feature constitutes by itself a strong indication

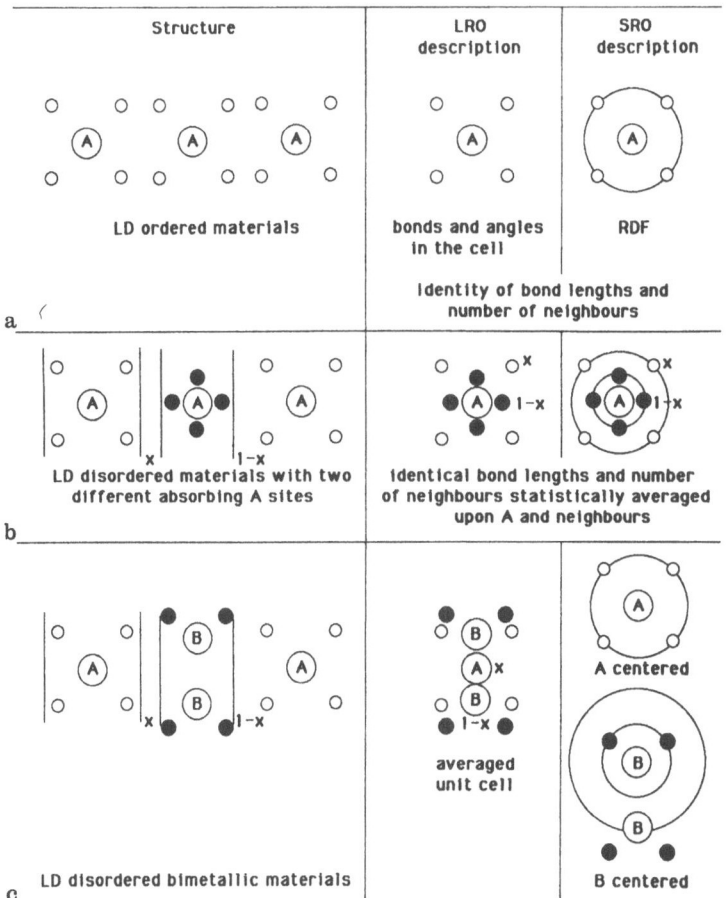

Fig. 9 a–c. Schematic comparison of SRO and LRO pictures in some ordered and disordered materials (see text)

for the existence of partial disorder in the structure of the solid solution under study. Indeed, when the occupation of the impurity site is only statistical, XRD data contain information averaged over the totality of similar sites in the sample (independently of their occupation by impurity or diluting atoms), while EXAFS data contain information related only to this site when occupied by an impurity atom (Fig. 9c). On the contrary, when a solid system with an ordered distribution of impurity atom is investigated, both XRD and EXAFS techniques lead qualitatively and quantitatively to the same information. In such cases, of course, coincidence between SRO and LRO description is expected (Fig. 9a, b).

Turning back to the above mentioned solid solutions, the comparison between EXAFS and XRD results allows to establish and in several cases to solve such statistical disorder problems. Particularly, when the dilute species occupies one type of sites only, data analysis allows a detailed description of the local arrangement around impurity (coordination number, distances to the nearest and next nearest neighbour atoms, magnitude of distortion of the site, etc.) and provides information of outstanding importance in the appreciation of the bulk properties.

3 Order and Disorder in Unidimensional Compounds

3.1 Ribbon Structure and Site Distortion in Copper Bromanilate Chains

We begin with a case which can be considered as a nice example of structure elucidation in LD materials using only room temperature data but combining two edges results, namely copper and bromine [4].

The first structural and magnetic properties of copper(II) chloranilate appeared in 1960 [38]. This compound behaves as a $S = 1/2$ antiferromagnetic chain with a J value around -24 cm^{-1}. The coupling constant J is defined by the phenomenological Heisenberg Hamiltonian:

$$\hat{H} = -J \sum_{i=1}^{N-1} \hat{S}_i \cdot \hat{S}_{i+1} \tag{4}$$

$\hat{S}_i$ and $\hat{S}_{i+1}$ are the quantum spin operators related to spins i and i + 1 respectively The explanation of such an important coupling between paramagnetic ions with polyatomic ligands is not a priori evident and, in the absence of XRD data, two structures were proposed (Fig. 10). Another magnetic feature appears puzzling: the fixation of two ammonia molecules (or two basic ligands L) per copper(II) ion induces an important reduction of the J coupling constant (-3.8 cm^{-1} in the ammonia adduct $Cu(C_6O_4X_2)$ $(NH_3)_2$ with X = Cl.

Bromanilato compounds (X = Br) present the same magnetic properties and the same lack of unambiguous structural determination. Nevertheless, their structure can be elucidated by carrying EXAFS experiments at two readily available edges (Copper at 9000 eV and Bromine at 13,000 eV). The modulus of the R-space spectra are displayed in Fig. 11, whereas Tables 1 and 2 give the corresponding fitting results.

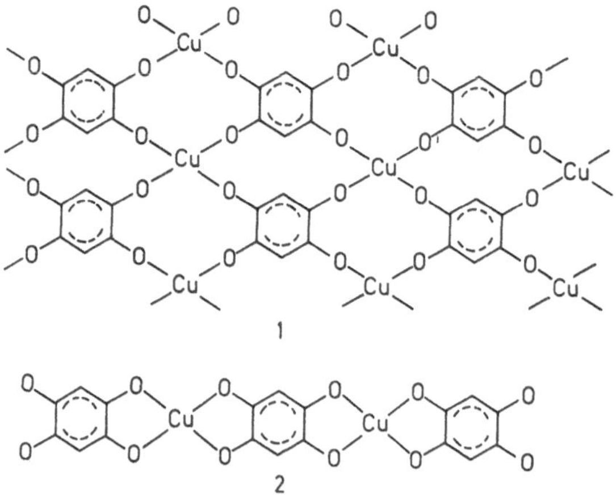

Fig. 10. Two proposed structures for chloranilato and bromanilato copper (II) complexes

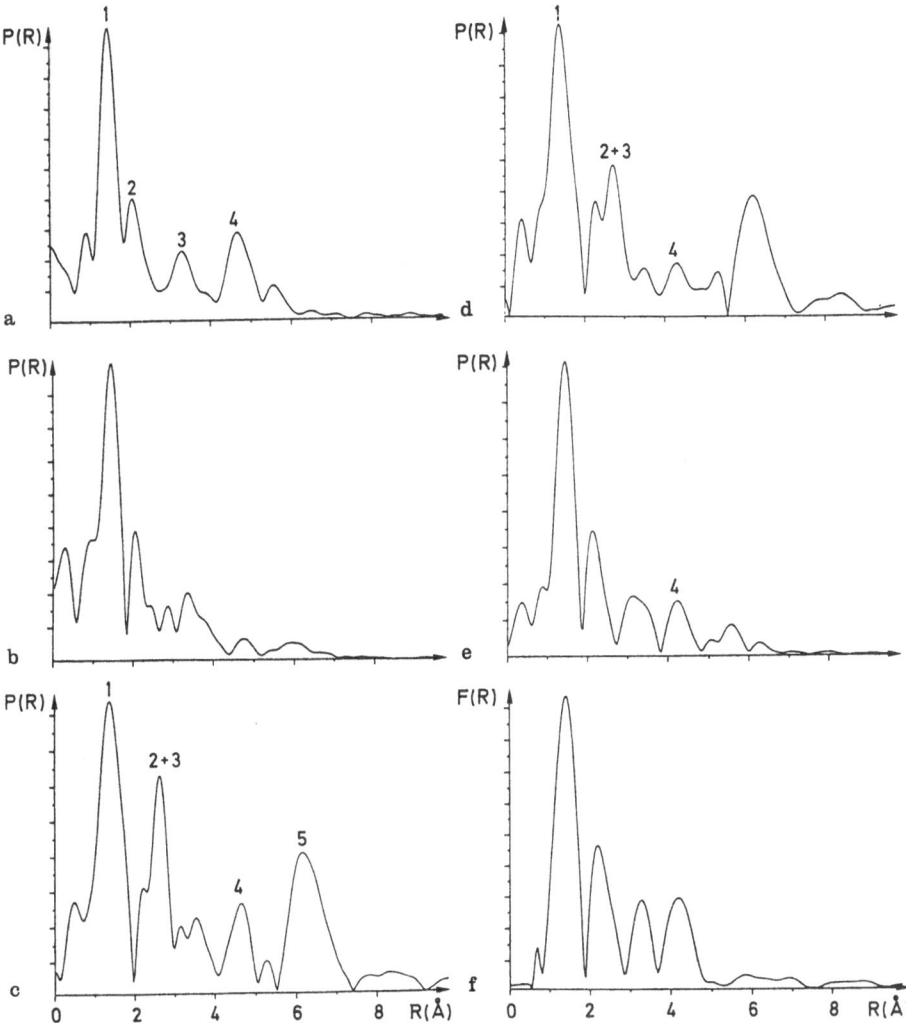

Fig. 11 a–f. Fourier transform spectra of chloranilato and bromanilato derivatives at copper and bromine edges.
(a) $Cu(C_6O_4Br_2)$; (b) $Cu(C_6O_4Br_2)$ $(NH_3)_2$; (c) $Cu(C_6O_4Br_2)$; (d) $Cu(C_6O_4Br_2)$ $(NH_3)_2$;
(e) $Cu(C_6O_4Cl_2)$, 295 K; (f) $Cu(C_6O_4Cl_2)$, 30 K.
The adsorption edge is italic.

Three distances are unambiguously determined at first sight: Cu—O = 1.95 ± 0.02 Å, Cu—C = 2.67 ± 0.04 Å both with small Debye-Waller factors and Br—C = 1.86 ± 0.02 Å. In addition, it is possible to attribute peak 4 in spectra a and c in Fig. 11 to Cu—Br or Br—Cu at a distance around 5 Å: the peak is intense (heavy atom peak); it is at the same distance at both edges; in chloranilato derivatives the corresponding peak is, as expected, slightly displaced towards shorter distances and less intense (Fig. 11 e); such an intense intermolecular heavy atom peak at such a distance is unlikely at room temperature. From these data alone, it is therefore possible to de-

Table 1. Copper chloranilate and bromanilate chains at the Copper edge: two-shell fitting results

	E_0/eV	R/Å	$2\sigma^2$/Å^2	ϱ/%
$Cu(C_2O_4) \cdot {}^1/_3\, H_2O$				
Cu—O	9002.79	1.97	0.019	
Cu—C	8992.53	2.64	0.027	0.57
$Cu(C_2O_4)\,(NH_3)_2 \cdot 2\,H_2O$				
fitted as: Cu—O	8994.80	1.98	0.026	
Cu—C	9001.17	2.71	0.046	0.97
$Cu(C_6O_4Cl_2)$				
Cu—O	9002.29	1.95	0.017	
Cu—C	8992.23	2.71	0.010	0.36
$Cu(C_6O_4Cl_2)\,(NH_3)_2$				
fitted as: Cu—O	8995.25	1.97	0.020	
Cu—C	8999.41	2.65	0.029	0.68
$Cu(C_6O_4Br_2)$				
Cu—O	9002.81	1.95	0.011	
Cu—C	8990.43	2.67	0.010	1.5
$Cu(C_6O_4Br_2)\,(NH_3)_2$				
fitted as: Cu—O	8993.55	1.98	0.021	
Cu—C	9000.83	2.65	0.048	1.48

Table 2. Copper bromanilate: fitting results for shells involving bromine atoms[a]

	E_0/eV	R/Å	$2\sigma^2$/Å^2	ϱ/%
$Cu(C_6O_4Br_2)$				
Br—C2	13483.2	1.86	0.011	6.7
Br—(C1 + O1)	13477.5	2.84	0.012	1.3
		3.11	0.021	
Br—Cu	13476.2	4.99	0.017	2
Br—Br	13475.3	6.55	0.010	12.5
$Cu(C_6O_4Br_2)\,(NH_3)_2$				
Br—C2	13484.12	1.87	0.012	8.5
	13477.42	2.84	0.009	
Br—(C1 + O1 + O2)		3.14	0.025	0.59
		3.17	0.018	
$Cu(C_6O_4Br_2)$				
Cu—Br	8984.25	5.04	0.015	4

[a] The absorbing atom is italic

monstrate with simple geometrical means that Cu(II) must lie in the plane and that a layer structure is incompatible with the determined Cu—O and Cu—C distances.

EXAFS supports therefore the planar ribbon structure shown in Fig. 12a for $Cu(C_6O_4X_2)$. Our conclusion is confirmed by the 30 K spectrum of copper chloranilate, with its intense peak 4 (Fig. 11f)[38c].

EXAFS spectroscopy, as well as magnetic properties turns to be very sensitive to the fixation of two ammonia molecules (Fig. 11b, d): increased complexity of the spectrum, particularly for Peaks 2 and 3, at both Cu and Br edges, reduced intensity of the Cu—Br Peak 4. These observations can be related simply to a site distortion of the square CuO_4 entity, leading to a copper surroundings with 4 short bonds (2 Cu—N

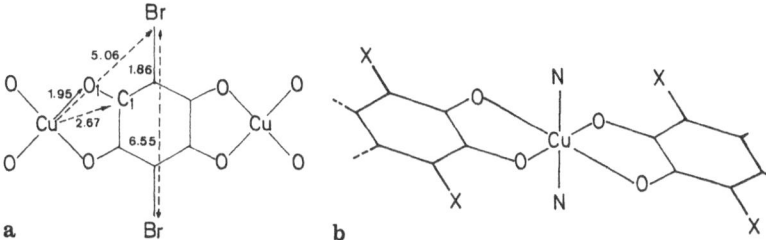

Fig. 12a and b. Proposed structure from EXAFS data: a) ribbon structure for bromanilate chain; b) distortion of the copper site in the diammino adduct

and 2 Cu—O) and two long Cu—O bonds (Fig. 12b). This situation was already known from XRD in $Cu(C_2O_4)$ $(NH_3)_2 \cdot 2$ H_2O [39]. The quantitative treatment of the data confirms this interpretation, in particular with the enhanced values of σ in all the ammonia adducts. This distortion can be then interpreted as the sign of an orbital reversal of the magnetic orbitals, which explains the reduced J constant by a poorer overlap through the bromanilato bridge [4]. Two other points related to our discussion in Sect. 2.3.2 may be raised: first, the EXAFS determined Cu—Br (or Br—Cu) distance appears too weak compared to geometrical evaluations (5.06 Å). This is naturally explained by the Cu—O_1—Br multiple scattering effect appearing clearly in Fig. 12a; secondly, the large Peak 5 at 6.55 Å in spectra c and d at Br edge is quite unusual; it does not vary with the distortion of the copper site: it can be assigned to an intramolecular Br—Br distance (6.58 Å by geometry) with, once more, an enhanced intensity due to a focusing effect through the alignment Br—C..C—Br (Fig. 12a).

3.2 Copper(II) Oxalate

3.2.1 Introduction

The structure and the magnetic properties of copper oxalate, $CuC_2O_4 \cdot 1/3$ H_2O has been discussed for many years. This compound cannot be obtained as single crystals suitable for XRD structure determination. It is not isostructural with the other metallic bivalent oxalates. In order to explain the 1D antiferromagnetic behaviour, various structures were proposed either using powder diffraction patterns [40] or assuming magnetic [41] or structural [42] analogies. Such structures are shown in Fig. 13. The first hypothesis (Fig. 13a) is based on a similarity with the copper(II) acetate dimer: copper oxalate and copper acetate antiferromagnetic coupling constants are large and it is tempting to correlate this property with a short distance between the two paramagnetic ions [41a]. The second one is the structure solved by powder diffraction data [40]. The third one assumes an analogy with $Na_2Cu(C_2O_4)_2 \cdot 2$ H_2O, the structure of which was determined by single crystals XRD [42a]. The second and third models differs essentially from the first one by their ribbon structure: in both, the copper(II) ions is bischelated by $(C_2O_4)^{2-}$ (Fig. 13b) and builds 1D planar ribbons. They differ by the packing of the ribbons in the crystal. In the structure proposed by

121

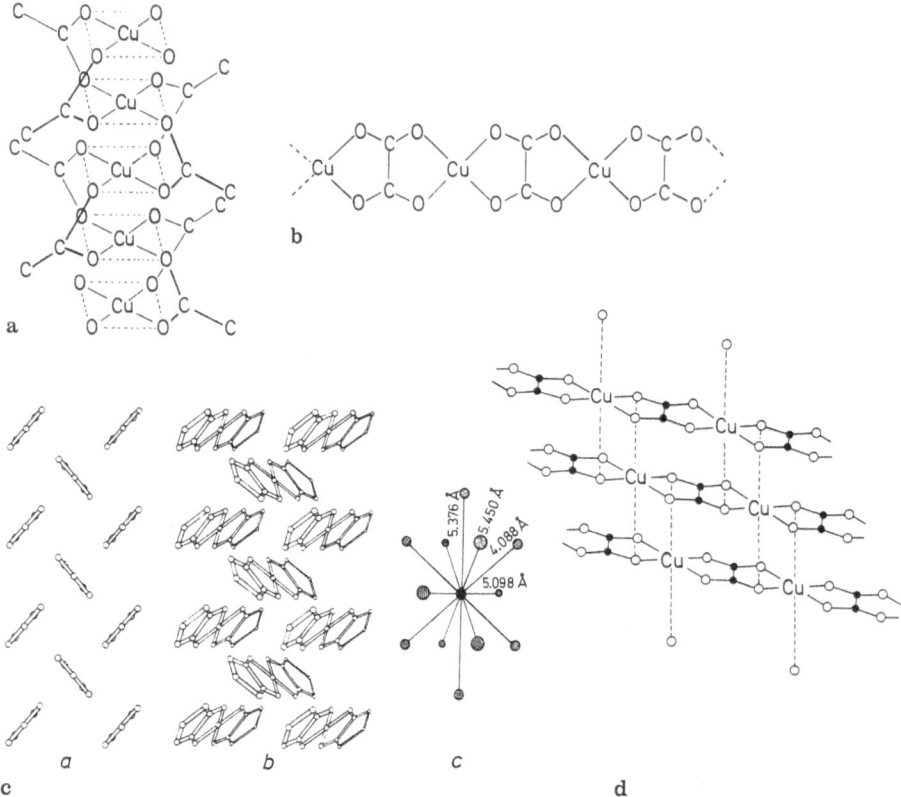

Fig. 13a–d. Proposed structures for copper (II) oxalate; **a)** copper acetate analog (41); **b)** ribbon structure; **c)** 3D structure after powder diffraction (40); **d)** 3D structure by extrapolation of XRD data of $Na_2Cu(C_2O_4)_2, 2 H_2O$ (42a)

Schmittler (Fig. 13c), adjacent ribbon planes are almost perpendicular and each copper "sees" four copper neighbours at 4.09 Å. In the structure proposed by Gleizes (Fig. 13d), the ribbons are parallel and each copper has 2 Cu neighbours at about 3.5 Å as in $Na_2Cu(C_2O_4)_2$.

3.2.2 The Ribbon Structure: Room Temperature Data

As the knowledge of CuC_2O_4 structure is necessary to explain its magnetic properties, an EXAFS analysis of this compound was undertaken in order to settle the correct model, since it is typically the kind of problem that EXAFS can solve in 3D organized compounds.

The first EXAFS study of copper oxalate is based upon room temperature data only [3]. The comparison of CuC_2O_4 EXAFS spectrum with those of model compounds

---

Fig. 14a–j. Fourier transform in the R-space of $Cu(C_2O_4)$, 1/3 H_2O at 300 K **a)** and 30 K **b)**; $Na_2Cu(C_2O_4)_2$, 2 H_2O at 300 K **c)** and 30 K **d)**; $K_2Cu(C_2O_4)_2$, 2 H_2O at 300 K **e)** and 30 K **f)**; $Cu_2(CH_3COO)_4$, pyrazine at 300 K **h)** and 30 K **g)**; $Ni(C_2O_4)$, 2 H_2O at 300 K **i)** and 30 K **j)**

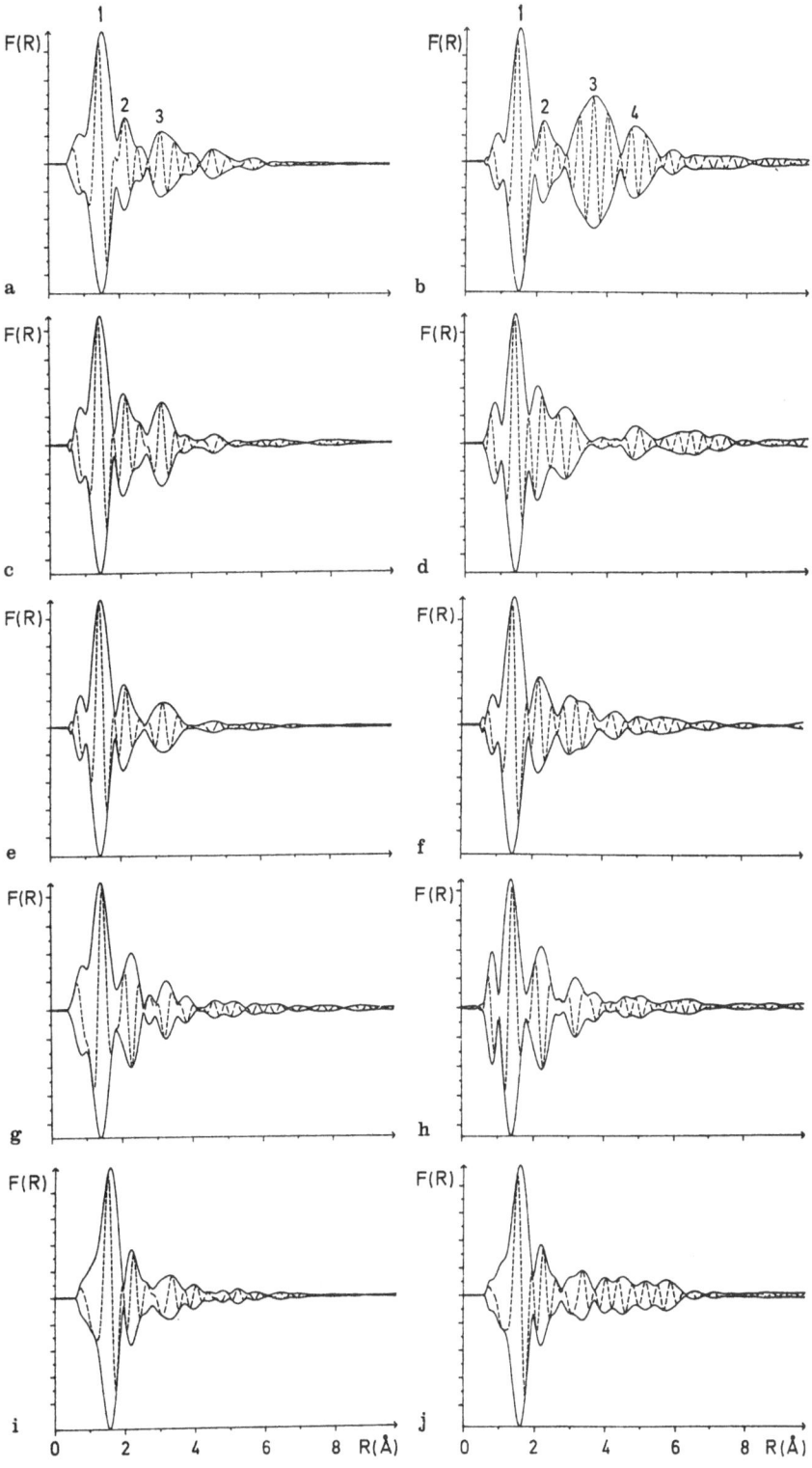

(Fig. 14a, c, e, g) proves that the second shell is composed of 4 Cu—C distances and cannot be attributed to Cu—Cu signals. This result rules out the copper acetate model and the ribbon structure is demonstrated to be the most likely (Fig. 13b).

3.2.3 Towards 3D Structure with EXAFS: Low Temperature Data

In the original work [3], the nature of the third peak (Fig. 14a) is not identified and no Cu—Cu contribution is evidenced, confirming that molecular packing discussion cannot be performed at room temperature.

The next step of the study is therefore the use of both 300 K and 30 K spectra of $CuC_2O_4 \cdot 1/3\,H_2O$ (Fig. 14a, b). These spectra are compared to those of $Na_2Cu(C_2O_4)_2 \cdot 2\,H_2O$ (Fig. 14c, d) and $K_2Cu(C_2O_4)_2 \cdot 2\,H_2O$ (Fig. 14e, f), which are two copper(II) bis-oxalato monomers. Their structures are shown in Fig. 15a, b. The $Cu_2(CH_3CO_2)_4$ pyrazine chain (Fig. 14g, h) is used as a model for the Cu—Cu contribution (one copper at about 2.58 Å) and $NiC_2O_4 \cdot 2\,H_2O$ (Fig. 14i, j) as a model of ribbon structure with a known molecular packing. A short account was published [44] and a detailed analysis is to be published in Inorganic Chemistry.

Qualitative Approach

Models Compounds

$M(II)(C_2O_4) \cdot 2\,H_2O$

$Fe(II)C_2O_4 \cdot 2\,H_2O$ is the only bivalent metallic oxalate whose XRD structure has been solved [45]. It is nevertheless isostructural with the Ni(II) and with the Zn(II) analogs. In the nickel derivative for example, the square planar NiO_4 site within the planar ribbons is completed with two axial water ligands. The shortest Ni—Ni intermolecular contribution, computed from X-ray powder diffraction data of Ref. [46] corresponds to 4 Ni at 5.20 Å. The room temperature EXAFS spectrum (Fig. 14i, j) present 3 peaks. Peaks 1 and 2 are unambiguously identified as Ni—O and Ni—C contributions. The third peak will be discussed later on. Upon lowering the temperature towards 30 K, these first peaks are almost invariant while new contributions appear at larger distances. Therefore, the three first peaks are certainly due to intramolecular contributions while the temperature dependent contributions probably corresponds to intermolecular Ni—Ni distances.

The 300 K and 30 K spectra of $FeC_2O_4 \cdot 2\,H_2O$ [3] and $ZnC_2O_4 \cdot 2\,H_2O$ (Fig. 18c) show, as expected, the same behaviour as $NiC_2O_4 \cdot 2\,H_2O$.

These new data allow also a better understanding of the third peak origin, in Fig. 14i, j. As a matter of fact, both radial position (about 3.5 Å) and amplitude (unexpectedly high) suggest that we are dealing with a Ni—O' contribution. O' is the second oxygen atom of a carboxylate group, not directly bound to a given metal (Fig. 15a). Indeed, the only intramolecular contribution at about 3.5 Å corresponds to Ni—O' distance. Furthermore, as the Ni—C—O' atoms are aligned, the observed enhancement of the peak under discussion can be accounted for by a strong multiple scattering effect (Fig. 15a). The effect is not only responsible for the two fold increase of amplitude of the peak with respect to the contribution of four oxygen atoms at $R \gtrsim 3.5$ Å but modifies also the phase shift and the corresponding distance. The

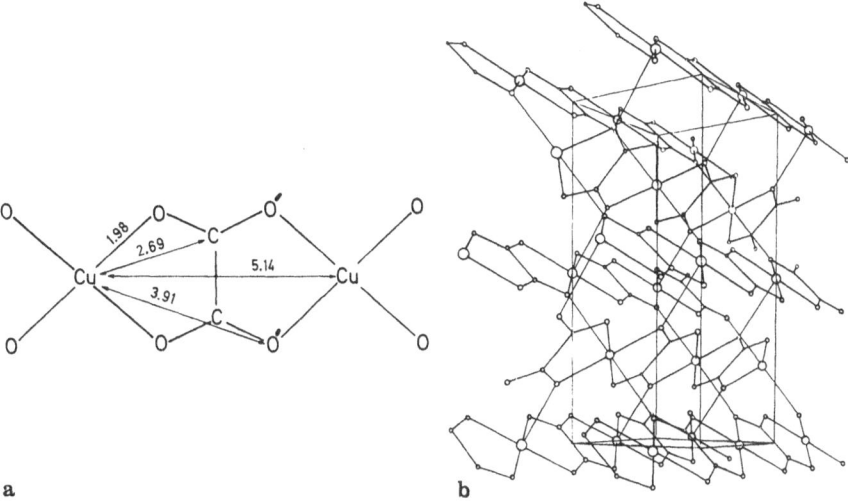

Fig. 15a and b. EXAFS determined copper oxalate structure: **a)** ribbon structure of the chains; **b)** 3D packing of the ribbons

Ni—O' distance should therefore be estimated larger than 3.5 Å. A similar effect was also observed in $K_2Pt(C_2O_4)_2$ [27].

This analysis about $NiC_2O_4 \cdot 2\,H_2O$ throws light into the more general problem of the chelation of metallic ions by oxalato anion. Two points deserved to be underlined:

(1) The three peaks structure (M—O around 2 Å, M—C around 2.7 Å, M—O' at about 4 Å with a multiple scattering effect, is a signature of the chelating binding of the metal ion to the oxalato ligand (Fig. 14a, c, e, i).

(2) Extra signals at low temperature are mainly due to metal-metal intermolecular contributions.

Under the light of the two previous statements, the discussion and analysis of EXAFS data for metallic bis-oxalate become almost straightforward as exemplified in the following.

$Na_2Cu(C_2O_4)_2 \cdot 2\,H_2O$ and $K_2Cu(C_2O_4)_2 \cdot 2\,H_2O$

The latter point can be illustrated by the discussion of the EXAFS data of $Na_2Cu(C_2O_4)_2 \cdot 2\,H_2O$ and $K_2Cu(C_2O_4)_2 \cdot 2\,H_2O$. The crystal packings are significantly different from each other; in the sodium salt the molecular complexes stack with R(Cu—Cu) = 3.58 Å [42a] whereas in the potassium salt, the first Cu—Cu distance is larger than 5.5 Å [43] Fig. 16. The room temperature EXAFS spectra of the two compounds (Fig. 14c, e) corroborate point A. Both spectra are characteristic of the metal-oxalato chelating mode identical to the one in the MC_2O_4 series. The three characteristic peaks (M—O, M—C, M—O') are present and fairly independent of the molecular packing. The differences between the EXAFS spectra appear only at 30 K. For the potassium salt, the three first peaks are almost temperature independent, whereas in the sodium salt the contribution of 2 Cu at 3.58 Å considerably affects the spectrum in the 3–4 Å range (Fig. 14d): the raising of the Cu—Cu contribution at 30 K results in an apparent shift of the third peak (Fig. 14c, d). Since the C_2O_4

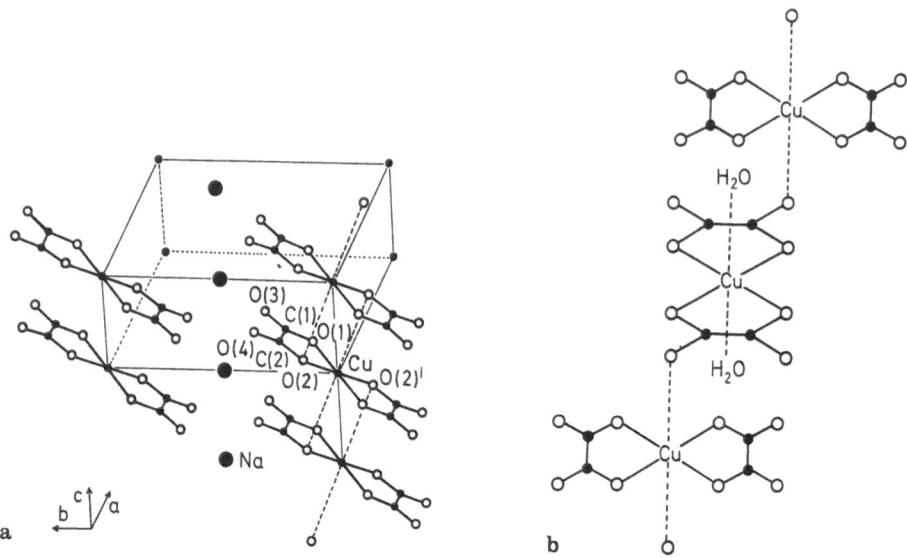

Fig. 16a and b. Crystallographic structures of **a)** $Na_2 Cu(C_2O_4)_2$, $2 H_2O$ (42A) and **b)** $K_2 Cu(C_2O_4)_2$, $2 H_2O$ (43)

chelation cannot be destroyed by cooling the sample, the Cu—O' signal should remain at 30 K, as in NiC_2O_4 and $K_2CuC_2O_4$. The explanation of this apparent shift is that Cu—Cu and Cu—O' signals are partially cancelled by a destructive interference effect, as discussed in Sect. 2.3.3.

Copper Oxalate

The evolution of the CuC_2O_4 spectrum from room temperature to 30 K is the most spectacular in the series.

The first two peaks, already assigned to Cu—O at 1.98 Å and to Cu—C at 2.66 Å [3], are temperature independent. The assignment of the second peak to the contribution of the oxalato carbon atoms only is confirmed by the 30 K spectrum. Moreover, the third room temperature peak, which was not identified in the first paper [3] can be assigned to Cu—O'. At low temperature, an important new signal appears in the 4 Å range, hiding the Cu—O' third peak of the room temperature spectrum, and a fourth peak raises at R > 5 Å. Comparing the CuC_2O_4 temperature evolution to the one of NiC_2O_4, it is clear that the CuC_2O_4 metal-metal intermolecular contribution involves more than 2 Cu at about 4 Å and more than 4 Cu in the 5–6 Å range. Among the models discussed in the introduction, the Schmittler's structure appears to be the most likely, with 4 Cu at 4.09 Å, and the other Cu—Cu distances displayed Fig. 13c.

Therefore, a simple qualitative study of the temperature variation of EXAFS spectra can be used to rule out molecular packing models and to select the more probable one. A quantitative approach allows then to extract quantitative information by curve fitting.

Quantitative Approach

The results of the fits of the filtered experimental spectra to the theoretical formula (1) are reported in Table 3 and Fig. 17. Phase shifts $\varphi(k)$ and amplitude functions $A(k)$ were extracted from the spectra of model compounds with known crystal structures: $Cu-O$, $Cu-C$, and $Cu-O'$ from $Na_2Cu(C_2O_4)_2 \cdot 2\,H_2O$ at 300 K and $Cu-Cu$ from $Cu_2(CH_3COO)_4 \cdot$ pyrazine at 30 K. These preliminary fits are also presented in Table 3

Table 3. Copper oxalate and models compounds: fitting results for intramolecular and intermolecular distances (see text)

	N	σ	R/Å	ΔE_0	Scale	$\varrho/\%$
Cu(Ac)$_2$ pyz 30 K						
Cu–O	4	0.01	1.96	−6.2	0.7	0.12
Cu–Cu	1.02	0.005	2.58	−1.9	0.7	0.5
Cu–Cu	0.5	0.00009	2.58	−1.5	0.7	2.6
Cu–Cu	2	0.1	2.58	−1.13	0.7	2.18
Cu(Ac)$_2$ pyz 300 K						
Cu–O	4	0.04	1.96	−6.5	1.09	0.2
Cu–Cu	0.8	0.02	2.56	−1.04	1.09	0.15
Na$_2$Cu(C$_2$O$_4$)$_2$ 300 K						
Cu–O	4	0.04	1.93	0.05	0.99	0.02
Cu–C	3.6	0.03	2.69	0.04	0.99	0.1
Cu–O′	4.6	0.05	3.9	−0.7	0.99	0.03
K$_2$Cu(C$_2$O$_4$)$_2$ 300 K						
Cu–O	4	0.01	1.93	−0.3	0.99	0.4
Cu–C	3.8	0.05	2.68	0.63	0.99	0.3
Cu–O′	2.8	0.026	3.93	0.6	0.99	0.6
CuC$_2$O$_4$ 300 K						
Cu–O	4	0.05	1.97	2.8	0.99	0.05
Cu–C	5.2	0.09	2.69	6.9	0.99	0.04
Cu–O′	3.3	0.001	3.94	4.4	0.99	0.06
K$_2$CuC$_2$O$_4$ 30 K						
Cu–O	4	0.004	1.93	1.74	1.05	0.2
Cu–C	3.75	0.023	2.72	8.1	1.05	0.2
Cu–O	2.8	0.008	3.89	−0.8	1.05	1.5
Na$_2$Cu(C$_2$O$_4$)$_2$ 30 K						
Cu–O	4	0.001	1.92	−1.6	1.04	0.84
Cu–C	6.2	0.005	2.67	−6.4	1.04	1.03
Cu–O′	4	0.02	3.9	0	1.04	
Cu–Cu	2.1	0.004	3.5	−12.6	1.04	2.21
Cu–Cu	1	0.00006	3.5	−9.8	1.04	3.2
Cu–Cu	4	0.1	3.5	−13.2	1.04	2.9
30 K–300 K	2.5	0.003	3.51	−9.6	1.04	2.4
CuC$_2$O$_4$ 30 K						
Cu–O	4	0.03	1.98	2.7	0.98	0.18
Cu–C	4.7	0.08	2.71	9.8	0.98	0.18
Cu–O′	4	0.006	3.9	−1.2	0.98	
Cu–Cu	4.73	0.006	4.13	3.5	0.98	3.8
Cu–Cu	2	0.00001	4.13	3.5	0.98	7.8
Cu–Cu	6	0.03	4.13	3.5	0.98	4.5
30 K–300 K	4.21	0.001	4.13	2.9	0.98	15

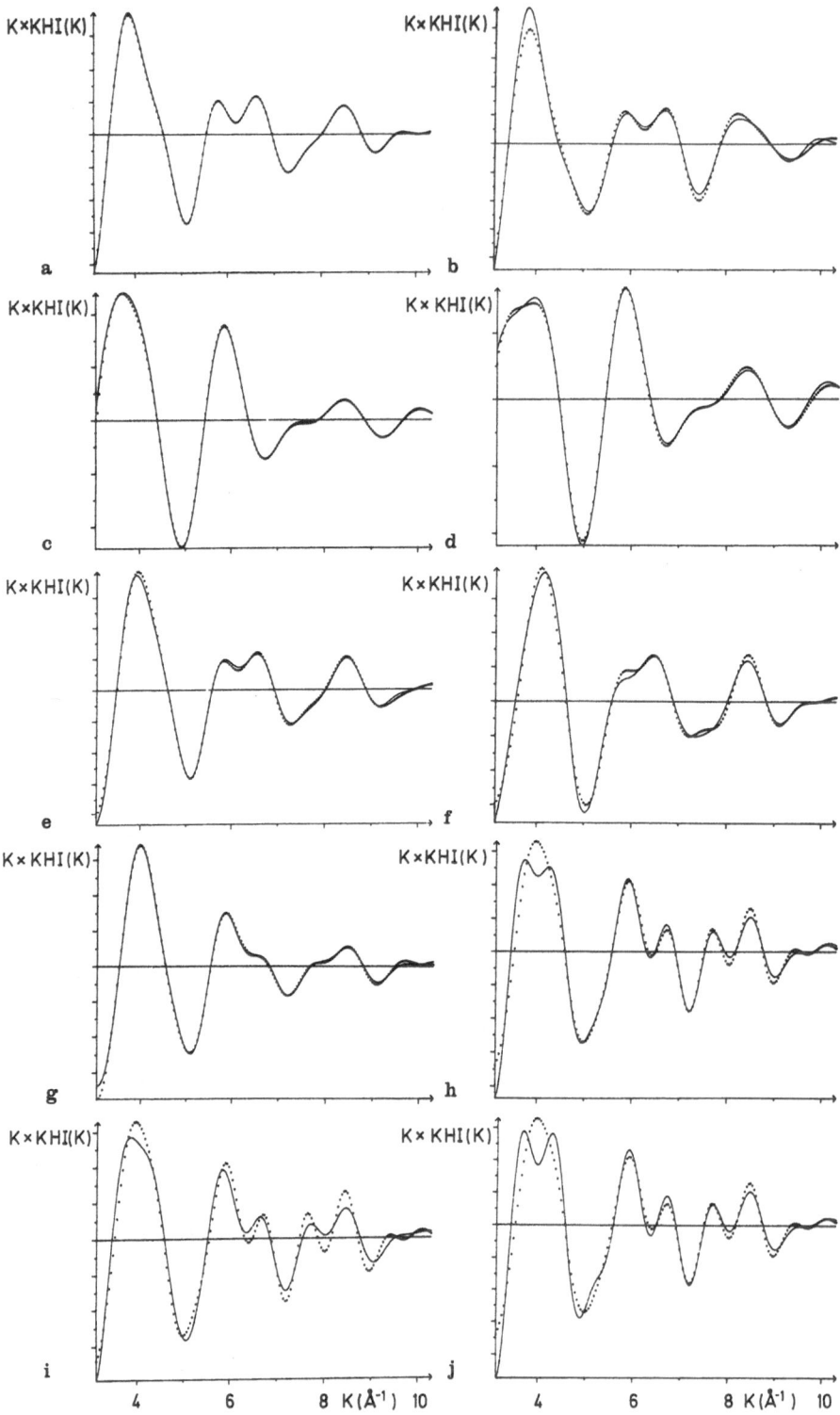

and Fig. 17. All the other fits were obtained in the same manner: the first Cu—O shell was filtered and fitted assuming a known coordination number [4]. This fit provides a scaling factor.

— then the first and the second shells are filtered together and a two shells fit is performed with all the first shell parameters settled to their previous values, in particular the scaling factor.

— the same operation is repeated for the more remote shells, up to complete filtering of the spectrum. When a copper shell overlaps another shell, the Cu—Cu contribution is previously fitted from the 30 K–300 K difference spectrum, and the result introduced in the final fit as a starting point.

Contrary to single shell filtering and fitting, such a step by step multishells fitting process presents the advantage to constrain the parameters of remote shells to be evaluated relatively to the previous ones. The same kind of treatment was used also in the fitting of lamellar MPS_3 in order to determine the relative Debye-Waller variation of the second shell (see Sect. 4.2).

Even when the overall amplitude is incorrect — due to insufficient control of sample thickness or homogeneity, or poor X-ray harmonic rejection —, it is still possible to determine the ratio of successive number of neighbours within 20% error.

For example, in $Na_2Cu(C_2O_4)_2 \cdot 2H_2O$ at 30 K, the best fit is obtained with 4 Cu—O at 1.92 Å and 2.1 Cu—Cu at 3.50 Å (crystallographic value: 2 Cu—Cu at 3.58 Å). Any attempt to fix the number of Cu—Cu contribution to 1 or 4 leads to a worse fit with abnormal σ values (Fig. 17i, j). With this treatment, the number of Cu—Cu contributions in the unknown $Cu(C_2O_4)$ structure is evaluated to 4.7 ± 1. Nevertheless, the quantitative treatment of Peak 4 (Fig. 14b) fails since it represents a too complex Cu—Cu distribution.

Our quantitative results confirms the qualitative interpretation above and agree well with the X-ray powder diffraction conclusions. EPR results are consistent. also with the presence of two orientations of the chains in the crystal [41b]. All these experimental results converge in the structure shown in Fig. 15b.

To conclude this section, we can say that comparison of room and low temperature EXAFS spectra may give a qualitative and semiquantitative radial image of the 3D packing. The conjunction of EXAFS and X-ray powder diffraction provides the best available 3D picture of the structure, EXAFS giving the more accurate short range order bond lengths and powder diffraction giving long range order.

3.3 Bimetallic Chains

Up to now, we described unimetallic chains. Some years ago, a new perspective was opened in the field of 1D magnetic systems by the synthesis and the study of the magnetic properties of an ordered bimetallic Cu(II)—Mn(II) chain which revealed

---

◀ **Fig. 17a–j.** Four shells fits of oxalato derivatives: $Na_2 Cu(C_2O_4)_2$, $2H_2O$ at 300 K **a)** and 30 K **b)**; $Cu_2(CH_3COO)_4$, pyrazine at 300 K **c)** and 30 K **d)**; $K_2Cu(C_2O_4)_2$, $2H_2O$ at 300 K **e)** and 30 K **f)**; $Cu(C_2O_4)$, 1/3 H_2O with the best fit at 300 K **g)**; $Cu(C_2O_4)$, 1/3 H_2O at 30 K, best fit **h)**, with 2 copper atoms included in shell 3 **i)** or with 6 copper atoms included in shell 3 **j)**

Alain Michalowicz et al.

1D ferrimagnetic behaviour [6]. The interest for this new field is rapidly raising [47-48].

We report here some examples of EXAFS analysis about badly crystallized Cu—Ni, Cu—Zn and Cu—Mn oxalato derivatives and Cu—Mn dithiooxalato chains.

3.3.1 Oxalato Chains

The reaction of bis-oxalato Cu(II) with Ni(II), Zn(II), Mn(II) yields light blue microcrystalline powders. Attempts to grow single crystals have systematically failed. The resulting Cu—Mn and Cu—Ni compounds do not present the properties expected

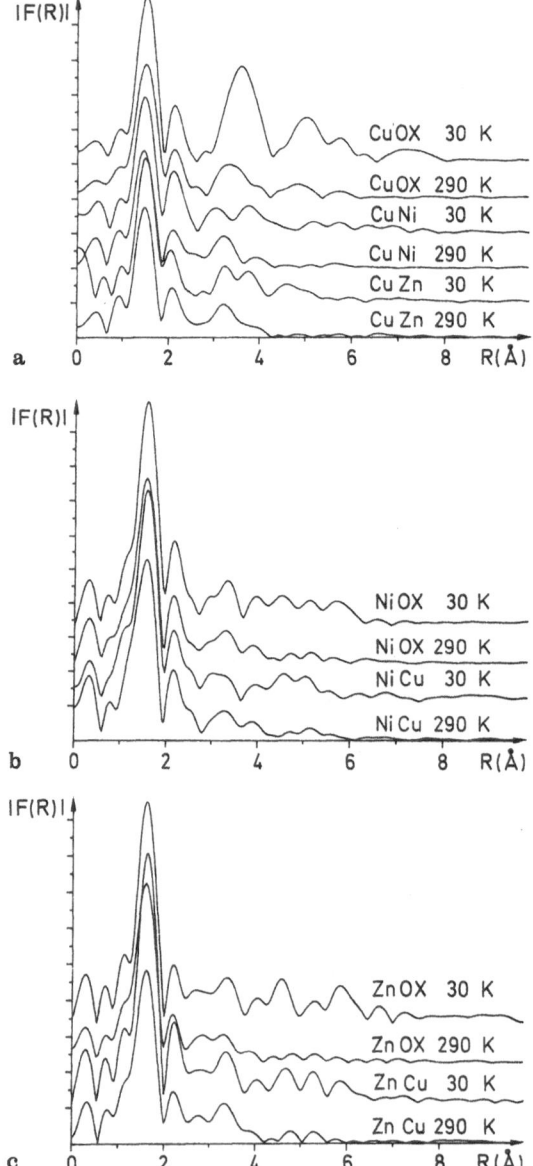

Fig. 18a–c. Modulus of the Fourier transform in a series of uni or bimetallic oxalates; **a)** at the copper edge; **b)** at the nickel edge; **c)** at the zinc edge

130

Table 4. Bimetallic oxalato chains: first-shell metal-oxygen fitting results

Compd	T/K	E_0/eV	R/Å	$2\sigma^2$/Å^2	ϱ/%
	Copper Edge				
$Cu(C_2O_4) \cdot {}^1/_3\,H_2O$	290	8998	1.98	0.012	1.0
	30	8998	1.97	0.005	2.0
$CuNi(C_2O_4)_2 \cdot 4\,H_2O$	290	8998	1.96	0.008	3.1
	30	9000	1.96	0.008	3.4
$CuZn(C_2O_4)_2 \cdot 4\,H_2O$	290	9000	1.95	0.005	2.5
	30	8999	1.96	0.002	2.2
	Nickel Edge				
$Ni(C_2O_4) \cdot 2\,H_2O$	290	8350	2.04	0.008	1.4
	30	8351	2.04	0.005	1.6
$NiCu(C_2O_4)_2 \cdot 4\,H_2O$	290	8353	2.03	0.010	1.9
	30	8352	2.04	0.006	1.2
	Zinc Edge				
$Zn(C_2O_4) \cdot 2\,H_2O$	290	9680	2.08	0.011	2.1
	30	9680	2.09	0.006	2.1
$ZnCu(C_2O_4)_2 \cdot 4\,H_2O$	290	9680	2.08	0.011	2.5
	30	9680	2.09	0.005	4.1

for ferrimagnetic materials [6], whereas the Cu—Zn derivative displays the properties of a very weakly coupled Cu(II) complex, as expected for an ordered bimetallic chain with a diamagnetic ion alternating with a Cu(II) one. The R-space spectra of Cu—Ni, Cu—Zn are shown in Fig. 18 together with those of unimetallic chains of copper nickel and zinc. Quantitative results are given in Table 4. The number of oxygen nearest neighbours around nickel or zinc is 1.5 times larger than around copper. This agrees with four nearest neighbours around Cu and 6 around Ni and Zn.

The M—O, M—C, M—O′ shells are determined as previously. If the stoichiometry of the compounds is taken into account, the bridging network shown above is only compatible with a chain structure. The X-ray powder pattern excludes the presence of copper oxalate and strongly suggests an ordered structure deriving from the structure of nickel oxalate. A 1/1 mixture of the pure microcrystalline powders can also be ruled out. Consequently, a bimetallic chain structure appears to be the most likely (Fig. 19).

We also performed room and low temperature experiments on the Cu—Mn deri-

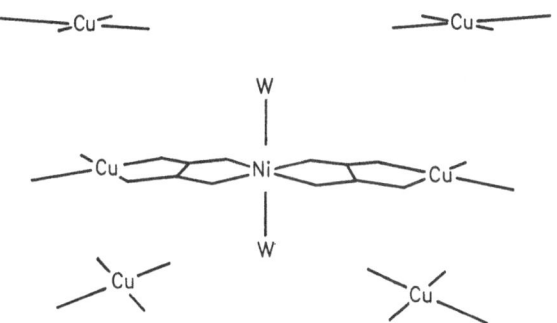

Fig. 19. EXAFS proposed structure for bimetallic copper-nickel oxalate

vative. In this case, and even though some Cu—Mn pairs have been detected by electronic spectroscopy, EXAFS at the copper edge, at 30 K, show evidently the very intense peak encountered in the Cu oxalate spectrum. The Mn edge spectrum is identical to the one of $Mn(C_2O_4) \cdot 2\,H_2O$. It seems therefore possible to conclude here that in the studied samples, the Cu—Mn bimetallic oxalate contain a certain amount of Cu oxalate and Mn oxalate. The usefulness of EXAFS here is to reject the bimetallic ordered chain view of the Cu—Mn derivatives synthesized sofar.

3.3.2 Dithiooxalato Chains

The problem encountered here is rather different. The Cu(II)—Mn(II) dithiooxalato chain (CuMnDTO $\cdot$ 7.5 H_2O) is obtained at room temperature under the form of single crystals and the structure of the pure product is fully characterized, without any disorder (Fig. 20). It is the first 1D compound to present the ferrimagnetic behaviour shown in Fig. 21 curve 1. However, the partially dehydrated compound CuMn—$(C_2O_2S_2)_2 \cdot 3\,H_2O$, obtained at higher temperature, presents no longer ferrimagnetic properties (Fig. 21, curve 2) and its structure is not known. It exists intermediate hydrated materials where the magnetic properties are also very different from the fully hydrated material (Fig. 21, curve 3).

The loss of 4.5 water molecules has striking consequences upon magnetic properties, where the ferrimagnetic behaviour is lost, as well as upon EXAFS spectra. The RDF at Cu and Mn edges at low temperature are displayed in Fig. 22. A qualitative discussion is sufficient to show that the loss of the water molecules affect the surroundings of both copper and manganese ions. At the Mn edge the molecular environment is modified as shown by the decrease of the first three peaks; the peaks at large distances (at 5 Å and above) are no longer present, indicating both intrachain and interchain disorders without significant ordered bringing together of heavy atoms belonging to different chains. Clearly, the whole structure is modified, and the loss of ferrimagnetic properties upon dehydration cannot be attributed only to the closeness of neighbouring chains (quenching ferrimagnetism by interchain interaction).

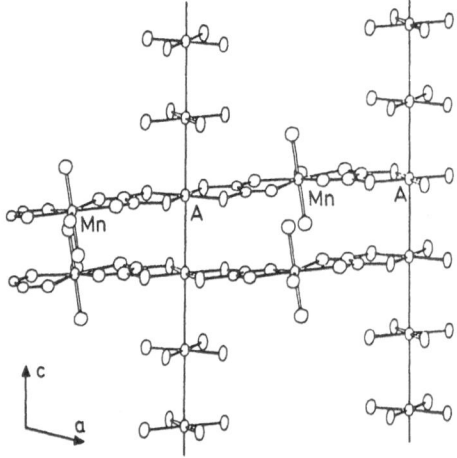

Fig. 20. XRD structure of the ferrimagnetic chain CuMnDTO

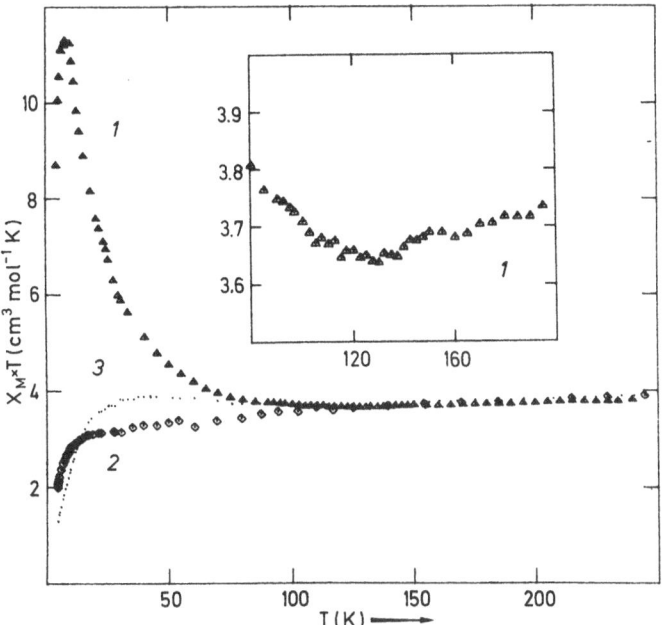

Fig. 21. Thermal variation of the molar susceptibility in bimetallic chains under the form $XmX_M \cdot T$ vs T: 1) pure CuMnDTO with 7.5 water molecules; 2) CuMnDTO with 3 water molecules; 3) partially dehydrated material

Even if the evolution is less spectacular, the spectra at copper edge confirm that dehydration induces the onset of disorder. In particular, the fourth peak at about 4 A in Fig. 22a attributable to the Cu—S distances within the stack of CuS_4 units (Fig. 20) shifts towards longer distances in the 3 H_2O compound. This may be attributed to a disorder within these stacks. Once more no ordered bringing together of metallic ions belonging to different chains is observed.

All these observations are consistent with the hypothesis of deep disorder in both the chain and the 3D structures, the first one being sufficient to explain the loss of the ferrimagnetic properties. The existence of intermediate hydrated materials tends to support the idea of a fragile material, losing progressively water molecules initially located within the interchains space and coordinated to the Mn(II) ion. Thermal differential and gravimetric analyses confirm these views [6c].

3.3.3 Doped TMMC

We want to conclude this section about 1D materials in showing another example of the care with which disorder problems should be handled. The structure of tetra-methylammonium manganese chloride (TMMC) which is the archetype of 1D anti-ferromagnets and its copper(II) analog, which is an example of the 1D ferromagnets, are known. The consequences of doping TMMC by Cu(II) or by Cd(II) has been studied. The magnetic measurements were interpreted in the frame of a random distribution of Cu(II) in the Mn(II) array [49c-e]. Testing this hypothesis by studying RDF at both Mn and Cu edge was tempting: if Cu(II) is randomly distributed, Mn and Cu will see mainly Mn nearest neighbours; on the contrary, if linear clusters of Cu(II) ions are formed in the chain, Mn will mainly see Mn, whereas Cu will see mainly Cu as nearest neighbours. The spectra at 30 K are shown in Fig. 23a and b. Pure TMMC displays a nice example of multiple scattering effect on the next nearest neighbour (the peak 4 at more than 6 Å in Fig. 23a presents an enhanced intensity and a phase shift, leading to a Mn—Mn distance between next nearest neighbours which is not apparently equal to two times the Mn—Mn nearest neighbours distance).

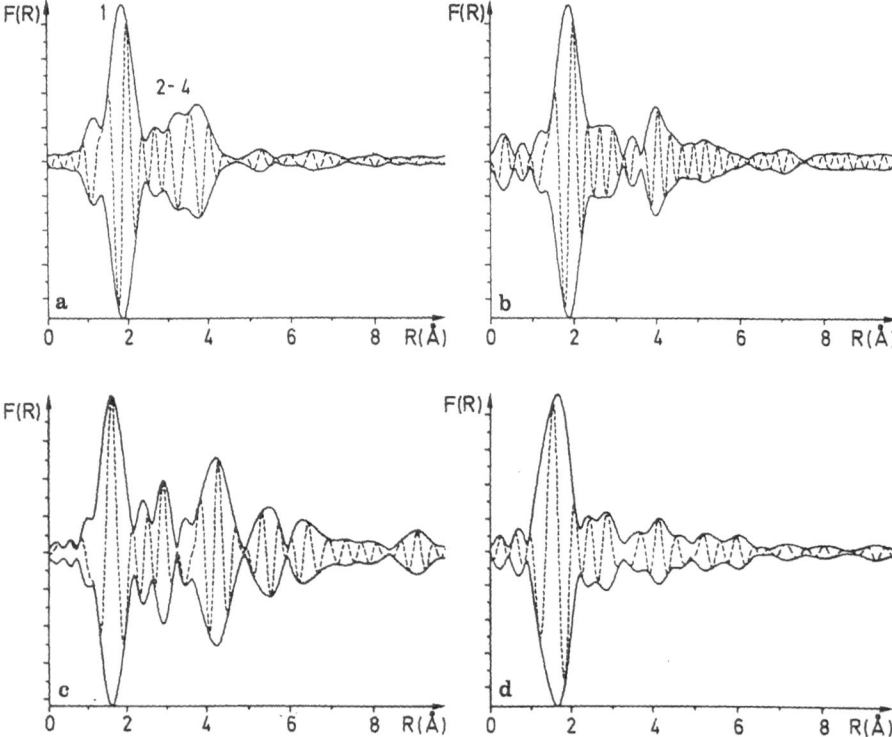

Fig. 22a–d. RDF Fourier of bimetallic CuMnDTO at 30 K: at copper edge in CuMnDTO, 7.5 H_2O (**a**) and in CuMnDTO, 3 H_2O (**b**); at manganese edge in CuMnDTO, 7.5 H_2O (**c**) and in CuMnDTO, 3 H_2O (**d**)

This is particularly evident from the spectra including the Mn—Mn phase correction, following the Beni and Lee criteria (Fig. 23d). The copper doped material (TMMC/Cu) displays a slight distortion of peak 3 (Fig. 23c) at the Mn edge and an important one at the copper edge on the same peak, which is not surprising with a Jahn-Teller ion. But, a quantitative analysis of peak 2, which corresponds to the nearest metallic neighbour is difficult since it involves a mixing of two metal-metal signals with unknown ratio, too close amplitude and phase factors, and therefore large correlation between the fitted parameters. This does not allow to put in evidence safely any clustering effect of Cu(II) ions. A detailed study of Cd(II) doped materials is underway [50].

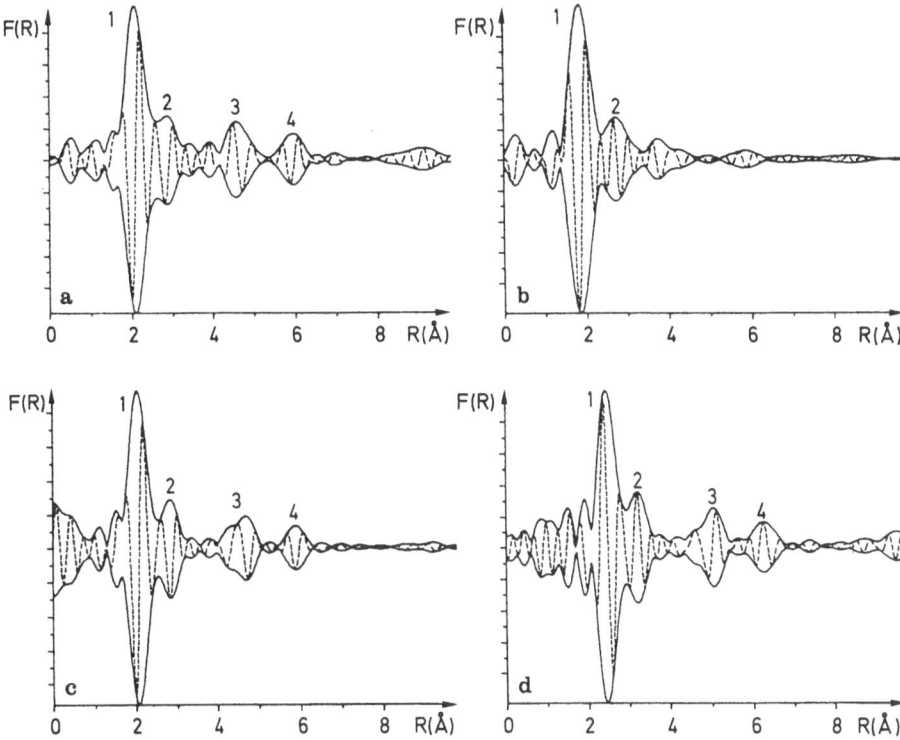

Fig. 23 a–d. RDF of pure TMMC without **a**) or with Mn—Mn phase correction **d**); TMMC doped with 20 % of copper(II) ions at manganese edge **b**) and at copper edge **c**)

4 Bidimensional Compounds: The MPS₃ Materials

4.1 Introduction

There has been an increasing interest over the past decade in pure and intercalated layered systems of the MPS_3 family, where M(II) is a transition metal ion [51, 52]. Indeed, these insulating materials associate an unusual chemical reactivity (intercalation through electron transfer as well as through cation exchange [12]) to a variety of physical properties [53] (optical, magnetic, electrical), sometimes tunable, and to a large structural versatility which is illustrated by the existence of materials such as $In(III)_{2/3} \square_{1/3} PS_3$ [54] ($\square$ stands for a metal vacancy), $Cr(III)_{1/2} Cu(I)_{1/2} PS_3$ [55] or even mixed valence layers $V(II)_{0.33} V(III)_{0.44} \square_{0.23} PS_3$ [56].

Although the structure of several pure MPS_3 have been fully solved by classical XRD, EXAFS spectroscopy has recently proved to be of invaluable help in characterizing structural arrangement and disorder in two typical situations encountered in these systems:

(i) materials having a low cristallinity, particularly those obtained upon reaction in the solid state at room temperature;

(ii) materials which can be obtained as monocrystals, but which present some disorder so that XRD only gives an averaged view.

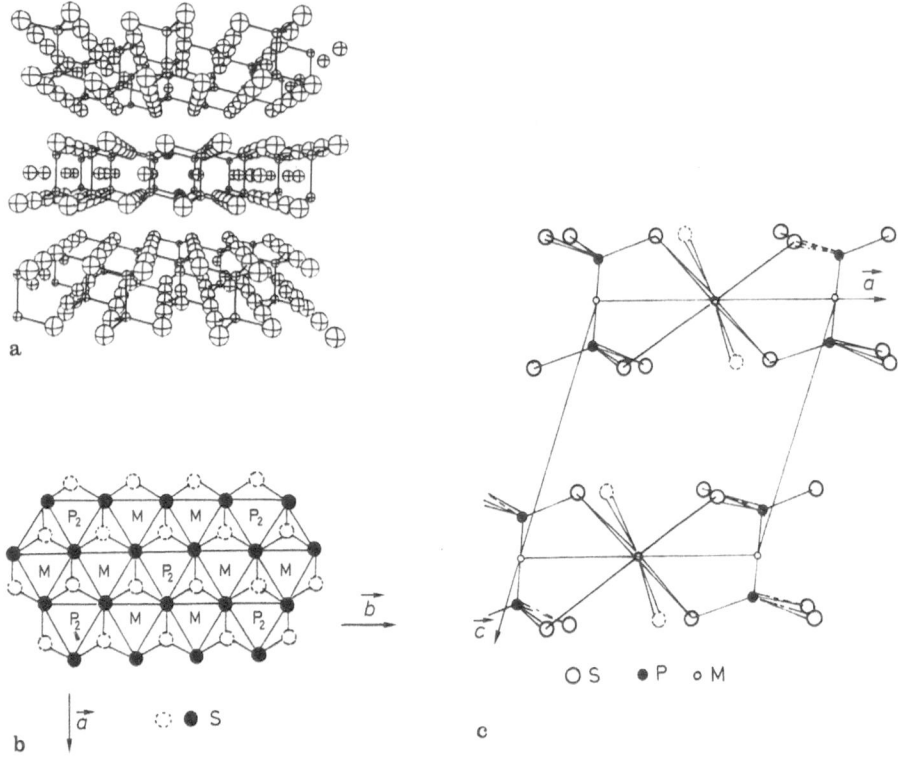

Fig. 24a–c. Perspective view of MPS$_3$ along the b axis displaying the Van der Waals gap between adjacent layers, after Piacentini (66) **a**); structural models of a MPS$_3$ layer showing the intralayer arrangement **b**) and the interlayer stacking **c**)

Some examples illustrating these situations are described hereunder. The crystal structure of a representative MPS$_3$ compound is sketched in Fig. 24.

The structure is related to that of CdCl$_2$, with M(II) ions and P—P pairs occupying the Cd positions and sulfur atoms (close packed) occupying the chloride positions. This structure is very similar to that encountered in the wide class of the layered dichalcogenides MX$_2$. Each MPS$_3$ layer can also be described as a polymetallic complex, which consists of an array of M(II) cations coordinated to the sulfur atoms of (P$_2$S$_6$)$^{4-}$ bridging ligands. This coordination chemist's view turns to be very useful to understand the reactivity of these peculiar materials.

4.2 Partial Disorder Into MPS$_3$ Layers Upon Intercalation: EXAFS Detection and Analysis

MPS$_3$ spontaneously reacts at room temperature with aqueous solutions of a wide variety of salts such as KCl, Co(C$_5$H$_5$)$_2$Cl or various ammonium chlorides. Intercalates Mn$_{(1-x)}$PS$_3$(G)$_{2x}$(H$_2$O)$_y$ are formed, where the positive charge of the guest cations G is counterbalanced by the removal of an equivalent amount of intralamellar Mn^{2+}

cations (which are expelled into the solution) [12]. These intercalates keep some LRO, as they exhibit quite sharp (hkl) XRD lines, which can be indexed in the same type of monoclinic cell as pure $MnPS_3$, simply expanded in the c direction (Fig. 24c).

However, the intercalation process induces modifications in the lattice since the magnetic properties of the intercalates are very different [58] from those of pure MPS_3: whereas the latter orders antiferromagnetically at low temperature, a spontaneous magnetization takes place in the intercalates (Fig. 25).

As crystals suitable for structure determination could not be obtained, EXAFS study of $MnPS_3$ and of several intercalates was performed to compare the SRO around manganese in these systems [7].

The modulus of the RDF of pure $MnPS_3$ and of the intercalates $Mn_{0.89}PS_3$((methyl-tris-octyl)N)$_{0.22}$ $(H_2O)_{\approx 0.3}$, $Mn_{0.89}PS_3$(n-octylNH_3)$_{0.22}$ $(H_2O)_{\approx 0.3}$, $Mn_{0.8}PS_3K_{0.4}$ $(H_2O)_{\approx 1}$ and $Mn_{0.83}PS_3[Co(C_5H_5)_2]_{(0.34)}(H_2O)_{\approx 0.3}$ (noted I, II, III, IV respectively) is presented in Fig. 26.

As expected from XRD data obtained with single crystals, the two main peaks on the curve of $MnPS_3$ correspond to the Mn—S (peak A) and to the unresolved Mn—Mn and Mn—P distances (peak B). Two main facts are obvious from Fig. 26. Firstly, the amplitude of the second peak (B) of intercalates II, III, IV is dramatically reduced with respect to that of pure $MnPS_3$. Secondly, the decrease of peak B is much less important

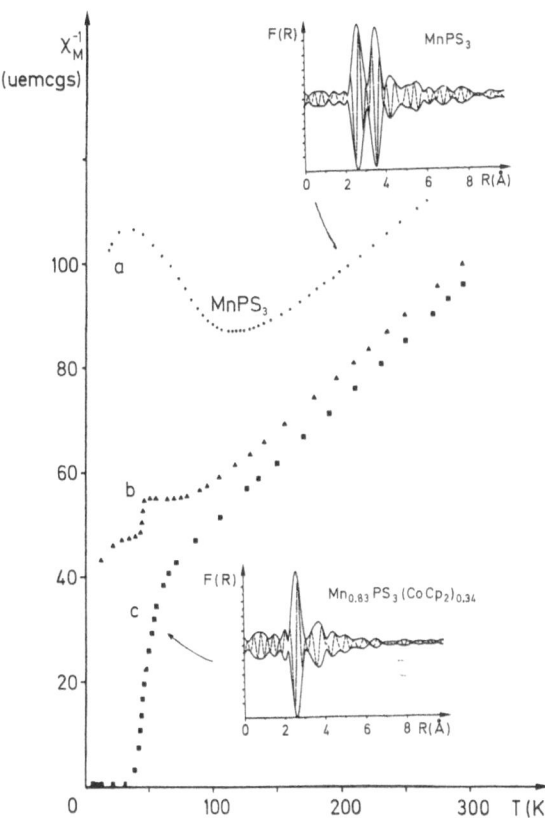

Fig. 25. Temperature dependence of the reciprocal magnetic molar susceptibility of $MnPS_3$ (a), $Mn_{0.89}PS_3$ (methyl-tris-octyl-N)$_{0.22}$ (b), $Mn_{0.83}PS_3$ $(CoCp_2)_{0.34}$ (c)

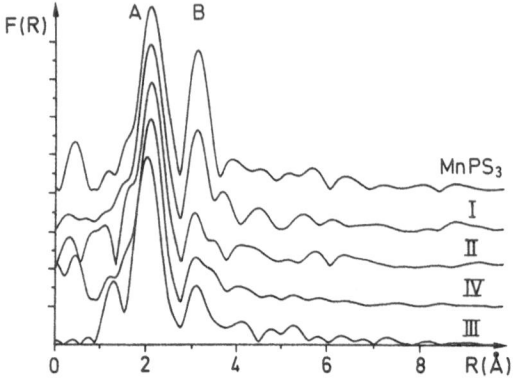

Fig. 26. RDF around Mn in MnPS$_3$ and four intercalates: (I) Mn$_{0.89}$ PS$_3$ (methyl-tris-octyl-N)$_{0.22}$; (II) Mn$_{0.89}$ PS$_3$ (n-octyl-NH$_3$)$_{0.22}$; (III) Mn$_{0.80}$ PS$_3$ K$_{0.40}$; (IV) Mn$_{0.83}$ PS$_3$ [Co(C$_5$H$_5$)$_2$]$_{0.34}$

in the case of intercalate I, which contains the very bulky methyl tris octylammonium cation. Quantitative data have been extracted from the experimental results, in particular the distances between the manganese ions and their first and second neighbours, and the mean deviation value σ on these distances. Results obtained at 300 K are gathered in Table 5.

Complementary results at low temperature on the cobalticinium intercalated compound show that the disorder induced by intercalation is mainly structural rather than thermal [10, 59].

A weak variation of σ is observed for the first coordination sphere on going from MnPS$_3$ to the intercalates. In contrast, the variation of σ for the second coordination shells is significant and the amplitude of its increase can be used to characterize

Table 5. MnPS$_3$ and intercalated derivatives: structural parameters extracted from EXAFS spectra. The numbers of neighbours are fixed

Compounds		Bond	Distances R/Å	Dispersion σ/Å
MnPS$_3$		Mn—S	2.58	0.08
		Mn—Mn	3.50	0.08
		Mn—P	3.66	0.08
Mn$_{0.89}$PS$_3$(methyltrisoctylN)$_{0.22}$	(I)	Mn—S	2.58	0.10
		Mn—Mn	3.49	0.11
		Mn—P	3.70	0.11
Mn$_{0.89}$PS$_3$(n-octylNH$_3$)$_{0.22}$	(II)	Mn—S	2.57	0.10
		Mn—Mn	3.50	0.15
		Mn—P	3.70	0.15
Mn$_{0.80}$PS$_3$K$_{0.40}$	(III)	Mn—S	2.57	0.09
		Mn—Mn	3.52	0.15
		Mn—P	3.73	0.15
Mn$_{0.83}$PS$_3$(Co(C$_5$H$_5$)$_2$)$_{0.34}$	(IV)	Mn—S	2.57	0.10
		Mn—Mn	3.51	0.16
		Mn—P	3.71	0.16

quantitatively the local disorder around the Mn(II) cations: it can be considered as a parameter measuring the disorder.

The fact that less disorder is induced by the methyl(tris-octyl) ammonium cation has been interpreted in term of a very low charge/volume ratio for the cation with respect to the other ones studied [59]. Two final remarks can be presented:

(i) the increase of disorder qualitatively parallels the extent to which the magnetic properties are modified: whereas no spontaneous magnetization occurs in MPS_3, a small one appears in the methyl tris octyl ammonium intercalate and strong ones in the three other intercalates. Disorder is probably associated to a differenciation of neighbouring manganese sites, which produces the observed weak ferromagnetism;

(ii) To our knowledge, very little work has been performed to study the influence of intercalation over the disorder of other host lattices. A study of the influence of rubidium insertion into $NbSe_2$ (60a) has led to the conclusion that only very little effects occurred: only slight flattening of the layer and very small variation of the bond distances were observed.

4.3 Disorder Induced Into MPS_3 Layers Upon Substitution: Resolution of Structural Disorder in $Mn_{(1-x)}Cu_{2x}PS_3$ Systems Through Complementary EXAFS and XRD Analysis

As exposed in the previous paragraph, MPS_3 systems can lead to series of intercalated compounds presenting relatively important concentrations of intralamellar metallic vacancies. At the present stage, no evidence for any superstructure was ever found in the X-Ray powder diffraction patterns of these $M_{(1-x)}\square_x PS_3 (G)_{2x} (H_2O)_y$ intercalates and no order is expected either on the intralamellar vacancies or on the interlamellar cations G^+. During the course of our search for ordered intercalated systems, several lamellar compounds were prepared, of general formula $M_{(1-x)}M'_{2x}PS_3$;

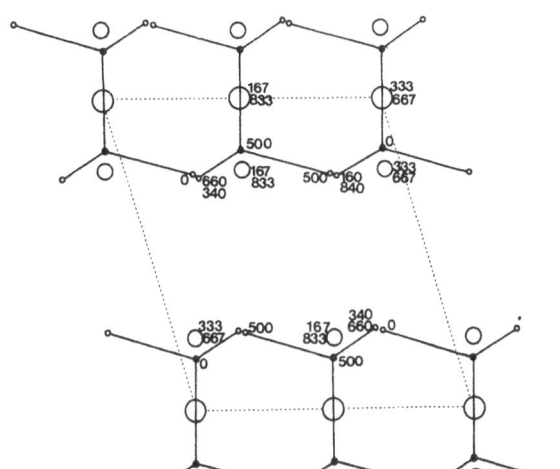

Fig. 27. Projection of the structure of $Mn_{(1-x)} Cu_{(2x)} PS_3$ (x = 0.13) along [010]. Small, medium and large open circles indicates sulfur, copper and manganese positions, respectively. Solid circles indicates phosphorus positions. Numbers represent relative position (x 1000) along the b direction

M(II) = Mn and Cd; M'(I) = Cu or Ag monocations were supposed to be distributed in an ordered manner over the interlamellar sites. In the particular case of $Mn_{0.87}$—$Cu_{0.26}PS_3$, suitable crystals (0.30 · 0.25 · 0.05 mm) were grown and a complete XRD analysis could be carried out at room temperature. Also low and room temperature EXAFS spectra were recorded at both manganese and copper K edges [8].

The apparent discrepancies between the results of the two studies gave rise to an interesting discussion which led to a deeper insight in the structure of the compound. We report the arguments hereunder.

XRD Analysis

Single crystal Weissenberg studies shows that $Mn_{0.87}Cu_{0.26}PS_3$ system is monoclinic and belongs to the B 2/m, B 2 or B m space group. Furthermore, some spots present a broad and diffuse character and no superstructure can be found. The cell parameters a = 6.090(4) Å, b = 10.539(8) Å, c = 6.815(4) Å, and β = 107.11(3) · are closely related to those of the $FePS_3$ pure system [57a] and the refinement of the structure resulted relatively easy to perform and led to the XRD picture given as a projection in Fig. 27.

This unambiguously reveals that the studied system has a layered structure built on the same ABC stacking scheme that $FePS_3$, i.e. an "empty" layer (or Van der Waals gap) alternating with a "filled" layer containing phosphorus pairs and manganese and

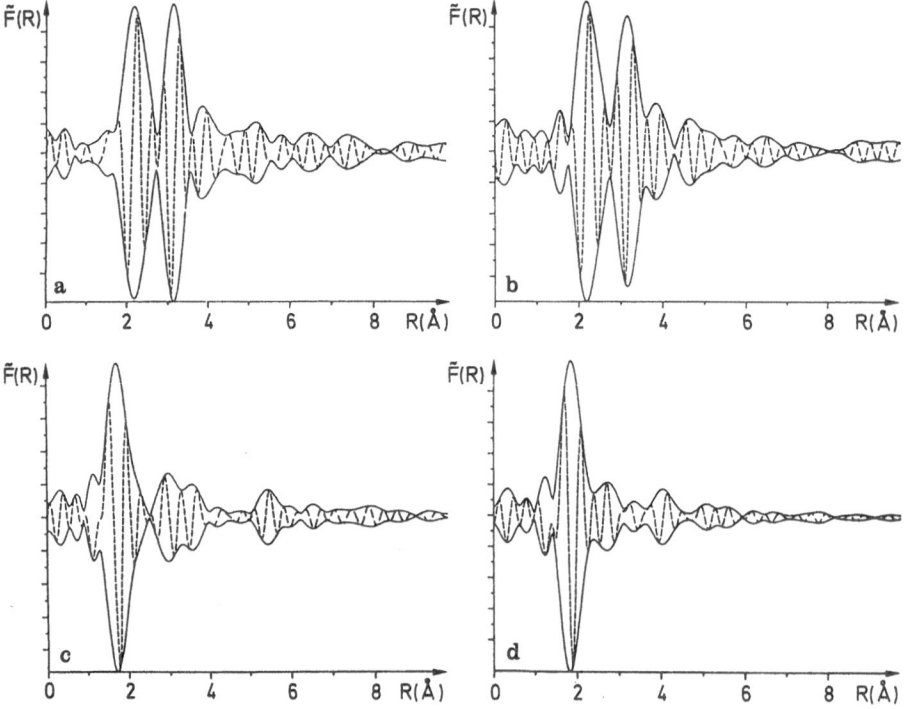

Fig. 28a–d. RDF at 15 K at manganese edge in $MnPS_3$ **a**) and in $Mn_{(1-x)}Cu_{(2x)}PS_3$ **b**) and at copper edge in $Mn_{(1-x)}Cu_{(2x)}PS_3$ and in $K_2Cu(C_2O_4)_2$ **d**)

copper atoms. According to this analysis, each monoclinic cell is made up of two $[P_2S_6]$ and two $[Mn_{(1-x)}Cu_{2x}S_6]$ octahedra, these last entities being simultaneously occupied by 0.85 ± 0.02 Mn in central positions and $2 (0.12 \pm 0.01)$ Cu atoms located on two eccentric positions close to the sulfur layer planes. It is obvious however that the same octahedron cannot accomodate both the two copper and the manganese atoms even with the above reported occupancy probabilities. This XRD picture rather reflects the disordered distribution of $85 \pm 2\%$ of $[MnS_6]$ octahedra, very similar to those encountered in the $MnPS_3$ or $FePS_3$ structure, together with $12 \pm 1\%$ of a new type of $[S_3Cu \ldots CuS_3]$ octahedra.

The calculated Mn—S distances are close to (although slightly larger than) the corresponding distance (2.60 Å) found in $MnPS_3$ [10] but the Cu—S distances are abnormally short compared to values estimated from the literature on Cu(I)-sulfur compound [61]. Finally, the P—S bond length agrees with the values usually encountered in the MPS_3 family.

EXAFS Analysis

Several qualitative conclusions are readily accessible from the EXAFS radial distribution functions gathered in Fig. 28.

(i) The $Mn_{0.87}Cu_{0.26}PS_3$ spectra at the manganese edge seem very similar to those of $MnPS_3$. The observed peaks and temperature dependence are analyzed in terms of Mn—S distances (first peak) and of a mixture of Mn—Mn and Mn—P distances (second peak). As this latter peak is extremely sensitive to any change in the local structure around the Mn centre (a sensitivity well illustrated by the dramatic reduction of this peak upon intercalation, see Sect. 4.2), its constancy demonstrates that the partial substitution of some Mn(II) ions by Cu(I) pairs does not significantly affect the symmetry of the remaining manganese sites.

(ii) The RDF at the copper edge (Fig. 28c) differ strongly from those obtained at the manganese edge. However, the intense first peak compare well (modulus and imaginary part) with the Cu—S peak (2.28 Å) of the model compound $K_2Cu(S_2C_2O_2)_2$ (Fig. 28d) and a first estimation of 2.19–2.20 Å can be proposed for the corresponding distances in $Mn_{0.87}Cu_{0.26}PS_3$. Another striking feature of these spectra is the strong temperature dependence of the complex structure observed above 2.5 Å and not reproduced here.

Table 6. $Mn_{0.87}Cu_{0.26}PS_3$ and $MnPS_3$ Three-shell fits of EXAFS spectra at Mn edge

Compound	T/K	Mn—S		Mn—P		(a) Mn—Mn		(b) $\varrho/\%$
		σ/Å	R/Å	σ/Å	R/Å	σ/Å	R/Å	
$MnPS_3$	300	0.08	2.58	0.08	3.66	0.08	3.50	1.8
	15	0.05	2.60	0.02	3.66	0.02	3.49	1.6
$Mn_{0.87}Cu_{0.26}PS_3$	300	0.08	2.59	0.08	3.66	0.08	3.51	1.5
	15	0.05	2.61	0.02	3.65	0.02	3.51	1.3

141

Table 7. $Mn_{0.87}Cu_{0.26}PS_3$ and model compound: single shell fits of EXAFS spectra at Cu edge

Compound	T/K	Cu—S		$\varrho/\%$
		$\sigma/\text{Å}$	R/Å	
$K_2Cu(S_2C_2O_2)_2$	300	0.07	2.28[a]	1.5
	15	0.05	2.28	1.9
$Mn_{0.87}Cu_{0.26}PS_3$	300	0.08	2.21	2.
	15	0.06	2.22	2.5

[a] d XRD (Cu—S) = 2.28 Å

(iii) Clearly, Mn and Cu ions are located at the centre of very different types of sites and any occupancy of the Mn(II) vacancies (at the centre of S_6 octahedra) by a significant amount of Cu(I) can be definitively ruled out.

The least square fitting results are summarized in Tables 6 and 7. Only the first shell was fitted for spectra recorded at the copper edge, while, as previously done on MPS_3 — see Sect. 3.1.2 —, a two-steps fit over three shells has been performed and has proved satisfactory for the manganese K edge spectra.

In order to prevent from any error due to a fit on a false minimum, several attempts using different E_0 starting values and R (Cu—S) values were carried out. Besides Mn—S distances indicative of the existence of a manganese site very similar indeed to the manganese site in the model compound $MnPS_3$, the Cu—S fitted distance is found to be 2.22 Å i.e. very close to the first qualitative estimate but extremely different from the crystallographic value of 2.07 Å proposed after the above XRD analysis (Table 8). The origin of such an enormous and surprising discrepancy was investigated with very careful consideration at each stage of the experimental procedures and data analysis. No step, neither in XRD nor in EXAFS, could be incriminated to account for this difference as well as for other serious divergences (on the Cu . . . Cu distances for example), lying between XRD and EXAFS data. The solution of this apparently puzzling problem arose finally from the following basic considerations:

(i) observing inconsistencies between sets of data obtained by methods based on fundamentally different probing modes is not surprising at all, as shown in Sect. 2. Indeed, radial distribution function R(r) and all subsequent structural information provided by XRD experiments are averaged over all pairs of atoms in the investigated sample of $Mn_{0.87}Cu_{0.23}PS_3$ and lead to a LRO description. This gives an unrealistic picture of the structure with indistinguishable $[Mn_{(1-x)}Cu_{2x}S_6]$ octahedra interpreted upon further considerations by means of a disordered ditribu-

Table 8. $Mn_{0.87}Cu_{0.26}PS_3$ Metal-sulfur distances as obtained by EXAFS and XRD. The EXAFS values are those obtained at 15 K since the thermal vibration at 300 K is responsible for a systematic error of (−0.01) to (−0.02) Å

M—S	XRD distances/Å	Exafs distances/Å
Cu—S	2.075 ± 0.007	2.22 ± 0.02
Mn—S	2.643 ± 0.009	2.61 ± 0.02

tion of [MnS$_6$] and [S$_3$Cu ... CuS$_3$] octahedra. In contrast, EXAFS allows to access to individual R(Mn(r)) and R(Cu(r)) RDF and leads directly to a SRO description with the detailed and separated local environments seen by Mn and Cu atoms respectively;

(ii) there is no a priori reasons for expecting identical sizes in the S$_6$ skeletons of the [MS$_6$] and [S$_3$Cu ... CuS$_3$] octahedra. As a matter of fact, the XRD experiment averages not only the "atomic" content of each octahedron but also its dimensions, yielding finally "indistinguishable" entities where "distinguishable" ones do exist. Fig. 29 compares the dimensions and the atomic arrangements of the two octahedra discerned by EXAFS with those averaged by XRD.

The presence of two Cu(I) ions in 13 % of the octahedra (Fig. 29 c) induces an expansion of these latter with respect to the 87 % of more usual [MnS$_6$] entities (Fig. 29 b). This relative dilatation is responsible for the weak but significant increase (0.03 Å) observed on the Mn—S distances upon averaging the centre to apex distances over all the octahedra. Correspondingly, as 87 % of the octahedra are smaller than the [S$_3$Cu ... CuS$_3$] ones, upon averaging, the XRD Cu—S distances drop dramatically from their local EXAFS value of 2.22 Å (Fig. 29 c) to the abnormally short value of 2.075 Å [30]. Turning now to the problem of the Cu ... Cu distances which are also underestimated by XRD analysis, a simple calculation based on an homothetic correspondence between XRD averaged (Fig. 29 a) and EXAFS isolated octahedra (Fig. 29 c) gives a value of $3.003 \cdot 2.22/2.075 = 3.21$ Å which compares well with the EXAFS estimation.

Clearly, the separation between the two kinds of octahedra with their true (un-averaged) dimensions and all the subsequent quantitative considerations, is allowed only by the EXAFS data, collected at both the Mn and Cu edges, whereas the unambiguous location of the Cu(I) in the Mn$_{0.87}$PS$_3$ framework is given only by the XRD picture. It is necessary to add two sets of data harvested by the means of two methods to understand the apparent mismatch between XRD and EXAFS data for the manganese and copper environments in the Mn$_{0.87}$Cu$_{0.26}$PS$_3$ compound and to obtain a

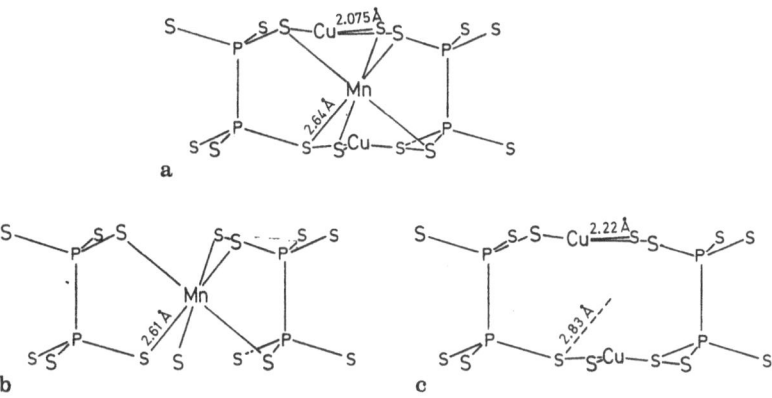

Fig. 29 a–c. Schematic representations of the Mn and Cu sites in Mn$_{(1-x)}$Cu$_{(2x)}$PS$_3$ (x = 0.13): a) XRD-averaged site; b) manganese site; c) copper site. The centre to sulfur distance of 2.83 Å is calculated with the assumption that the three types of sites, [MnS6], XRD-averaged octahedron and [S$_3$Cu ... CuS$_3$] are homothetic

"true" image of the reality. Still better, we consider now, that such discrepancies observed between raw XRD and EXAFS can constitute, after a thorough critical check of the validity of the experimental and analysis procedures, the signature of a partial disorder among XRD averaged sites. As disorder exists in a wide variety of low dimensional coordination solids such a result deserves to be emphasized.

Therefore, the complementary nature of the EXAFS and XRD techniques should be highly profitable for getting detailed insight into the structure of such disordered solids if (i) the disorder occurs over a reasonable, detectable, number of sites; (ii) each kind of site is associated with a particular type of atom and local structure and (iii) EXAFS measurements can be performed at the corresponding edges.

Further works illustrating the power of the EXAFS and the XRD techniques when used in conjunction on disordered layered systems of the MPS_3 family were recently reported (11) or are in progress [62].

4.4 EXAFS Site Selectivity as a Tool for Characterizing Heterometallic $Mn_{(1-x)}Ni_xPS_3$ Layers Obtained by Mild Chemistry

The ability of $MnPS_3$ to react under mild conditions leads to the possibility of facile syntheses in the solid state at room temperature; this unfortunately yields microcrystalline powders which cannot be fully characterized by X-Ray powder diffraction. In such cases, EXAFS can bring complementary information which may be

Table 9. Structural parameters extracted from EXAFS spectra. In $Mn_{1-x}Ni_xPS_3$ at nickel edge, two different compositions of the second shell were tested: 3 (P—P) and 3 Mn (a), 3 (P—P) and 3 Ni (b). Fit (b) represents the case of a mechnical $MnPS_3$—$NiPS_3$ mixture. Since the fit yields a similar residue, the actual environment of Ni in $Mn_{1-x}Ni_xPS_3$ cannot be solved by EXAFS alone, but the mechanical mixture possibility has been ruled out by XRD.

Compounds	Bond	Distance R/Å	Dispersion σ/Å	Residual factors
$MnPS_3$	Mn—S	2.59	0.08	
	Mn—Mn	3.49	0.07	1.7%
	Mn—P	3.66	0.07	
$Mn_{1-x}PS_3K_{2x}$	Mn—S	2.57	0.09	
	Mn—Mn	3.52	0.15	3.4%
	Mn—P	3.73	0.15	
$Mn_{1-x}Ni_xPS_3$	Mn—S	2.58	0.09	
	Mn—Mn	3.45	0.07	1.2%
	Mn—P	3.69	0.07	
$Mn_{1-x}Ni_xPS_3$	Ni—S	2.42	0.1	
	Ni—P	3.60(a)	0.08	0.82(a)%
		3.62(b)		
	Ni—Mn(a)	3.38	0.08	
	Ni—Ni(b)	3.34	0.08	0.95(b)%
$NiPS_3$	Ni—S	2.41	0.09	
	Ni—Ni	3.35	0.06	1.7%
	Ni—P	3.55	0.06	

essential to the description of the obtained material, as illustrated in the following example.

The potassium ions inserted in the $Mn_{(1-x)}PS_3K_{2x}(H_2O)_{\approx 1}$ intercalate are solvated by a monolayer of water molecules and are therefore highly mobile. Treating this intercalate with an aqueous $NiCl_2$ solution results in the replacement of the K^+ ions by $[Ni(H_2O)_6]^{2+}$ species. Upon drying, the nickel get rid of the aqua ligands and an anhydrous $Mn_{(1-x)}Ni_xPS_3$ ($x = 0.72$) compound is formed, in which the layers are back in Van der Waals contact (according to the value of the measured interlamellar distance $\simeq 6.45$ Å). X-Ray powder diffraction shows that the sample is monophasic: the parameters of $Mn_{0.72}Ni_{0.28}PS_3$ are intermediate between those of $MnPS_3$ and those of $NiPS_3$ (Table 9), a fact which rules out the possibility that the mixed material might be a mechanical mixture of $MnPS_3$ and $NiPS_3$, and which is further confirmed by the study of its magnetic properties [9].

The structural problem which then arises is to determine on which sites (intra or inter lamellar) the nickel ions are located. EXAFS gave a definite answer to this question by comparison of the local environment of manganese and nickel in the mixed material with their respective environment in pure $MnPS_3$ and pure $NiPS_3$.

The RDF of $MnPS_3$, $Mn_{(1-x)}PS_3K_{2x}(H_2O)$, $NiPS_3$, $Mn_{(1-x)}Ni_xPS_3$ (at both Mn and Ni edges) are shown in Fig. 30.

Comparing the curves of the potassium intercalate and of $Mn_{(1-x)}Ni_xPS_3$ at the manganese edge immediately shows that the second peak, which is very small in the potassium intercalate (because of the disorder discussed in Sect. 4.2, reappears with a strong amplitude in the mixed anhydrous material. Therefore, a decrease of the local disorder around the Mn(II) ions accompanies the chemical process. Furthermore, careful study of the spectra shows that the environment of manganese in the mixed $Mn_{(1-x)}Ni_xPS_3$ compound is quasi identical to that of Mn in pure $MnPS_3$. Similarly, comparison of the radial distribution curves of $NiPS_3$ and $Mn_{(1-x)}Ni_xPS_3$ at the nickel edge shows that the surroundings of nickel is quasi identical in both cases. Therefore, Ni(II) ions as well as Mn(II) ions in $Mn_{(1-x)}Ni_xPS_3$ are located on intramolecular sites, at the centres of the sulfur octahedra. If manganese or nickel ions were located in the Van der Waals gap, they would not have a second neighbour shell consisting of three metallic ions and three P—P pairs, and the second shell peak in the mixed material would not be identical to that in the pure ones. Quantitative structural data extracted from the experimental spectra are gathered in Table 9.

We wish to emphasize here that these data provide a second example of the way in which EXAFS site selectivity can be used to obtain local distances in a material in which X-ray data would only lead to an average picture.

5 Conclusion

We would like to draw up three comments to conclude.

(1) Properly handled, EXAFS can be of invaluable help in solving order and disorder structural problems in low-dimensional materials.

(2) Up to date, exploration of a given compound at different edges and low temperature data have not been used systematically. The availability of inexpensive liquid nitrogen cryostats and of X-ray spectrometers in an expanded energy range, specially

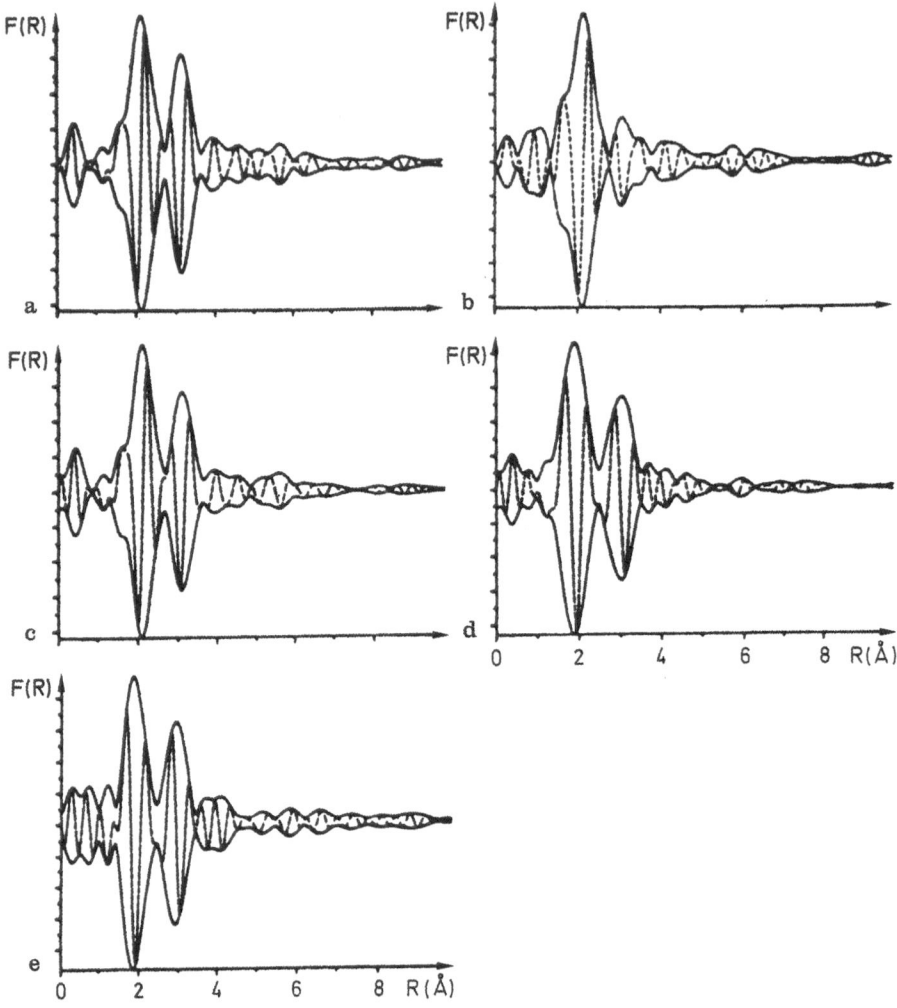

Bild 30 a–e. RDF around the absorbing metallic atom at 300 K: **a)** $MnPS_3$; **b)** $Mn_{(1-x)}PS_3 K_{(2x)} (H_2O)$: **c)** $Mn_{(1-x)} Ni_{(x)} PS_3$; **d)** $Mn_{(1-x)} Ni_{(x)}$; **e)** $NiPS_3$ (absorbing atoms are italic)

in the soft X-ray part where a ligand-centered view (O, P, Cl, S) can be expected, would allow new useful insights to solve structural problems in materials. We got promising preliminary results in this direction at sulfur and phosphorus edges in the MPS_3 series.

(3) We have said as manybody, before to become synchrotron radiation users, that nothing was better that single crystal X-ray diffraction crystallographic structure. Our story tells that when order and disorder problems are involved, this settlement should be tempered and that SRO techniques have an irreplaceable interest. Indeed, instead of opposing SRO and LRO pictures, we show that it is the comparison of both views which is fruitful in low-dimensional materials, where anisotropy makes the problem easier to solve. We guess that a more systematic combined LRO and SRO

approach of 3D solids, though more difficult, would give new and sometimes surprising results when applied to some materials considered as perfectly ordered.

Appendix: Data Collection and Analysis

The X-Ray absorption spectra presented here were recorded at LURE, the french synchrotron radiation facility, on the DCI EXAFS I, II or III spectrometers described by Raoux et al. [63], and J. Goulon et al. [64].

The spectra were recorded within a period of several years, but with analogous operating storage conditions: 1.72 eV energy, 100–250 mA current intensity.

We used channel-cut Si 220 or Si 400, or double crystal Si 311 monochromators, and air ionization chambers detectors. Most of the spectra were recorded with a step of 2 eV on a range of 800 to 1000 eV.

The samples were well-pounded microcrystalline powders uniformly spread between two close-fitting non absorbing ribbon tapes of controlled thickness.

The low temperature measurements used an helium continuous-flow cryostat, allowing work at fixed temperature down to 15 K. Most of the experiments were performed at 30 K ± 5 K.

The data analysis was performed with use of the classical EXAFS formula (1) within the plane-wave approximation. The amplitude and phase shifts of Teo et al. [65] were used except when otherwise specified in the text. When tabulated and experimental parameters are used on the same data, slight differences appear in the final results (Tables 1 and 3 for CuC_2O_4 for example).

The classical analysis way was followed: extraction of EXAFS signal $k*\chi(k)$, Fourier transform of $k^3 \cdot \chi(k)$ in the R space, filtering of one or more shells, fitting of the filtered inverse Fourier transformed signal in the k space. The figures represent either the modulus of the Fourier transform or the imaginary part (dotted line) and the modulus (full line).

The quantity minimized in the fit is the weighted residual factor R or ϱ:

$$R = \sum_i w(k) \, (\chi_{exp}^i - \chi_{calcd}^i)^2 \Big/ \sum_i \chi_{exp}^{i2} \cdot w(k)$$

$w(k)$ is the weighting factor: $w(k) = 0$ for $k < 3\,\text{Å}^{-1}$ and $= k^5$ for $k > 3\,\text{Å}^{-1}$. The programs were written by Michalowicz after a first version by Goulon. All these FORTRAN 77 programs are available in free-access at the Paris-Sud informatique centre.

6 Acknowledgement

We are most grateful to Dr Marin and his colleagues of the Laboratoire de l'Accelerateur Lineaire and LURE for the operation of the storage ring DCI, to Drs Lagarde, A. Sadoc, H. Dexpert for the access to the EXAFS I and III spectrometers, to Dr Goulon and R. Cortes for the access to the EXAFS II spectrometer and to J. Goulon for his help in initial programming, to Drs C. Dupas, J. J. Girerd, A. Gleizes, M. Julve, H. Mercier for kindly supplying samples and for many enjoyable discussions and to Prof. R. Fourme and O. Kahn for their constant support and interest.

7 References

1. See e.g.:
   a. Bernasconi J, Schneider T (Ed.), (1981) Physics in one-dimension, Springer, Berlin
   b. Miller JS (Ed.), (1981–1983) Extended linear chain compounds, Plenum, New York, 3 vol.
   c. Gatteschi D, Kahn O, Willett RD (Ed.), (1985) Magneto-structural correlations in exchange-coupled systems, NATO ASI Series, C140, Reidel, Dordrecht
   d. Delhaës P, Drillon M (Ed.), 1987, organic and inorganic low dimensional crystalline NATO ARW Series, Reidel Dordrecht under press
2. See e.g.:
   a. Wittingham MS, Jacobson AJ (Ed.), (1982) Intercalation Chemistry, Academic Press, New York
   b. Levy FA (Ed.), (1979) Physics and Chemistry of materials with layered structures, Vol. 1–6, Reidel, Dordrecht
3. Michalowicz A, Girerd JJ, Goulon J, (1979) Inorg. Chem. *18*: 3004
4. Verdaguer M, Michalowicz A, Girerd JJ, Alberding N, Kahn O, (1980) ibid. *19*: 3271
5. Verdaguer M, Julve M, Michalowicz A, Kahn O, (1983) ibid. *22*: 2624
6. a. Gleizes A, Verdaguer M, (1984) J. Am. Chem. Soc. *106*: 3727
   b. Verdaguer M, Gleizes A, Renard JP, (1984) J. Seiden Phys. Rev. *29*: 5144
   c. Gleizes A, Audiere JP, Verdaguer M, work in progress
7. Michalowicz A, Clement R, (1982) Inorg. Chem. *21*: 3872
8. Mathey Y, Michalowicz A, Toffoli P, Vlaic G, (1984) ibid. *23*: 897
9. Clement R, Michalowicz A, (1984) Rev. Chim. Min. *21*: 426
10. Clement R, Mercier H, Michalowicz A, (1985) ibid. *22*: 135
11. Mathey Y, Mercier H, Michalowicz A, Leblanc A, (1985) J. Phys. Chem. *46*: 1025
12. Clement R, Garnier O, Jegoudez J, (1986) Inorg. Chem. *25*: 1404
13. Sayers DE, Stern EA, Lytle FW, (1971) Phys. Rev. Lett. *27*: 1207
14. Teo BK, Joy DC (Ed.), (1981) EXAFS Spectroscopy, Plenum, New York
15. Winick H, Doniach S (Ed.), (1980) Synchrotron radiation research, Plenum Press
16. a. Bianconi A (Ed.), 1982 EXAFS and Near Edge Structures, Springer Series Chem. Phys. 27, Berlin
    b. Hodgson KO, Hedman B, Penner-Hahn JE (Ed.), (1984) EXAFS and Near Edge Structure III, Springer Berlin
17. Penner-Hahn JE, Hodgson KO (Ed.), (1985) EXAFS III, Springer Berlin
18. Proc. Conf.: Progress in X-ray studies by synchrotron radiation, Strasbourg, 1–4 April 1985
19. a. See e.g. papers in Part VI (amorphous materials and glasses) in Ref. 16b, p. 267sq
    b. See e.g. in Ref. 18 papers by Gaskell PH, Babanov YA, Sadoc A, Maurerand M, Petiau J
20. a. Eisenberger P, Brown GS, (1979) Sol. Stat. Comm. *29*: 481
    b. Crozier ED, Seary AJ, (1980) Can. J. Phys. *58*: 1388
    c. Fontaine A in Ref. 18, C4
21. a. Catlow CRA et al., (1985) Nature (London) *312*: 601
    b. Greaves GN, Hatton PD, (1985) Chem. Brit. 371
22. a. Doniach S, Eisenberger P, Hodgson KO in Ref. 15, p. 425
    b. See e.g. Part III (Biological systems) in Ref. 16b, p. 95sq
23. Ref. 14, a) p. 17; b) p. 18; c) p. 22
24. Michalowicz A, Huet J, Gaudemer A, (1982) Nouv. J. Chim. *6*: 79
25. Poncet JL, Guilard R, Friant P, Goulon-Ginet C, Goulon J, (1984) ibid. *8*: 583
26. Eisenberger P, Lengeler B, (1980) Phys. Rev. B *22*: 3551
27. a. Teo BK, (1980) J. Am. Chem. Soc. *103*: 3990
    b. Co MS, Hendrikson WA, Hodgson, KO, Doniach S, (1983) ibid. *105*: 1144
28. a. Teo BK, Eisenberger P, Kincaid BM, (1978) ibid. *100*: 1735
    b. Goulon J, Friant P, Goulon-Ginet C, Coutsolelos A, Guilard R, (1984) Chem. Phys. *83*: 367
29. Murata T, Lagarde P et al. in Ref. 16, p. 271 and in Ref. 17, p. 432
30. Craievitch A, Dartyge E et al. in Ref. 16, p. 274
31. Raoux D, Fontaine A et al., (1981) Phys. Rev. B *24*: 5547
32. Marcus M, Tsai CL, (1984) Solid State Comm. *52*: 551

33. Mikkelsen JC, Boyce JB, (1983) Phys. Rev. B *28*: 7130
34. Marboeuf A, Dexpert H et al. private communication
35. de Novion CH, Landesman JP, (1985) Pure Appl. Chem. *57*: 1391
36. Krill G, Godart C, (to be published) Solid State Comm.
37. Malzfeldt W, Niemann W, Rabe P, Haensel R in Ref. 17, p. 445
38. a. Kanda S, (1960) Bull. Chem. Soc. Jpn. Pure Chem. Sect. *81*: 3147
    b. (1962) ibid. *83*: 283
    c. Note added in proof: the planar bischelating configuration found by EXAFS has just been confirmed in a binuclear iodanilato derivative. See Tinti F, Verdaguer M, Kahn O, Savariault JM (1987) inorg. chem. *26*, 2380
39. a. Garaj J, Lanfelderova M, Lundgren G, Gazo J, (1972) Collect. Czech. Chem. Comm. *37*: 3181
    b. Girerd JJ, Kahn O, Verdaguer M, (1980) Inorg. Chem. *19*: 274
40. a. Schmittler H, (1968) Monatsber. Deutsch. Akad. Wiss. Berlin *10*: 581
    b. Fichtner-Schmittler H, (1984) Cryst. Res. Technol. *19*: 1225
41. Dubicki L, Harris C, Kokot E, Martin RL, (1966) Inorg. Chem. *5*: 93
    b. Mc Gregor KT, Soos ZG, (1976) ibid. *15*: 2159
42. a. Gleizes A, Maury F, Galy J, (1979) ibid. *18*: 3004
    b. Maury F, (1980) These 3eme cycle, Universite de Toulouse
43. Viswamitra, (1962) Z. Krist. *117*: 437
44. Michalowicz A, Fourme R, (1981) Acta Cryst. *A37*: C307
45. a. Mazzi C, Garavelli F, (1957) Period. Mineral. *26*: 2–3
    b. Caric S, (1959) Bull. Soc. Fr. Mineral. Cristallogr. *82*: 50
46. Dubernat J, Pezerat H, (1974) J. App. Cryst. *7*: 387
47. a. Drillon M, Coronado E, Beltran D, Georges R, (1983) Chem. Phys. *79*: 449
    b. Coronado E, Ph.D. Thesis, Valencia, Spain
48. a. Kahn O, Pei Y, Sletten J, Verdaguer M, Renard JP (1986) J. Am. Chem. Soc. 108, 7428–7430
    b. Note added in proof: see also contributions by Darriet J, Drillon M, Georges R, Kahn O, Landee C, Palacio F, Pei Y, Renard JP, Rey P, Verdaguer M in Ref. 4d
49. a. Dingle R, Lines ME, Holt SL, (1969) Phys. Rev. *187*: 643
    b. Dupas C, (1978) These, Orsay
    c. Dupas C, Renard JP, Seiden J, Cheikh-Rouhou A, (1982) Phys. Rev. B *25*: 3261
    d. Richards PM, (1974) Phys. Rev. B *10*: 805; (1976) ibid. *13*: 458
    e. Clement S, Dupas C, Renard JP, Cheikh-Rouhou A., (1982) J. Physique *43*: 767
50. Clement S, Dupas C, Verdaguer M in Ref. 18, 3(4)p10
51. a. Klingen W, (1969) Thesis, Hohenheim
    b. Klingen W, Ott R, Hahn H, (1973) Z. Anorg. Allg. Chem. *396*: 271
52. Johnson JW, (1982) in: Intercalation Chemistry, Whittingbottom MS, Jacobson AJ (Ed.), Acad. Press, New York
53. Brec R, Sleich DM, Ouvrard G, Louisy A, Rouxel J, (1979) Inorg. Chem. *18*: 1814
54. Soled S, Wold A, (1976) Mat. Res. Bull. *11*: 657
55. Colombet P, Leblanc A, Danot M, Rouxel J, (1982) J. Solid State Chem. *41*: 171
56. Ouvrard G, Freour R, Brec R, Rouxel J, (1985) Mat. Res. Bull. *20*: 1053
57. a. Klingen W, Eulenberger G, Hahn H, (1973) Z. Anorg. Allg. Chem. *401*: 97
    b. Ouvrard G, Brec R, Rouxel J (to be published) Mat. Res. Bull
    c. Brec R, Ouvrard G, Louisy A, Rouxel J, (1982) Ann. Chim. Fr. *5*: 499
58. Clement R, Audiere JP, Renard JP, (1982) Rev. Chim. Min. *19*: 560
59. Michalowicz A, Clement R (to be published) J. Inclus. Chem.
60. a. Bourdillon AJ, Pettifer RF, Marseglia EA, (1980) Physica B + C (Amsterdam) *99*: 64
    b. Kauzlarich SM, Teo BK, Averill BA, (1986) Inorg. Chem. *25*: 1209
61. a. Brown DB, Zubieta JA, (1980) ibid. *19*: 1945
    b. Eller PG, (1971) J. Chem. Soc., Chem. Comm. 105
    c. Siiman O, (1981) Inorg. Chem. *20*: 2285
62. Mathey Y, Colombet P, Michalowicz A (to be published)
63. Raoux D, Petiau J, Bondot P, Calas G, Fontaine A, Lagarde P, Levitz P, Loupias G, Sadoc A, (1980), Rev. Phys. Appl. *15*: 1079
64. Goulon J, Lemonnier M, Cortes R, Retournard A, Raoux D, (1983) Nucl. Instrum. Meth. *208*:
65. Teo BK, Lee PA, Simmons AL, Eisenberger P, Kincaid BM, (1977) J. Am. Chem. Soc. *99*, 3854
66. Piacentini M et al., (1982) Chem. Phys. *65*: 289

Resonant X-Ray Scattering in Biological Structure Research

Heinrich B. Stuhrmann

Institut für Physikalische Chemie Johannes-Gutenberg-Universität, Mainz, FRG

Table of Contents

The use of anomalous X-ray scattering of light elements like sulfur and phosphorus is of particular interest in biological structure reasearch. These elements serve as native labels in proteins, nucleic acids and membranes. Their medium scattering power is drastically changed at their K absorption edges at wavelengths between 5 and 6 Å where X-ray absorption excludes the use of open air diffractometers. The construction of a new diffractometer tunable to wavelengths between 1.2 and 7 Å is presented. First results of anomalous scattering from sulfur in bacteriorhodopsin near the K absorption edge have been obtained recently. Their possible impact on crystallography will be considered. A comparison with nuclear spin dependent neutron scattering is given.

Present Adress: GKSS-WS, 2054 Geesthacht, FRG

1 Introduction

Resonant (or anomalous) X-ray scattering is a non-destructive labelling technique of X-ray structure analysis. Heavy atom derivatives are used for the determination of protein structures and other biomolecules. These crystals would also lend themselves to direct structure determination by resonant X-ray diffraction. This technique is discussed in many textbooks of protein crystallography [1, 2]. A review of X-ray resonant scattering of non-crystalline macromolecular structures is given in [3]. This acticle will concentrate on recent developments leading to the use of native label atoms, e.g. sulfur in proteins and phosphorus in nucleic acids and polar head groups of lipids.

Resonant scattering is highly selective. The chemical element of interest will give rise to strong anomalous dispersion only in a few narrow energy bands very close to the absorption edges. Gold, for instance, shows strong dispersion of X-ray scattering at its K edge ($\lambda = 0.153$ Å), at its L_3 edge ($\lambda = 1.040$ Å) and at its M_5 edge ($\lambda = 5.584$ Å). The dispersion at the K and L_3 edges is most easily measured with open air diffractometers. The X-ray spectrum near the M_5 edge is strongly absorbed by air. This absorption edge would not be an easy candidate for anomalous X-ray scattering experiments.

With light elements like sulfur and phosphorus we do not have the choice demonstrated in the case of gold. The only absorption edges which could be of any use in anomalous X-ray diffraction work are the K absorption edges, and these are at wavelengths near 5.0185 Å and 5.784 Å, respectively. The use of anomalous scattering of light elements for macromolecular structure determination needs a new concept for the construction of the diffractometer and in particular of the sample environment. The considerations concerning the technique start with the relation between the total cross-section and the cross-section of resonant X-ray scattering.

2 Scattering by a Single Atom

When the wavelength of the incident X-ray beam is close to the absorption edge of an atom, the atomic scattering factor becomes complex to a greater extent:

$$f = f_0 + f' + if'' \tag{1}$$

f_0 is the short-wavelength limit of the scattering amplitude. Following the convention of X-ray diffraction, f is given in electrons, e.g. $f_0(0)$ of iron is 26. The scattering length of one electron is 2.8×10^{-12} mm. f' and f'' exhibit a wavelength dependence (dispersion). The absorption edge for X-rays represents the threshold frequency above which an inner electron can be ejected into the continuum. The resonance absorption, which is also known as photoelectric absorption, is an inelastic channel. The optical theorem for X-rays is then

$$\sigma_{photoelectric} = 2\lambda \frac{e^2}{mc^2} f'' \tag{2}$$

Since $\sigma_{photoelectric}$ exists only in the short wavelength region of the absorption edge, so also does f"(0). The dispersion of f"(0) is fully controlled by $\sigma_{photoelectric}$. For the K edge, $\sigma_{photoelectric}$ is represented fairly well by the empirical formula

$$\sigma_{photoelectric} = \left(\frac{\omega_k}{\omega}\right)^3 \sigma_{photoelectric}(\omega_k) \ (\omega > \omega_k) \tag{3}$$

Since for photons $\omega = 2\pi c/\lambda$, f"(0) should vary with frequency roughly as $1/\omega^2$. Figure 1 shows that the dispersion of f"(0) is slightly more complicated, because ionization of p and d·electrons will give rise to additional abrupt changes of f.

In fact, the schematic representation of the absorption edge in Fig. 1 has to be replaced by a more structured dispersion profile at higher frequency resolution, as is shown in Fig. 2.

The absorption coefficient μ (in mm^{-1}) and density 1 Mg/m^3 is

$$\mu = \frac{2N_L}{M}\lambda\frac{e^2}{mc^2}f" = \frac{337.1}{M}\lambda f" \tag{4}$$

($N = 6.02 \times 10^{23}$, M = atomic weight, $e^2/mc^2 = 2.8 \times 10^{-12}$ mm).

Of the numerous empirical formulae used for the computation of X-ray absorption coefficients we just mention

$$\mu = 0.016\lambda^3 Z^{3.94}/M \qquad \lambda < \lambda_K . \tag{5a}$$

$$\mu = 0.000\,529\lambda^3 Z^{4.3}/M \qquad \lambda > \lambda_K \tag{5b}$$

Z is the atomic number.

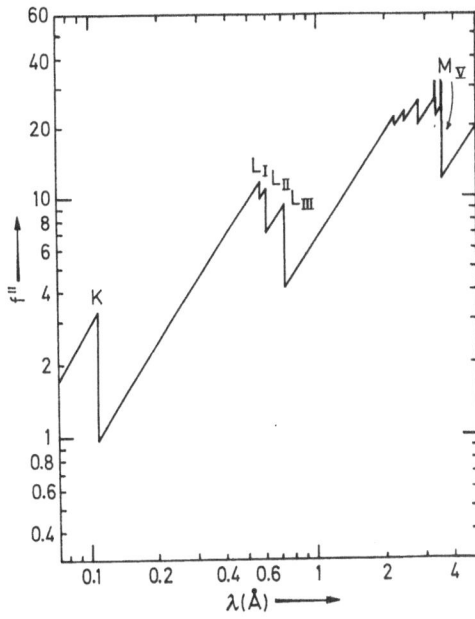

Fig. 1. Dispersion of the imaginary part f" of the atomic form factor of uranium. f" (in units of electrons) reaches nearly two-thirds of the non-resonant atomic form factor $f_0 = 92$. The schematic representation has been taken from The International Tables of Crystallography IV. The dispersion of f" shows strong peaks at the M_{IV} and M_V absorption edges [4]

153

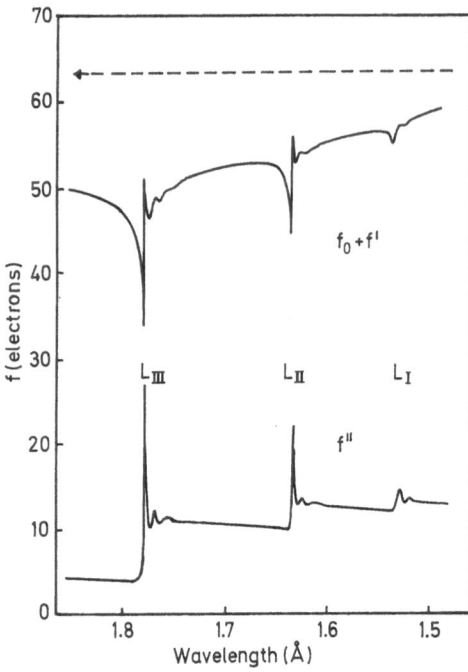

Fig. 2. Dispersion of $f_0 + f'$ of europium in Eu(PhAcAc) at the three absorption edges, after [5]. The f' values were calculated by using the Kramers-Kronig relation, Eq. (7). The absolute scale (in electrons) relies on values of f" and f' which were taken from The International Tables of Crystallography IV

On crossing the X-ray K absorption edge from lower to higher frequencies the absorption coefficient μ will increase by the factor $63.868/Z^{0.6207}$. Since the spatial distribution of the core electrons is confined to a very small volume near the nucleus, f" (and f') shows a relatively weak dependence on the scattering angle. Few experiments have been done to confirm theoretical predictions, e.g. on Barium [6] with MoK$_0$ radiation where

$$f'' = 2.40(5) - 0.59(11) \sin^2 \theta/\lambda^2 \tag{6}$$

The measurement of the total cross-section determines completely the dispersion of f". The dispersion of the real part is not independent of that of f", because there is a general relationship between them, which is known as the Kramers-Kronig relation:

$$f'(\omega) = \frac{2}{\pi} \int_{\omega=0}^{\infty} \frac{\omega' f''(\omega')}{\omega^2 - \omega'^2} \, d\omega' \tag{7}$$

Thus a knowledge of $f''(\omega)$ over a sufficiently wide frequency region permits the evaluation of $f'(\omega)$. With X-rays the frequency interval of strong anomalous dispersion may be very narrow. Then Eq. (7) simplifies to

$$f'(\omega) = -\frac{1}{\pi} \int_{\varepsilon=-\delta}^{\delta} \frac{f''(\omega + \varepsilon)}{\varepsilon} \, d\varepsilon \tag{8}$$

where δ is of the order of 0.01. Approximation of the integral by a sum of the integrand at −δ and +δ yields

$$-\frac{1}{\pi}\frac{f''(\omega+\delta)-f''(\omega-\delta)}{\delta}=-\frac{2}{\pi}\frac{\Delta f''(\omega)}{\Delta\omega} \tag{9}$$

The f′ dispersion looks like the first derivatives of f″.

The interplay of f′ and f″ becomes clearer when the amplitude f is represented in the plane of complex numbers (Argand diagram). As is shown in Fig. 3, f of iron in ferritin, an iron-storing protein, follows two-thirds of a circular line at wavelengths near its K absorption edge at λ = 1.743 Å [7]. In the case of nuclear resonance a nearly closed circle would be observed. Anomalous X-ray scattering at the L_3 edges of rare earth atoms very often shows a strongly enhanced peak at the absorption edge, yielding an Argand diagram very similar to nuclear resonance. The strong variation of the resonant amplitude of these elements has gained considerable importance in X-ray resonant diffraction [8,9].

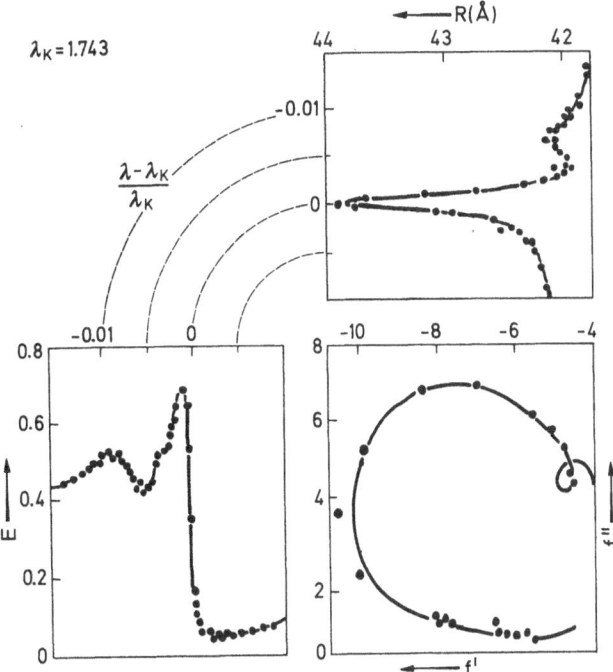

Fig. 3. Construction of the Argand diagram from the dispersion of f′ and f″. f′ is taken from the dispersion of apparent radius of gyration, R, of ferritin. f″ is proportional to the absorption of the ferriton solution at the K absorption edge of iron at λ_K = 1.743 Å

3 Scattering by Oriented Macromolecules

The description of resonante scattering starts from an assembly of M atoms, each having fixed coordinates r with respect to an origin. Of these, a smaller number N of atoms is assumed to have strong resonant scattering. The structure $\varrho(r)$ is then represented by

$$\varrho(r) = \sum_{n=1}^{M} f_{o,n} \delta(r - r_n) + (f' + if'') \sum_{n=1}^{N} \delta(r - r_n)$$

$$\equiv u(r) + (f' + if'') \, v(r) \tag{10}$$

It has been assumed that all strongly resonant scatters have the same dispersion, i.e. they represent the same chemical element and they are assumed to have the same chemical environment. With U(r) and V(r) as the real and imaginary parts of $\varrho(r)$ respectively, we obtain the scattering amplitude of the structure $\varrho(r)$ as

$$F(h) \sim \int [U(r) + iV(r)] \, e^{ih \cdot r} \, d^3r \tag{11}$$

where h is the scattering vector and $h = |h| = 4\pi \sin \theta/\lambda$. The multiplication of F(h) with its complex conjugate

$$F^*(h) \sim \int [U(r) - iV(r)] \, e^{-ih \cdot r} \, d^3r \tag{12}$$

yields the scattering intensity S(h)

$$S(h) = A(h) \, A^*(h) \tag{13}$$

$$= \iint [U(r) + iV(r)] \, [U(r') - iV(r')] \, e^{ih \cdot (r-r')} \, d^3r \, d^3r'$$

$$= \iint [U(r) \, U(r') + V(r) \, V(r')] \cos [h \cdot (r - r')] \, d^3r \, d^3r'$$

$$+ \iint [U(r) \, V(r') - V(r) \, U(r)] \sin [h \cdot (r - r')] \, d^3r \, d^3r'$$

Separating the resonant real part we obtain

$$S(h) = \iint u(r) \, u(r') \cos [h \cdot (r - r')] \, d^3r \, d^3r'$$

$$+ 2f' \iint u(r) \, (r') \cos [h \cdot (r - r')] \, d^3r \, d^3r'$$

$$+ (f'^2 + f''^2) \iint v(r) \, v(r') \cos [h \cdot (r - r')] \, d^3r \, d^3r'$$

$$+ f'' \iint [u(r) \, v(r') - v(r) \, u(r')] \sin [h \cdot (r - r')] \, d^3r \, d^3r'$$

$$= S_u(h) + f'S_{uv}(h) + (f'^2 + f''^2) \, S_v(h) + f''\varphi(h)$$

This is the dispersion of the geometrical structure factor in crystallography. The overall effect of resonant scattering is to cause the breakdown of Friedel's law so that the Bijvoet pairs of reflections S(h) and S(−h) are unequal (see e.g. [1]). The difference

$$S(h) - S(-h) = 2f''\varphi(h) \tag{15}$$

is used to determine the absolute configuration [10].

4 Scattering by Free Molecules

A first step towards the consideration of free, randomly oriented molecules (or microcrystals) is the superposition of $S(\mathbf{h})$ on $S(-\mathbf{h})$. Only the cosine terms in Eq. (12) are conserved. Introducing the wavelength dependence of f' and f'' explicitly, we obtain by integration of $S(\mathbf{h})$ over the solid angle [3]

$$
\begin{aligned}
I(\mathbf{h}, \lambda) = &\int S(\mathbf{h}, \lambda) \, d\Omega \\
= &\iint u(\mathbf{r}) \, u(\mathbf{r}') \sin h \, |\mathbf{r} - \mathbf{r}'|/|\mathbf{r} - \mathbf{r}'| \, d^3r \, d^3r' \\
&+ 2f'(\lambda) \iint u(\mathbf{r}) \, v(\mathbf{r}') \sin h \, |\mathbf{r} - \mathbf{r}'|/|\mathbf{r} - \mathbf{r}'| \, d^3r \, d^3r' \\
&+ [f'^2(\lambda) + f''(\lambda)] \iint v(\mathbf{r}) \, v(\mathbf{r}') \sin h \, |\mathbf{r} - \mathbf{r}'|/|\mathbf{r} - \mathbf{r}'| \, d^3r \, d^3r' \\
\equiv &\, I_u(\mathbf{h}) + f'(\lambda) \, I_{uv}(\mathbf{h}) + [f'^2(\lambda) + f''^2(\lambda)] \, I_v(\mathbf{h}) \quad (16)
\end{aligned}
$$

Macromolecules in solution as described by Eq. (10) give rise to three basic scattering functions. $I_u(\mathbf{h})$ is the off-resonance scattering, I_v originates from the structure of the resonant scatterers alone, and $I_{uv}(\mathbf{h})$ is a cross term, containing the influence of both $u(\mathbf{r})$ and $v(\mathbf{r})$ as their convolution. In many cases the resonant scattering terms in Eq. (16) may be much smaller than $I_u(\mathbf{h})$. The quadratic term in Eq. (16) can then be neglected and we obtain

$$
I(\mathbf{h}, \lambda) = I_u(\mathbf{h}) + f'(\lambda) \, I_{uv}(\mathbf{h}) \quad (17)
$$

This equation stresses the dominant role of f' in resonant scattering of non-crystalline structures. As can be deduced from Fig. 3, resonant scattering in this case will be observed very close to the absorption edge. This is quite different from diffraction by non-centrosymmetric unit cells where the f'' dispersion according to Eq. (14) allows resonant scattering measurements even at much shorter wavelengths than that of the absorption edge [11]. The ability to tune synchrotron radiation to the narrow regions of strong dispersions of f' near the absorption edge created for the first time the conditions necessary for successful measurements for resonant scattering from disordered systems, which is basically linked to $f'(\lambda) \, I_{uv}(\mathbf{h})$.

The analysis of the resonant solution scattering data demands a different representation of the Debye equation (16). If the macromolecular structure had a spherical appearance, then the formalism of isomorphous replacement in single crystal diffraction outlined in the preceding section would apply. This is not surprising as the rotation of a spherical structure could not be noticed anyway. In more complicated, asymmetric macromolecular structures it is the spherical average of the structure which must be subjected to the phase analysis described above. As this statement is less trivial, we shall extend the description of a macromolecular structure beyond its spherical average by introducing an expansion of $\varrho(\mathbf{r})$ as a series of spherical harmonics $Y_{lm}(\omega)$ [12, 13]:

$$
\varrho(\mathbf{r}) = \sum_{l=0}^{\infty} \sum_{m=-l}^{l} \varrho_{lm}(r) \, Y_{lm}(\omega) \quad (18)
$$

Heinrich B. Stuhrmann

where

$$\varrho_{lm}(r) = \int \varrho(\mathbf{r}) \, Y^*_{lm}(\omega) \, d\omega$$

ω is a unit vector in physical space.

The amplitude F(h) is again represented as a sum of partial waves:

$$F(\mathbf{h}) = \sum_{l=0}^{\infty} \sum_{m=-l}^{l} F_{lm}(h) \, Y_{lm}(\Omega) \tag{19}$$

where the radial function are uniquely related by Hankel transformations

$$F_{lm}(h) = \sqrt{\frac{2}{\pi}} \, i^l \int_{r=0}^{\infty} \varrho_{lm}(r) \, j_l(hr) \, r^2 \, dr \tag{20}$$

$$\varrho_{lm}(r) = \sqrt{\frac{2}{\pi}} \, (-i)^l \int_{h=0}^{\infty} F_{lm}(h) \, j_l(hr) \, h^2 \, dh \tag{21}$$

On averaging F(h) F*(h) with respect to all orientations (Ω) in reciprocal (or momentum) space we obtain

$$I(h) = 2\pi^2 \sum_{l=0}^{\infty} \sum_{m=-l}^{l} F_{lm}(h) \, F_{lm}(h) \tag{22}$$

This is the form of a scalar product, which can be written as $I = \langle F \mid F \rangle$. Each multiple $\varrho_{lm}(r) \, Y_{lm}(\omega)$ has its own scattering function $|F_{lm}(h)|^2$. In particular, the scattering of the average structure appears as the monopole scattering $|F_{00}(h)|^2$.

This representation of the scattering from randomly oriented particles is of considerable importance, as it opens up a general method of separating the various multipole contributions to I(h). In fact, Svergun have proposed an algorithm which allows one to distinguish between groups óf multipoles with the same index 1 [14]

$$\sum_{m=-l}^{l} \varrho_{lm}(r) \, Y_{lm}(\omega) \leftrightarrow \sum_{m=-l}^{l} |F_{lm}(h)|^2 \tag{23}$$

This means that there is a way to separate the monopole scattering $(1 = 1)$ from the scattering of the three dipoles p_x, P_y, P_z $(1 = 1; m = -1, 0, 1)$ and the five quadrupole functions $(1 = 2; m = -2, -1, 0, 1, 2)$ and increasingly larger groups of higher multipoles. To combine this mathematical analysis of I(h) with the method of resonant scattering we modify Eq. (18)

$$\varrho(\mathbf{r}) = \sum_{l=0}^{\infty} \sum_{m=-l}^{l} [u_{lm}(r) + (f' + if'') \, v_{lm}(r)] \, Y_{lm}(\omega) \tag{24}$$

$$F(\mathbf{h}) = \sum_{l=0}^{\infty} \sum_{m=-l}^{l} [A_{lm}(h) + (f' + if'') \, B_{lm}(h)] \, Y_{lm}(\Omega) \tag{25}$$

The dispersion of I(h) then reads

$$I(h) = \sum_{l=0}^{\infty} \sum_{m=-1}^{l} A_{lm}^2(h) + f'[A_{lm}(h) B_{lm}^*(h) + B_{lm}(h) A_{lm}^*(h)]$$

$$+ (f'^2 + f''^2) B_{lm}^2(h) \tag{26}$$

More explicitly and omitting (h) we obtain

$$
\begin{aligned}
I(h) = A_{00}A_{00} \quad &+ 2f'A_{00}B_{00} &&+ (f'^2 + f''^2) B_{00}B_{00} \\[4pt]
A_{10}A_{10} \quad &+ 2f'A_{10}B_{10} &&+ (f'^2 + f''^2) B_{10}B_{10} \\
A_{1-1}A_{1-1} \quad &+ f'(A_{1-1}B_{1-1}^* + B_{1-1}A_{1-1}^*) &&+ (f'^2 + f''^2) B_{11}B_{11}^* \\
A_{11}A_{11} \quad &+ f'(A_{11}B_{11}^* + B_{11}A_{11}^*) &&+ (f'^2 + f''^2) B_{11}B_{11}^* \\[4pt]
A_{20}A_{20} \quad &+ 2f'A_{20}B_{20} &&+ (f'^2 + f''^2) B_{20}B_{20} \\
A_{2-1}A_{2-1} \quad &+ f'(A_{2-1}B_{2-1}^* + B_{2-1}A_{2-1}^*) &&+ (f'^2 + f''^2) B_{21}B_{21}^* \\
A_{21}A_{21}^{\cdot} \quad &+ f'(A_{21}B_{21}^* + B_{21}A_{21}^*) &&+ (f'^2 + f''^2) B_{21}B_{21}^* \\
A_{2-2}A_{22} \quad &+ f'(A_{2-2}B_{2-2}^* + B_{2-2}A_{2-1}^*) &&+ (f'^2 + f''^2) B_{22}B_{2-2}^* \\
A_{22}A_{22} \quad &+ f'(A_{22}B_{22}^* + B_{22}A_{22}^*) &&+ (f'^2 + f''^2) B_{22}B_{22}^*
\end{aligned}
\tag{27}
$$

More blocks with $2l + 1$ lines may follow.

Methods from data analysis can now be used. Contrast variation determines the columns of Eq. (27), whereas groups of rows of $2l \times 1$ terms can be identified by the Svergun method [14].

For $l = 1$ the monopole structure can be determined completely if $B_{00}(h)$ is known. In fact, the phase problem of F_{00} reduces to the determination of the sign of F_{00}. This is usually not too difficult a task for B_{00}, as plausible arguments can be made concerning the corresponding radial mass distribution $v_{00}(r)$ of the resonant label atoms. Once, the signs of the sinusoidal function $B_{00}^2(h)$ are known for each peak, the phases of $A_{00}^2(h)$ can be determined directly by using the cross term.

If the structure is elongated the quadrupole terms (d-functions) have to be considered. Of the five terms three can be elimated by rotation of the structure through the Eulerian angles α, β and γ. Thus we are left with [15]

$$I2_u(h) = A2_1(h) A2_1(h) + A2_2(h) A2_2(h) \tag{28}$$

$$I2_{uv}(h) = A2_1(h) B2_1(h) + A2_2(h) B2_2(h) \tag{29}$$

$$I2_v(h) = B2_1(h) B2_1(h) + B2_2(h) B2_2(h) \tag{30}$$

I2 denotes the basic scattering functions for $l = 2$. It is also assumed that the complex radial functions have been converted to real functions A2 and B2 by forming appropriate linear combinations of the A_{2m} and B_{2m}. Again it is assumed that the spatial distribution of the resonant atoms is simple enough to allow the signs of $B2_1(h)$ and $B2_2(h)$ to be guessed. Then $A2_1(h)$ and $A2_2(h)$ can be found by the following geometrical construction, which has to be made for each h-interval. In the $(A2_1, A2_2)$-plane Eq. (28) represents a circle which is intersected by a straight line defined by Eq. (29). There will be two pairs $(A2_1, A2_2)$ as solutions of Eqs. (28–30). Approximating the radial function by just a few members of a series of polynomials a correlation between the solution in various h-intervals can be achieved. Laguerre polynomials are of particular interest because of their simple transformation properties [13, 15].

The evaluation of the three A_{1m} leads to a construction in three-dimensional space. The solutions are found as the intersection line of a plane with the surface of a sphere. With increasing index 1 of the multipoles the correlation between the resonant structure and the whole structure through the basic scattering function $I_{uv}(h)$ usually gets weaker and weaker.

As an important result of this joint use of resonant scattering and advanced multipole analysis we note that a partial structure $\sum_m u_{1m}(r) \, Y_{1m}(\omega)$ can be split into its constituents by the introduction of the known resonant label structures $v_{1m}(r) \, Y_{1m}(\omega)$. In the case of the quadrupole structure, one obtains the relative orientation of the main axes between the resonant structure and the total molecule. If a different resonant structure can be used, more information can be obtained, as the non-resonant structure is convoluted with another reference structure as a probe. This procedure should find considerable application in ribosome structure work [16].

Very often it is not possible to obtain all basic scattering functions with the same accuracy. In resonant scattering we are often left with the cross term $I_{uv}(h)$ only. This is still quite an acceptable situation, if the resolution to a monopole approximation of the structure is required, as $A_{00}(h)$ and $B_{00}(h)$ may be determined completely from the two remaining functions $I_u(h)$ and $I_{uv}(h)$. However, a straightforward method for the evaluation of higher multipoles in the sense of the above calculation then no longer exists. The analysis of resonant scattering in this case has to resort to models.

5 Instruments

Diffraction is a process controlled by a reciprocal law: The scattering angle is smaller, the larger the dimensions of the scattering object in comparison to the wavelength. Colloidal structures with characteristic dimensions of some 100 Å irradiated by X-rays of 1.5 Å wavelength will give rise to scattering into a cone with about 0.1° opening angle. The directional spread of the incident beam must be considerably smaller than that which can be achieved with common devices (powder camera, rotation camera, etc.). The design of an appropriate camera is essential for the experimental detection of any X-ray small-angle scattering. Collimation of a sufficiently narrow beam penetrating the sample is a basic prerequisite for recording at very small angles.

5.1 Collimators and Monochromators

The easiest way to produse a collimated beam is to combine two narrow pinhole diaphragms in a line. For reasons of intensity, however, line-shaped primary beams (i.e. with a band-like dross-section) have been preferred in nearly all fields of X-ray small-angle scattering research [17]. A wavelength of 1.54 Å (Cu—K_α radiation) is used most frequently.

Small-angle scattering experiments using X-ray synchrotron radiation have to take into account the unique features of this radiation source. The spectrum of synchrotron radiation does not show any discrete lines as does the spectrum from X-ray tubes; it is continuous. It exhibits high brilliance, directionality and polarisation in a wide range of wavelengths extending from the visible region to hard X-rays. Synchrotron radiation is emitted in the plane defined by the closed path of the electrons circulating at a speed very close to that of light. The off-plane radiation vanishes within small fractions of 1° [18, 19]. The restriction of the emitted light to the plane of the orbit facilitates the construction of X-ray scattering instruments considerably. The continuous spectrum introduces a new element into the design of diffractometers, because very efficient monochromators become necessary.

X-ray small-angle instruments using synchrotron radiation look quite different from common laboratory instruments. They always use crystal monochromators at least and their resolution relies on rather long distances between collimators. The instruments are larger. Very often much effort is put into a focussing system. The following arrangements are presently in use:

a. Bent crystal monochromator. The beam is deflected by the angle 2θ in the horizontal plane, yielding the wavelengths

$$n\lambda = 2d \sin \theta \tag{31}$$

The 111 plane of germanium (2d = 6.53 Å) or silicon (2d = 6.32 Å) is often used as the 222 reflection is forbidden. The monochromatic beam is convergent, resulting in a line at the focus of the bent crystal. This simple setup can be used because the incident white spectrum does not provide wavelengths of the third- and higher-order harmonics [20].

b. Bent mirror — bent crystal monochromator. (Fig. 4) Grazing incidence of the beam on the horizontal mirror cuts off the short wavelength part of the X-ray spectrum. With quartz the range of reflected wavelengths starts of roughly

$$\lambda(\text{Å}) = 0.38\theta \quad (\text{mrad}) \tag{32}$$

Using a bent mirror the cross-section of the premonochromated beam converges towards a horizontal line. The selection of a narrow wavelength band is achieved as described under *a*. The X-ray photon flux in the point-like focus (typically of the order of 0.5×0.5 mm) can amount to about 10^{11} photons/second. This setup is used for time-resolved measurements [20].

c. Bent mirror — double crystal monochromator. This system provides rapid tunability to a desired wavelength [21, 22]. The mirror acts as described in *b*. The parallel faces of the two crystals shift the beam in the vertical plane. This may be corrected for by independent rotation and displacement of the axis of the second crystal [23]. A very attractive alternative is to use a toroidal mirror, which concentrates the incident beam into a point-like focal spot [7]. It is used for resonant X-ray diffraction.

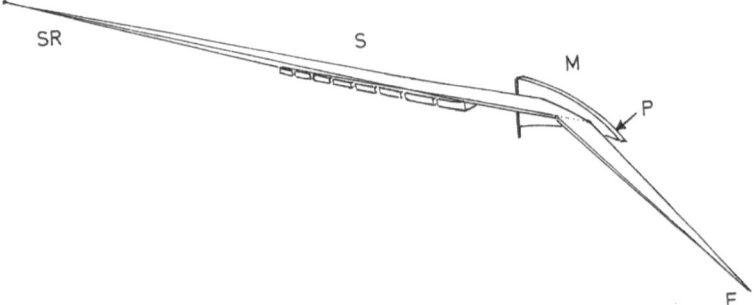

Fig. 4. Schematic diagram of the double focussing camera X13 of the EMBL Outstation at DESY Hamburg. Synchrotron radiation (SR) is focussed in the vertical plane by eight plane quartz mirrors (= S, 20 cm length each) at 20 m distance from the source point. A bent, triangular crystal monochromator (length: 5 cm) focusses the beam in the horizontal plane onto the sample at F. The focal distance can be varied from 3 to 6 m (i.e. the end of the optical bench) by a bending force at P and by realigning the mirrors. The convergence angle of the beam is larger than the divergence of the beam accepted by the optical components. The size of the beam is typically less than 1 mm^2. This kind of X-ray instrument is now used at many synchrotron radiation laboratories

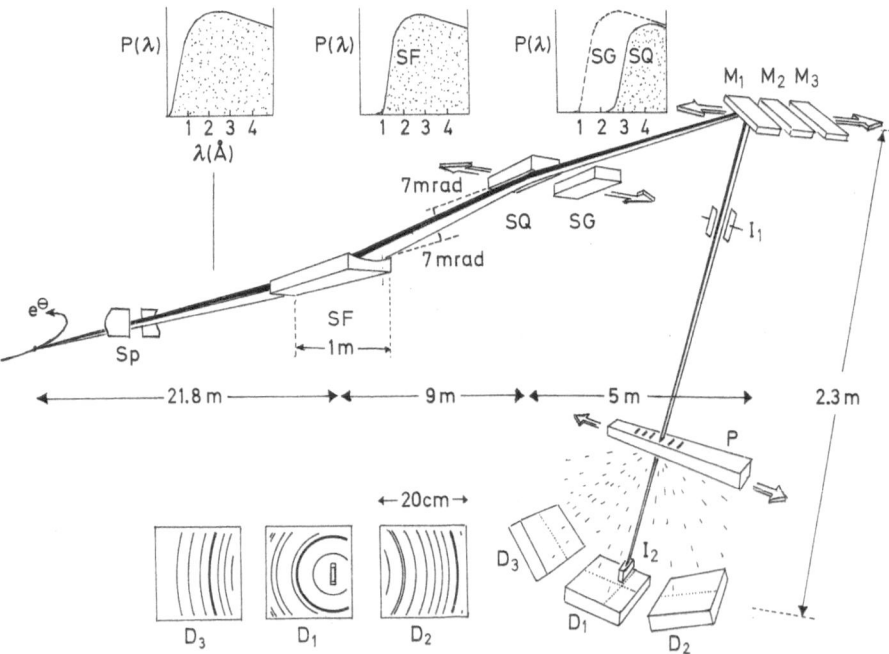

Fig. 5. Double mirror — single crystal optics for soft X-rays (d). SF = gold-coated double focussing mirror. SQ = plane quartz mirror. One half of it is coated with gold (SG). M1 M2, M3 are the crystal monochromators which can be used alternatively. P = sample exchanger. I1, I2 = ionization chambers. D1, D2, D3 are position-sensitive area counters. The upper inserts show the actual synchrotron radiation spectrum. The lower inserts show the diffraction pattern of bacteriorhodopsin as it would appear on the three detectors using 5 Å photons. This instrument is installed at beam A1 of HASYLAB

d. Double mirror — single crystal monochromator. This system (at DESY, Fig. 5) provides rapid tunability over a still wider range of wavelengths than described in *c*. With $\theta = 7$ mrad the first gold-coated mirror reflects wavelengths from about 1.2 Å onwards. The two halves of the second mirror surface have very different electron densities due to gold and quartz. The reflectivity of the latter starts at wavelengths around 3 Å. Lateral displacement of this mirror will tailor the spectrum to the needs of monochromatisation by the single crystal monochromator over a wide range of wavelengths extending from 1.2 to about 8 Å. As the reflectivity of crystals decreases significantly (about 0.15 at $\lambda = 6$ Å from 111 reflection of Ge [24]) only one crystal is used. This means that the scattering instrument has to be turned around the axis of the crystal monochromator. Rotation in the vertical plane is preferred in order to avoid losses due to the polarisation of synchrotron radiation from a bending-magnet section of the electron storage ring [25]. This instrument extends the use of resonant X-ray scattering experiments to the near soft X-ray spectrum (Fig. 6).

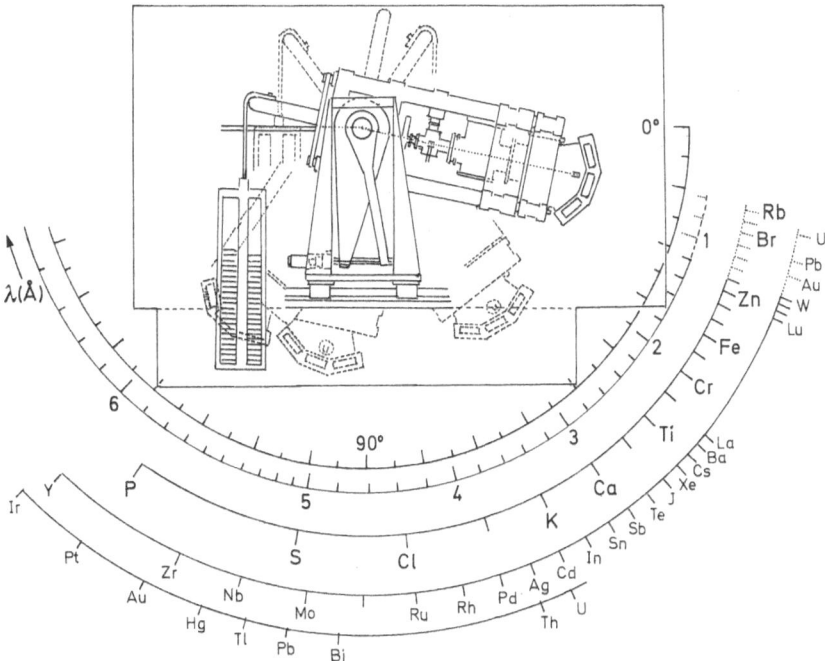

Fig. 6. The wavelength range of the diffractometer at beam line A1 of HASYLAB. Using the 111 plane of a germanium single crystal, wavelengths up to 6.4 Å can be reached by rotating the crystal by $\theta = 80°$ and the camera by $2\theta = 160°$. Thus the K absorption edges down to phosphorus (Z = 15), the L_3 absorption edges down to Yttrium (Z = 39) and the M_5 absorption edges down to iridium (Z = 78) can be used. Peplacing the germanium crystal by indium antimonide as the monochromator extends the wavelength range to 7.3 Å, hence including the K absorption edge of silicon (Z = 14) as well [26]

5.2 Sample Environment

Liquid samples are usually kept in thin-walled glass capillaries of appropriate diameter. The optimal sample thickness of aqueous solutions irradiated with

1.5 Å X-rays is 1 mm. It depends strongly on the wavelength. With 3 Å X-rays the thickness has to be adjusted to $1/2^3 = 1/8$ mm and at 6 Å it further decreases to $1/8 \times 8) = 1/64 = 0.016$ mm. Glass capillaries are then no longer useful. They have to be replaced by plastic windows a few micrometers thick.

As the beam path in small-angle instruments may have a length of several metres, absorption by air will reduce the intensity considerably. 1 m of air absorbs X-rays to nearly the same extent as 1 mm of water. Then only $1/e = 36\%$ of 1.5 Å X-rays are transmitted. Soft X-rays are strongly absorbed in air. The penetration depth of 6 Å X-rays in air is only 16 mm.

This is the reason why the X-ray beam path has to be filled with helium or evacuated, which reduces the scattering from gas. X-ray small-angle instruments are either put into a large vacuum vessel (camera) or they are made of a series of evacuated tubes with very thin windows at the entrace and exit of the beam. Large windows for the connection to area detectors are either made of beryllium or, more recently, of a thin plastic foil supported by a steel grid [23].

5.3 Detectors

In most cases X-rays are detected by xenon-filled proportional counters. In the last 15 years position-sensitive counters have come into use [27]. The spatial resolution of linear position-sensitive counters may reach 0.05 mm. Two-dimensional multiwire proportional counters (MWPC) with a sensitive area of 200×200 mm usually resolve 2 mm in the x and y directions [23]. The detector is based on the delay line method [27-29]. When an X-ray photon traverses the gas of the detector chamber it loses energy due to ionisation of gas molecules. In a $Xe-CO_2$ mixture this is 1 electron-hole pair for about 27 eV of energy loss. By means of thin anode wires at high voltage ($+4$ kV) one produces a region of high electric field in the gas. In the proximity of the wire the charges are amplified ($\times 10000$) in an avalanche-like fashion. On the cathode sides in front and behind the anode wire plane, one makes these charges propagate along the delay line. The differences in time of arrival at the respective ends of the delay line corresponds to the position at which the photon arrived. The area detectors we use are equipped with 300 ns total delay which means that about 3×10^5 positional determinations/second can be performed with less than 20% coincidence or dead time loss.

The electronics associated with the detector have to digitize the differences in time of arrival at the ends of the delay line. Classical methods have relied on time-to-pulse-height conversion followed by pulse height analysis. A considerable improvement in speed is achieved by direct time digitization [30]. The system is capable of handling about 3×10^5 events/s.

A further increase in speed of data taking is expected by segmentation of detectors, i.e. each detector segment has its own read-out system. A linear position-sensitive detector develop recently at EMBL consists of 128 anode wires, with each of them acting as an individual counter. The integral count rate is 10 MHz of statistical events. A similar development has been started for area counters. The new diffractometer for X-ray resonance scattering in HASYLAB is equipped with three area counters (Fig. 5). This system can handle nearly a million statical events per second [26].

High spatial resolution of about 0.1 mm on a smaller area of 80 mm diameter is achieved by a vidicon camera (e.g. from Westinghouse) [31]. The X-ray are converted to electrons on a screen (10 µm thick layer of Gd_2S_2O:Tb or 30 µm ZnS(Ag) from Proxitronic, Germany) located on the surface of a 7 mm thick and 80 mm wide fiber optics plate (Galileo, USA). The charge pattern is created on the silicon target of the vidicon and is read out in the fast scan standard technique (25 frames per second). This system is very convenient for handling very high count rates of 10^8 or so. For low intensity applications, which is the case in small-angle scattering studies, the charge may be integrated on the target for up to 999 frames before it is read out. The signal-to-noise ratio is of critical importance for such applications. The noise mainly results from the dark current and can be reduced by cooling the target.

The present modern positon-sensitive counters in general cope with the scattering intensity produced at the present synchrotron radiation sources. As an increase of the brilliance of synchrotron radiation by 1000 to 10000 is expected with planned electron storage rings, considerable effort has to go into the development of area detectors and their read-out systems.

In a typical experiment we want to record the scattering pattern as a function of 5–30 different wavelengths near an X-ray absorption edge. Using three position-sensitive area counters shown in Fig. 5, 65536 data items represent one scattering pattern. One scan typically produces $20 \times 65536 = 1310720$ data items in about one hour. On-line data reduction then becomes necessary.

6 Recent Advances in Contrast Variation

Neutron diffraction in H_2O/D_2O mixture is probably the most often used technique of contrast variation. The exchange of the mother liquor of protein crystals or of solutions does not present any problems in general. In a similar way heavy metal atoms are introduced, particularly into protein crystals, without any apparent change of the molecular structure. Both methods are presently extended by
— anomalous scattering of native labels, e.g. sulfur and phosphorus, and
— nuclear spin dependent coherent neutron scattering.

The use of native labels avoids the difficult task of finding isomorphous derivatives. Sulfur is present in the amino acids methionin and cystein. The latter very often form disulfide bridges between adjacent protein chains. It is an important constituent of rubber and it is also found in fossil fuels and many minerals. Phosphorus is present in ribonucleic acids and in polar head groups of membranes.

The isotopic exchange of hydrogen in subcellular structures has made considerable progress [16]. Labels of nearly any size can be introduced into any site of a large biological structure. This is the preparative basis for the use of nuclear spin dependent neutron scattering.

6.1 Anomalous Scattering of Sulfur

The use of 5 Å X-rays near the K absorption edge of sulfur has to overcome several technical difficulties, which are due to the strong absorption by nearly any kind of matter. Biological samples have to be very thin. The optimal thickness is about 30 μm. The detector still has to be efficient and the monochromator system should transmit only the fundamental wavelength (n = 1 in Eq. (31)). Using synchrotron radiation the quality of the premonochromator as shown in Fig. 5 is of utmost importance. In view of these problems a first attempt has been made with two-dimensional bacteriorhodopsin crystals [32]. The unit cell dimensions of the hexagonal arrangment of this membrane protein are 63 Å. A sample of appropriate thickness is easily prepared by sedimentation of the small membrane sheets onto the surface of a thin plastic foil. As the water content of the sample is very low it can be introduced into the vacuum of the instrument (Fig. 6) without major damage to the crystals. The choice of a crystalline structure offers several advantages:

— Possible contamination of the monochromatic beam by higher order wavelengths (λ/n, n = 3, 4, ... with Ge(111)) would lead to additional diffraction rings.

— The spatial resolution of the position-sensitive area counters can be characterised more readily.

— Finally, the successful experiment should reveal the spatial distribution of the seven sulfur atoms among the seven helical rods of this membrane protein. This structural information would help to find the N- and C-terminals of the the protein chain.

The performance of the imstrument at the beam line A1 of HASYLAB was quite satisfactory. Apart from the expected energy resolution of 1 eV (Fig. 7) at the K absorption edge of sulfur, E = 2470 eV, the main results were:

— The contamination by third-order harmonics is small. It does not affect the diffraction pattern of the 5 Å photons.

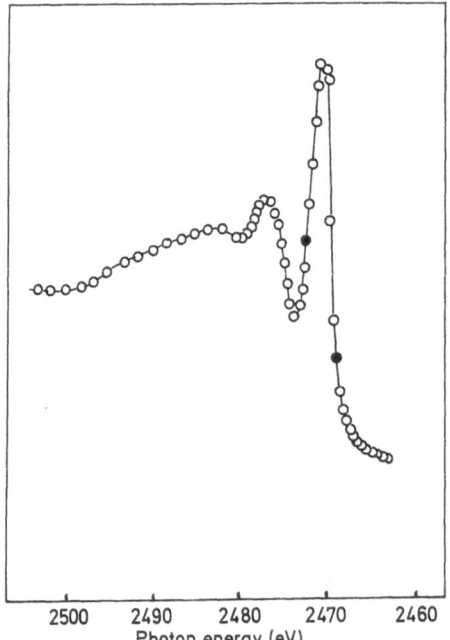

Fig. 7. The X-ray absorption spectrum of sulfur near the K edge at λ = 5.018 Å or at E = 2470 eV. The solid points indicate the two wavelengths which were used by Munk et al. [32]

2500   2490   2480   2470   2460
Photon energy (eV)

— The multiwire proportional counter filled with Ar/CO_2 (7/3) is still active. The spatial resolution increased slightly to about 4 mm (2 mm at $\lambda = 1.5$ Å).

The count rate was rather low: several thousand counts per second were measured. This is due to various plastic windows which add up to 100 μm thickness.

The diffraction pattern of bacteriorhodopsin was measured at two wavelengths where the change of X-ray absorption by sulfur (Fig. 7) is strong and hence a large difference in f' can be expected (compare Fig. 6).

Anomalous scattering of sulfur was found in the 1,1 reflection of the hexagonal bacteriorhodopsin lattice [32]. The contribution of anomalous scattering to the intensity of the other reflections falls within the error bars of the present data (Fig. 8).

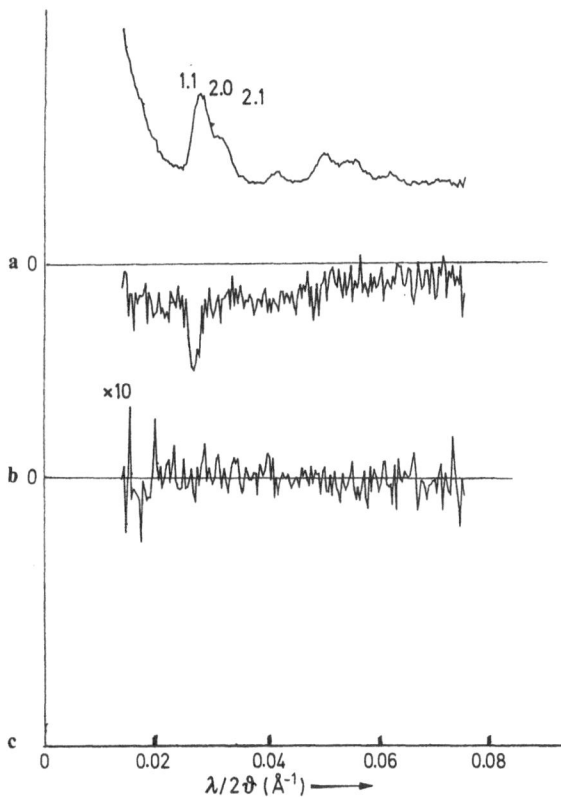

Fig. 8a–c. First measurement of the anomalous scattering of sulfur near the K absorption edge of sulfur [32]. **a** X-ray diffraction of bacteriorhodopsin at $\lambda = 5$ Å. **b** Anomalous X-ray diffraction obtained as difference between measurements at two wavelengths indicated in Fig. 7. **c** Difference between the intensity profiles taken at the same wavelength

6.2 Polarised Neutron Scattering by Dynamic Polarised Targets

The technique of polarised neutron scattering at dynamic polarised targets saw a renaissance in small-angle neutron scattering in 1985 [33]. Polarised targets as used in high-energy physics experiments may be of considerable interest in biological structure research [34, 35]. So far this promising technique has been facing difficulties in getting reasonable polarisation of target nuclei. Intense beams of polarised neutrons become available with the development of "super mirros" [36] around 1980. We report on some experiments which were carried out at the research reactor of GKSS, Geesthacht, in collaboration with ILL, Grenoble, and CERN, Geneva [33].

Spin contrast variation primarily applies to hydrogen nuclei [34, 35]. The basic scattering functions are obtained by scattering polarised neutrons from samples with nuclear spins polarised to a known degree P. In a schematic way we may write

$$I(h) = I_0(h) + \mathbf{n} \cdot \mathbf{P} I_{uv}(h) + P^2 I_v(h) \qquad (33)$$

where $\mathbf{n}$ and $\mathbf{P}$ describe the polarisation of the incident neutrons and the target nuclei [35, 37]. Using the super mirros of Schärpf the polarisation of the thermal neutron beam is nearly unity [36]. The polarisation direction of the neutron beam is easily inverted by a flat coil spin flipper. Measurements of the neutron scattering from a polarised target at opposite neutron spin directions yields $I_{uv}(h)$ as a difference pattern.

Among the nuclear spins, those of the protons are most easily polarised. Nevertheless, the proton spin polarisation reaches only $P(H) = 0.25\%$ in a field of 2.5·Tesla at T = 1 K in about one hour. The time to reach more favourable equilibria at lower temperatures increases dramatically to days and weeks. Proton spins are most easily aligned by dynamic nuclear polarisation (DNP), which is achieved by irradiating the sample with 4 mm microwaves at helium bath temperatures at 0.3 K in a magnetic field of 2.5 T in the presence of an organic radical (e. g.

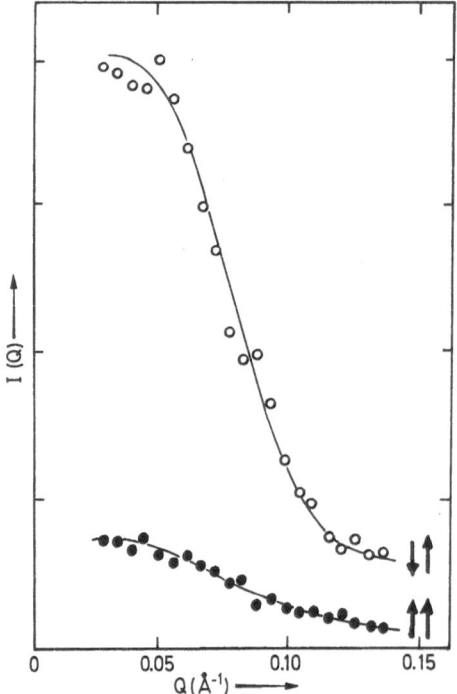

Fig. 9. Polarised neutron scattering by bovine serum albumin at $P(H) = 0.63$. Changing the polarisation direction of the neutron spins to the opposite direction leads to a large change of small-angle scattering [33]

Cr(V)) [35]. The polarisation of the protons is measured by continuous-wave nuclear magnetic resonance (NMR), calibrated by measurments of the NMR signal in thermal equilibrium with the helium bath around 1 K.

The proton spins of biological macromolecules in a deuterated solvent can be nearly as easily polarised as the best frozen spin target materials in high-energy physics research. Lysozyme in a mixture of D_2O and deuterated 1,2-propande-diol reached a polarisation of 77% after two hours of microwave irradiation at T = 0.3 K. Figure 9 shows neutron small-angle scattering of bovine serum albumin at P(H) = 63% for opposite spin directions of the incident neutron beam. The uncorrected data (no background subtraction) show a drastic change of small-angle scattering by a factor of four.

Several proteins, tRNA and ribosomes have been investigated in the same way. At T = 0.4 to 0.5 K proton polarisations around 50% were achieved. Higher nuclear polarisations are expected at lower temperatures.

Spin contrast variation will probably bear fruit in in situ structure determinations of arbitrary parts of biological structures. For this purpose a selected region of the macromolecular structure has to have the usual hydrogen isotope ^{1}H whereas the larger remainder is preferably completely deuterated. This technique of specific isotopic labelling is now well established with ribosomes and other subcellular particles which allow in vitro recombination of the constituents [16].

7 Comparison of Non-Destructive Labels

Although both nuclear spin dependent neutron scattering and resonant X-ray scattering are non-destructive, reversible, labelling techniques, their use in macromole-cular structure research is complementary.

Nuclear spin dependent neutron scattering relies on polarisation variation, i.e. on a change of the order of the nuclear spin system. Resonant (or anomalous) X-ray scattering is a concomitant feature of core electron ionisation (Eq. (2)). The dispersion of the resonant coherent scattering is very small compared to the rise of absorption (and fluorescence) due to electronic excitation. It typically amounts to 0.001 [3]. Usually, resonant X-ray scattering wins because of the f' dispersion of the cross term $I_{uv}(h)$ due to coupling of the resonant scattering amplitude with the usually much larger non-resonant amplitude [3].

The latter fact also applies to $I_{uv}(h)$ from spin contrast variation in neutron scattering. However, the variation of coherent scattering is relatively large and is comparable to the change of absorption. In fact, incoherent scattering is converted to coherent scattering, when the target is polarised completely and in the same direction as the incident polarised neutrons [34, 35, 37]. Spin contrast variation suffers much less from absorption corrections than does the measurement of resonant scattering.

Another important point is radiation damage. Thermal neutrons appear to destroy protein crystal structures much less than X-rays do. This is so at room temperature. Recently, it has been found that cooling of ribosome crystals to liquid nitrogen temperature makes the crystal structure extremely resistant to the intense synchrotron

X-rays [38]. Low temperatures appear to overcome radiation damage problems in general. The production of frozen crystals of biomolecules is of interest for the preparation of polarised targets which are always used at temperatures below 1 K.

The new techniques reported here are the result of collaborations with CERN in Geneva, ILL in Grenoble, EMBL in Heidelberg and HASYLAB at DESY, Hamburg. The construction of the diffractometer for the use of soft X-rays at HASYLAB and the polarised target project at GKSS are supported by the Bundesministerium für Forschung und Technologie (Grants No. 05 353 FAI and 01-I-21E05).

8 References

1. Woolfson MM (1970) An Introduction to X-ray Crystallography. Cambridge University Press, Cambridge
2. Rameseshan S, Abrahams SC (1974) Anomalous Scattering. Munksgaard, Copenhagen
3. Stuhrmann HB (1985) Adv. Polym. Sci. *67*: 123
4. Lerr Del Grande N, Oliver AJ (1981) UCRL — 85683 Preprint
5. Lye C, Phillips JC, Kaplan D, Doniach S, Hodgson KO (1980) Proc. Natl. Acad. Sci. USA *77*: 5884
6. Schäfer GF, Fischer K (1983) Z. Kristallographie *62*: 273
7. Stuhrmann HB (1980) Acta Crystallogr. A *36*: 996
8. Phillips JC, Hodgson KO (1980) Single Crystal X-ray Diffraction and Anomalous Scattering Using Synchrotron Radiation. In: Winick H, Doniach S (eds) Synchrotron Radiation Research. Plenum Press, New York
9. Templeton DH, Templeton LK, Phillips JC, Hodgson KO (1980) Acta Crystallogr. A *36*: 436
10. Okaya Y, Saito Y, Pepinsky R (1955) Phys. Rev. *98*: 1857
11. Hendrickson WA, Teeter M (1981) Nature *290*: 107
12. Guiner A, Fournet G (1955) Small Angle Scattering of X-rays. Wiley, New York
13. Stuhrmann HB (1970) Acta Crystallogr. A *26*: 297
14. Svergun DI, Feigin LA, Schedrin BM (1982) Acta Crystallogr. A *38*: 827–836
15. Stuhrmann HB (1970) Z. Phys. Chem., N.F. *72*: 177
16. Nierhaus KH, Lietzke R, May RP, Novotny V, Schulz H, Simpson K, Wurmbach P, Stuhrmann HB (1983) Proc. Natl. Acad. Sci. USA *80*: 2889
17. Kratky O, Stabinger H (1984) Colloid Polym. Sci. *262*: 345
18. Winick H (1980) Properties of Synchrotron Radiation. In: Winick H, Doniach S (eds) Synchrotron Radiation Research. Plenum Press, New York
19. Materlik G (1982) Properties of Synchrotron Radiation. In: Stuhrmann HB (ed) Uses of Synchrotron Radiation in Biology. Academic Press, London
20. Koch MHJ, Stuhrmann HB, Vachette P, Tardieu A (1982) Small-Angle X-ray Scttattering of Solutions. In: Stuhrmann (ed) Uses of Synchrotron Radiation in Biology. Academic Press, London
21. Bordas J (1984) Applications of X-ray Spectroscopy to Biophysical Problems. In: Stuhrmann (ed) Uses of Synchrotron Radiation in Biology. Academic Press, London
22. Hermes C, Gilberg E, Koch MHJ (1984) Nucl. Instrum. Methods Phys. Res. *222*: 207
23. Stuhrmann HB, Gabriel A (1983) J. Appl. Crystallogr. *16*: 563
24. Materlik G private communication
25. Stuhrmann H (1982) Makromol. Chem. *183*: 2501
26. Stuhrmann H, Bartels KS, Boulin C, Dauvergne F, Gabriel A, Goerigk G, Munk B (1986) Annual Report 1986 of HASYLAB at DESY, pp. 387–388
27. Gabriel A (1977) Rev. Sci. Instrum. *48*: 1303
28. Hendrix J (1984) IEEE Trans. Nucl. Sci. *NS-31*: 281

29. Bordas J, Mandelkow E (1983) Time Resolved X-ray Scattering from Solutions Using Synchrotron Radiation. In Shadafi RI, Fernandez SM (eds) Fast Methods in Physical Biochemistry and Cell Biology. Elsevier Science Publishers, Amsterdam
30. Bordas J, Koch MHJ, Clout PN, Dorrington E, Boulin C, Gabriel A (1980) J. Phys. E *13*: 938
31. Prieske W, Riekel C, Koch MHJ, Zachmann HG (1983) Nucl. Instrum. Methods *208*: 435
32. Munk B, Goerigk G, Stuhrmann HB, Büldt B, Plöhn HJ (1986) Annual Report 1986 of HASYLAB at DESY, pp. 354–355
33. Knop W, Nierhaus KH, Novotny V, Niinikoski TO, Krumpolc M, Rieubland JM, Rijllart A, Schärpf O, Schick HJ, Stuhrmann HB, Wagner R (1986) Helv. Phys. Acta *59*: 741
34. Abragam A, Bacchella GL, Coustham J, Glättli H, Fourmont M, Malinovski A., Meriel P, Pinot M, Roubeau P (1982) J. de Phys. *43*: 373
35. Stuhrmann HB, Schärpf O, Niinikoski TO, Rieubland M, Rijllart A, (1986) Eur. Biophys. J. *14*: 1
36. Schärpf O (1982) AIP Conf. Proc. *89*: 182
37. Abragam A, Goldman M (1982) Nuclear Magnetism: Order and Disorder. Clarendon Press, Oxford
38. Kratky C (1987) Short communication at the AGKr Meeting at Berlin

X-Ray Studies on Biological Membranes Using Synchrotron Radiation*

Peter Laggner

Institut für Röntgenfeinstrukturforschung der Österreichischen Akademie der Wissenschaften und des Forschungszentrums Graz Steyrergasse 17, A-8010 Graz, Austria

Table of Contents

---

* dedicated to Professor Otto Kratky on the occasion of his eightyfifth birthday.

Peter Laggner

Current views on the structure and function of biological membranes imply a great variety of dynamical aspects. These range from the dynamic phase behaviour of phospholipids and conformational variations of membrane proteins during membrane-associated processes, to the interactions between membranes in fusion and pore formation. The structural description of these processes, many of which imply the possibility of intermediate structures, calls for fast time-resolved diffraction methods, i.e. the cinematographic approach. The present article gives first an overview on the specific problems and the theories for their solution in X-ray diffraction on membranes, and then reviews the present possibilities of time-resolved structural studies using synchrotron radiation on phospholipid model systems and functional membranes (sarcoplasmic reticulum membrane). It is shown that such studies bear great promise in entering the millisecond time domain and bridging the existing gap between static structural information and the wealth of dynamic data derived from spectroscopic methods.

1 Diffraction Approaches to Membrane Structure

Biological membranes are supramolecular systems consisting of various types of phospholipids and different proteins as the major constituents; in many cases they also contain other lipidic constituents such as cholesterol, and carbohydrates attached either to protein or lipids. The most general structural feature of membranes is the phospholipid bilayer, but also other morphologies, such as spherical micelles or cylindrical rods can occur either transiently or permanently. The proteins may be either penetrating the bilayer ("integral" proteins) or be attached to the membrane surface ("peripheral" proteins). Of considerable structural and functional importance, in addition to their internal structure, is the interaction of membranes with their aqueous surroundings. From this complexity it is obvious that the question of membrane structure is among the most demanding problems of molecular biophysics.

Notwithstanding this complexity, the need for three-dimensional, structural information at the atomic level of resolution is central and indispensable to biomembrane science. X-ray, and to a lesser extent neutron-diffraction, as the most important sources for such information have, therefore, been widely used in this field (for reviews, see Refs. 1–4). The success of this approach, however, has generally been less spectacular than for instance in the cases of protein or nucleic acid structure. The reasons for this lie in the very nature of biological membranes: with few, notable exceptions (such as the purple membrane of halobacterium halobium, which can be viewed essentially as a two-dimensional crystal of bacteriorhodopsin with only little lipid, Refs. 5, 6, 25) biological membranes are characterized by highly complex and variable molecular compositions, and by the structural dynamics, "fluidity", which is in many cases essential for enzymatic, or other, functions of membranes. As a reflection of this most natural membranes do not crystallize, and a full, three-dimensional atomic structure analysis seems out of reach.

Despite these limitations, there are ways to obtain highly useful structural information on membranes by diffraction techniques, which cannot be gained using other methods. In at least three important biological systems, namely nerve myelin (for reviews, see Refs. 3, 7), retinal rod membranes (see, for example Refs. 8, 9), and chloroplasts [2, 10], the membranes are naturally arranged in multilayer stacks with considerable degree of one-dimensional order in the direction vertical to the membrane planes. The diffraction pattern then consists of discrete Bragg-reflections spaced at integral multiples the reciprocal lamellar repeat distances. From the intensities of the reflections, structural information in terms of an electron density profile can be obtained. The majority of cell- and plasma membranes, however, does not exist in such stacked arrays but are obtained, upon isolation from the cells, in the form of large membrane vesicles. Artificial stacking can be obtained by centrifugal deposition onto a flat surface, a method which has been first applied with some success to erythrocyte membrane ghosts [11] and subsequently to several other types of natural membranes as e.g. sarcoplasmic reticulum membranes from muscle [12, 13], rat liver mitochondrial membranes [14], and intestinal brush border membranes [15].

However, also completely unoriented specimens of membrane preparations give a diffraction pattern, in this case a diffuse, continuous one, which allows important structural conclusions. Wilkins et al. [16] have first identified this pattern with scattering caused by the electron density profile of the individual membranes. This approach lacks in resolution since the diffuse pattern fades out already at relatively small scattering angles, but has the advantage that no drastic measures for orientation and pelleting have to be applied. It is also an advantage that the membranes are freely accessible to solvent, which allows structural changes upon specific reactions to be studied without preparative complications.

An alternative strategy, most widely used in membrane biophysics, is the study of isolated membrane components. A wealth of information has been obtained from X-ray diffraction on pure lipids either unhydrated or in aqueous environment, notably by the group of Luzzati and colleagues (for his own account of this work, see Refs. [17, 18]). This approach still continues to provide important basic insight into the thermotropic and lyotropic mesomorphism of biologically important lipids, an area of highest actual research activities. This domain begins also to be conquered by classical X-ray crystallography as is demonstrated by several examples of phospholipids for which the crystal structure has been analyzed to atomic resolution (for a review, see Ref. [19]).

More difficult, but also promising to be honoured with success proves the isolation of pure, integral membrane proteins and their structure analysis by diffraction techniques. One avenue is given by defined solubilization with detergents and the evaluation of the small-angle (particle) scattering pattern from dilute solution (for reviews on this method, see Refs. [20] and [21]). This has so far been attempted with bovine rhodopsin, the major protein component of retinal rod outer sement membranes [22, 23], with the Ca^{2+}-dependent ATPase from sarcoplasmic reticulum [24] and with the Na/K-ATPase of plasma membranes [25]. These studies have provided, for the first time, unambiguous data for the molecular shape and dimensions of two integral proteins. The other possibility, i.e. crystal structure analysis of membrane proteins has recently been successfully verified with the reaction center complex of photosystem II in chloroplast membranes [26].

2 Special Tasks for Synchrotron Radiation

In the context of this review it is necessary to evaluate specifically those problems of membrane structure research which carry an essential requirement for the characteristics of synchrotron radiation. In other words, what problems are there which cannot, or only with great difficulty, be solved with the much cheaper facilities of an ordinary X-ray diffraction laboratory?

The most obvious and also most demanding task which tries to capitalize on the enormous X-ray intensity of synchrotron radiation is the cinematographic, time-resolved analysis of structural changes associated with membrane-specific processes. With conventional X-ray sources, even considering high-power rotating anodes, the exposure times for obtaining a standard quality diffraction pattern from a stacked membrane sample do normally not reach below the order of minutes, which is too slow for enzymatic processes under biologically meaningful conditions. Synchrotron radiation, by present standards of performance, provides a gain factor of about 10^3–10^4 in useful intensity, and, therefore, the exposure times could theoretically reach the ms-range, a time-domain which becomes already very interesting biologically, thus bridging the present gap between the static, time-average structure information from conventional X-ray diffraction and spectroscopic methods, which are extensively used in this field, with their much faster time scale but often restricted information content.

One example, which demonstrates the feasibility of this approach, is a time-resolved study on sarcoplasmic reticulum membranes during active transport; this will be reviewed in some detail in the following paragraphs. Other important applications are the studies on the kinetics of lipid phase transitions, where only optical turbidity methods and certain spectroscopic relaxation techniques are able to provide useful information on the rates, however not on the actual structural processes and possible intermediates.

Another type of application for synchrotron radiation is based upon the possibility to continuously vary the wave-length and perform diffraction experiments close to the absorption edge of a specific element. This approach, termed "anomalous dispersion" or "resonant scattering" is in principle well suited to obtain spatial information on certain groups of atoms within a membrane assembly, notably bound metal ions, but has so far not found broader use in this field. For a review on this technique, see the chapter by Stuhrmann in this volume.

3 Outline of the Theory

The essentially lamellar nature of membranes bears certain theoretical aspects which are not explicitly stated in general treatments of diffraction analysis. Therefore, a brief recollection of the theory is warranted.

As indicated in the introduction, both the diffuse scattering from individual, randomly oriented membranes and the discrete diffration from ordered stacks of many membranes have to be considered. Conceptually, this involves the theories of diffuse scattering as well as crystallographic aspects. From an experimental point of view, it has become customary to distinguish between low-angle and wide-angle diffraction, since quite different instrumental requirements exist for these two domains. The low-angle domain extends from several hundred Å to about 10 Å (with a wave-length of 1.5 Å this corresponds to diffraction angles 2θ between about 10 minutes and 10 degrees); within these limits lies normally the resolution (i.e. the Bragg's spacing at the highest angle where diffraction is observed), of data pertaining to the electron density profile vertical to the membrane plane, although in come cases of very well-ordered specimens better resolution has been achieved. The wide-angle domain is centered around the 4 Å region (angles of about 22 degrees), where discrete reflections arising from an ordered lateral packing of the lipid hydrocarbon chains is often observed.

The following symbols will be used in this section:

λ	wavelength of X-rays;
2θ	angle between incident and diffracted beam;
$h = \dfrac{4\pi}{\lambda}\sin\theta$	angular argument; in the literature one frequently finds also the notation by s, which corresponds to the reciprocal Bragg's spacing and is related to h by: $h = 2\pi \cdot s$;

Peter Laggner

n	order of reflection $(= 0; 1, 2, ...)$;
I(h)	diffracted intensity at h;
F(h)	structure factor or amplitude at h;
r	real space distance of a volume element from an arbitrary origin;
$\varrho(r)$	electron density distribution relative to the solvent electron density;
$\gamma(x)$	autocorrelation function of $\varrho(r)$;

For comprehensive treatments of diffraction theory and experimentation, the reader is referred to monographs (Refs. [27, 28]).

3.1 Small-Angle Scattering from Unoriented Membranes

A random assembly of infinitely thin planes would give a diffuse scattering pattern, which is isotropic around the primary beam position in the plane of registration and which decays with the square of the scattering angle [29, 30] according to

$$I(h) = h^{-2} \tag{1}$$

This term is often called "Lorentz factor of the plane", in analogy to a geometrical correction made in crystallography. For thin leaflets with finite thickness, the total scattering intensity is given by

$$I(h) = I_t(h) \cdot h^{-2} \tag{2}$$

where $I_t(h)$ is the scattering of the electron density profile along a line perpendicular to the plane, often called "thickness factor" or, in parts of the membrane literature (c.f. Ref. [4] and references therein), the "corrected intensity". For a membrane with no regular structure over its surface, or with motional averaging in the membrane plane (a condition which is met in many cases), this thickness factor contains all the available structural information and is obtained by multiplying the measured intensity I(h) with h^2.

The corresponding structure factors (amplitudes) of the lamellae are, therefore, calculated from the observed intensities according to

$$|F_t(h)| = \sqrt{I_t(h)} = h \cdot \sqrt{I(h)} \tag{3}$$

These are related to the electron density distribution across the membrane by the Fourier transform pair

$$F_t(h) = c_1 \int_0^\infty \varrho(r) \exp(-ihr) \, dr \tag{4a}$$

and

$$\varrho(r) = c_2 \int_0^\infty F_t(h) \exp(-ihr) \, dh \tag{4b}$$

where c_1 and c_2 are normalization factors, which are omitted in the following for brevity.

178

If a center of symmetry is present, i.e. $\varrho(r) = \varrho(-r)$, which is frequently the case with bilayers (at least at low resolution), the phase factor $\exp(-ihr)$ reduces to the symmetric cosine term, by omitting the sine term in the identity

$$\int_0^\infty \varrho(r) \cdot \exp(-ihr)\ dr = \int_0^\infty \varrho(r) \cdot \cos(hr)\ dr - i \cdot \int_0^\infty \varrho(r) \sin(hr)\ dr$$

The sine transform becomes zero in this case since the sine function is antisymmetric. Therefore, one can write the Fourier transforms in the simpler cosine forms

$$.F_t(h) = \int_0^\infty \varrho(r) \cos(hr)\ dr \tag{5a}$$

and

$$\varrho(r) = \int_0^\infty F_t(h) \cos(hr)\ dh \tag{5b}$$

Equation 5b would allow the straightforward evaluation of the desired electron density profile $\varrho(r)$, were it not for the complication of the a priori unknown signs of the structure factors $F(h)$ which are only known in their absolute magnitude from the experiment (see Eq. 3). This is the wellknown phase problem. In the centrosymmetrical case the problem of finding the phase angle α reduces to the choice between a plus or minus for $F_t(h)$ since only the real (cosine) part of the structure factor is non-zero, according to

$$F_t(h) = |F_t(h)| \cos \alpha - i \cdot |F_t(h)| \sin \alpha$$

and $\sin \alpha = 0$, so that $\alpha = 0, \pi, 2\pi$ ...

An additional constraint is given by the fact that the signs can only change when the amplitude, and hence the intensity, goes through zero. The sign and magnitude of $F_t(h)$ at $h = 0$ is determined by the net electron density contrast between the membrane and the solvent, since at zero angle all diffracted waves are in phase and, therefore (from Eq. 5a)

$$F(o) = \int \varrho(r)\ dr \tag{5c}$$

Even with these constraints there remain 2^{m-1} possibilities to be chosen from, with m being the number of maxima in the corrected intensity curve.

With three or four maxima usually observable in the low-angle region (Fig. 1) this leaves between four and eight choices. Chemical knowledge on the possibility or impossibility of a given result often allows to further reduce this ambiguity. In particular the sign of the second, strong maximum has drastic effects on the result, so that a reasonable selection can be made in many cases.

For asymmetric $\varrho(r)$ profiles, which neither show zeros nor have simple periodicities in $F_t(h)$, this approach is not applicable. In this case the only source of information is

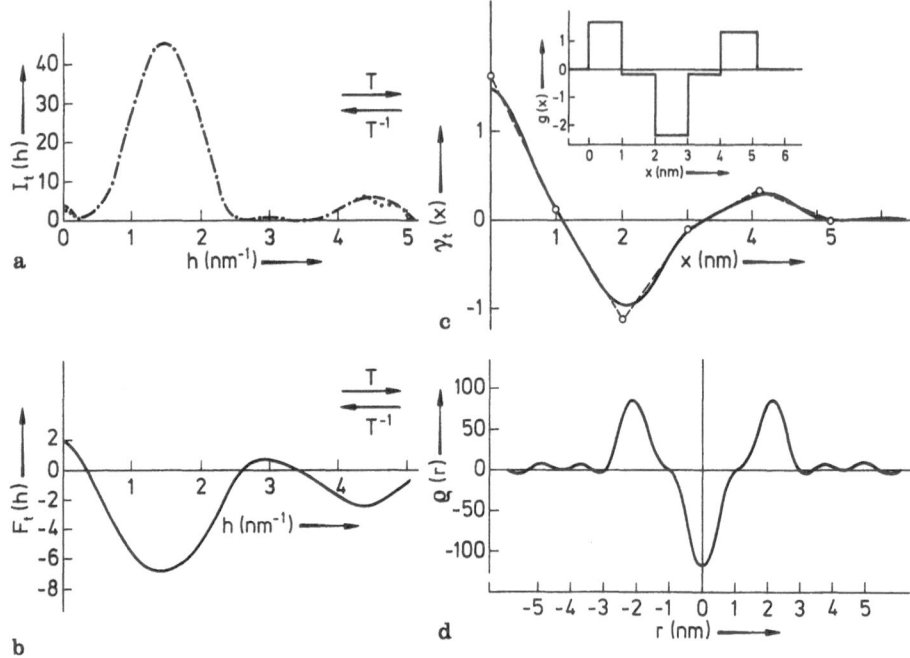

Fig. 1a–d. Small-angle scattering from a dilute, random dispersion of membranes (vesicles). **a**: corrected intensities or thickness factor obtained from the experimental intensity distribution I(h) by multiplication with h^2. **b**: Structure factor (amplitude function) with arbitrarily chosen signs $(+,-,+,-,)$. **c**: Autocorrelation function of the electron density $\varrho(x)$ profile across the membrane obtained by cosine transformation of $I_t(h)$ (Eq. 5a); the insert shows the profile obtained by deconvolution. **d**: Centrosymmetric electron density profile obtained by cosine transformation (Eq. 5b) of $F_t(h)$. From a study on lipoprotein X, an assembly of unilamellar vesicles (Ref. 84, with permission)

the autocorrelation function $\gamma_t(x)$, frequently also termed $Q(x)$, which is obtained by Fourier transformation of the thickness factor $I_t(h)$ according to

$$\gamma(x) = \int_0^\infty I_t(h) \cdot \cos hr \, dh \qquad (6)$$

This autocorrelation function is the convolution product (symbolized by $*$) of $\varrho(r)$ and is defined by

$$\gamma(x) = \varrho(r) * \varrho(-r) = 2 \int_{-0}^\infty \varrho(r) \cdot \varrho(r - x) \, dr \qquad (7)$$

It is a symmetric function and has its maximum at the origin, and determines the probability of finding an excess electron at a distance of x along a line perpendicular to the membrane plane, starting from any arbitrary origin within the membrane. One important piece of information can be directly obtained from $\gamma(x)$: the membrane thickness, from the point where $\gamma(x) \to 0$. A further, unambiguous analysis is generally not possible, however, the $\gamma(x)$ function is often better suited for trial-and-error modeling of structures than the intensity distribution itself since it is accessible to real-space imagination.

An important advantage of the autocorrelation function lies in the possibility of deconvolution which can lead to a unique solution, if the electron density profile is centrosymmetric [31]. Despite stringent requirements on the precision of the data to be analyzed this method in principle overcomes the phase problem. Special methods for this deconvolution have been developed by several authors [32-34], however most of them suffer from the fact that they are reliable only with exact input data and do not take account statistical errors. This has been overcome in more recently developed, numerical methods [35, 36], which allow for all systematic and statistical errors.

Completely different methods, involving the analytical expansion of the thickness factor $I_t(h)$ in the complex plane, have been proposed independently by King [37] and Mitsui [38]. These methods, too, can lead to unique solutions for $\varrho(r)$.

3.2 Small-Angle Diffraction from Stacked Membranes

The diffraction pattern of stacked membranes typically consists of a number (up to a dozen or more) regularly spaced sharp reflections, which can be indexed as the integral orders of Bragg diffraction lines from a one-dimensional crystal with the lattice periodicity along the stacking axis.

A useful quantity to be evaluated without further assumptions in the lamellar repeat distance d, which follows directly from Bragg's law

$$d = \frac{n \cdot \lambda}{2 \sin \theta} \tag{8}$$

In cases where the single membrane and the adjacent water layer form the repeating unit, the membrane thickness d_m can be estimated from the knowledge of the weight fraction of water, f_w, and the respective partial specific volumes of water and membrane, $\bar{v}_m$ and $\bar{v}_w$, according to

$$d_m = d \cdot \frac{\bar{v}_m(1 - f_w)}{\bar{v}_w \cdot f_w + \bar{v}_m(1 - f_w)} \tag{9}$$

Furthermore, if the membrane can be assumed to be a lipid bilayer, the molecular surface can be calculated by the formula

$$s = \frac{2\bar{v}_m \cdot M}{d_m \cdot N_L} \cdot 10^{24} \tag{10}$$

where M and N_L are the molecular weight and Loschmidt's number, respectively. It should be noted, however, that these are but crude estimations since they imply the assumption that membrane and water are separated by a discrete border plane which, of course, is unrealistic.

A more detailed analysis follows the evaluation of the intensities of the Bragg reflections. Similarly to the continuous scattering from single membranes, the intensities have first to be multiplied by the Lorentz factor, which for small angles is proportional to n^2. Generally, the structure factor is given by

$$|F_n| = \sin 2\theta_n \cdot \sqrt{I_n} \tag{11}$$

The structure factors are related to the electron density distribution of the unit cell in the stacking direction. A necessary condition for the analysis of such a system is the existence of centrosymmetric units: this can either originate from centro-symmetrical membranes, in which case the unit cell consists of the single membrane electron density profile plus two halves of the intermembrane water space; alternatively, a centrosymmetrical unit cell may arise from flattened vesicles so that each pair of adjacent membranes is arranged in mutual mirror-symmetry (Fig. 2).

The relation to the electron density distribution, $\varrho(r)$, of the unit cell is given by the Fourier series, rather than a transform as in Eq. 5b,

$$\varrho(r) = \sum_n F_n \cos (2\pi nr/d) \tag{12}$$

The electron density profile is, therefore, represented by the sum of cosine functions with their individual amplitudes given by the structure factors. Since the higher order reflections lead to smaller wave-lengths of the cosine functions, the resolution of details in $\varrho(r)$ increases as more of the higher orders are included in the sum. Again the resolution is determined by the Bragg's value of the highest order reflection. As

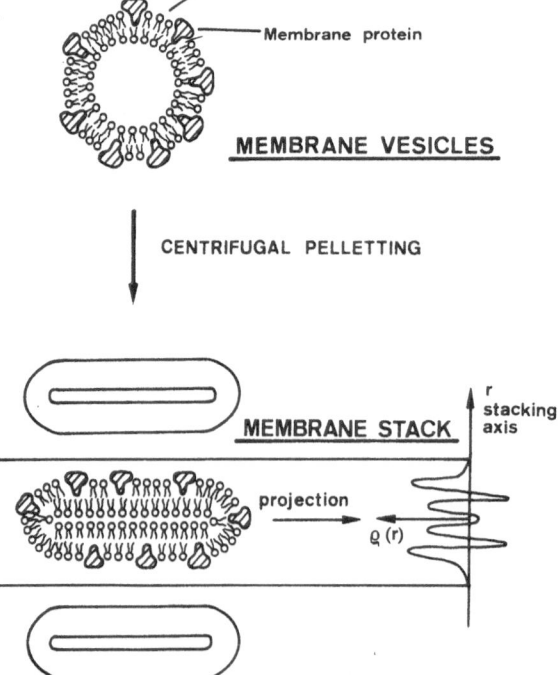

Fig. 2. Schematic representation of the centrosymmetric unit cell and its electron density distribution, obtained by stacking of asymmetric membrane vesicles

stated in Eq. 5c the first term $F_{(0)}$ in Eq. 12 is given by the mean value of $\varrho(r)$ relative to the solvent background.

The main problem is solving the Fourier series in Eq. 12 is again the indeterminany of the signs of F_n, i.e. the phase problem. An important aid to the correct assignment can be obtained from the continuous structure factors F_t determined from experiments with random, unstacked samples provided, of course, that the membrane structures are the same in both cases. This has been discussed in detail recently by Blaurock [4]. Many other methods have been described in the literature, most of them, however, have to rely on trial-and-error searches involving information from other sources [1, 3].

Another approach which also leads to the continuous structure factors of the membrane uses the fact that the regular stacking periodicity can be changed by varying the water content between the membranes. The structure factors from a series of swelling states should fall on the continuous $F_t(h)$ curve of the single membrane, and therefore the latter function can be found and, in particular, the zeros identified where phase changes should occur. An example for this approach is shown in Fig. 3. This swelling method relies on the assumption, that the membrane structure does not change with varying hydration. Discontinuous changes can be excluded directly from the diffraction data if, after proper scaling by the condition

$$\sum_n I_t(n) = d/d_{min} \tag{13}$$

where d_{min} is the minimum spacing in the swelling series [39], the intensities fall on a smooth curve.

Continuous structural changes can also be detected if the data points for the individual reflection orders fall on smooth lines, but these lines cannot be inter-

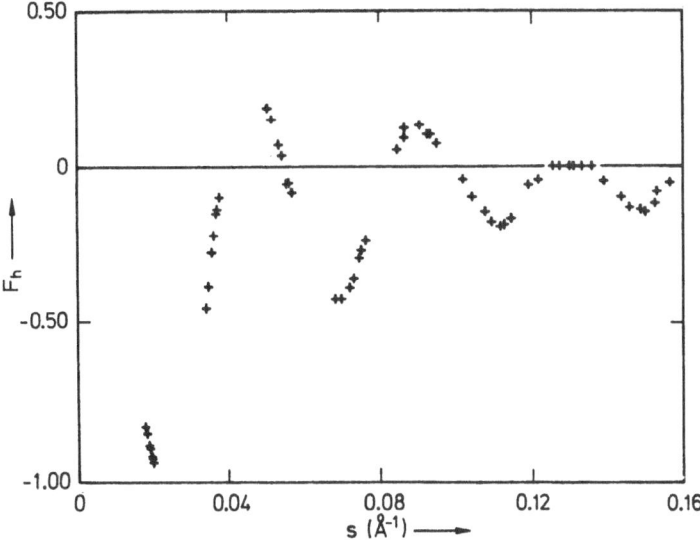

Fig. 3. Approach to the continuous structure factor of a membrane obtained by plotting the properly scaled discrete amplitude values of a membrane stack at different degrees of swelling. (From Ref. 41, with permission)

polated by one continuous function. The problem of possible structural changes may be seen as a weakness of the swelling approach but, nevertheless, the method continues to find use in this field.

There exist also methods which do not require a whole series of swelling experiments to evaluate the continuous structure factors. Shannon's sampling theorem from information theory [40, 41] implies that only two sets of discrete structure factors are sufficient. More specifically, this theory states that the complete continuous structure factor $F(s)$ can be reconstructed by Fourier interpolation from just one set of observed F_n

$$F(s) = \sum_{-n}^{n} F_n \frac{\sin \left[\pi(sd - n) \right]}{\pi(sd - n)} \tag{14}$$

with the chosen signs of F_n. Only the correct combination of signs for two independent sets of data (with different d) will lead to the same continuous transform. However, it has to be kept in mind that, again, this implies that the structure is the same in the two swelling states. Furthermore, the method is critically affected also by termination effects and by the precision of the data. In practice, therefore, the use of more than only two data sets is recommended.

The heavy atom labeling method (isomorphous exchange) which is a standard procedure for phase determination in protein crystal structure analysis has found relatively limited use in this field. This is due to the fact, that generally the data from membranes are of lower resolution, which means that the data are insufficient to specify the position of a single atom. Nevertheless, it has been shown that in favourable cases where good resolution can be achieved, the benefits of heavy atom labeling could be verified even leading to electron density information on an absolute scale. This was first demonstrated on the important membrane model system dimyristoyllecithin:cholesterol, with iodine or bromine bound to cholesterol in position C-26 [42].

In analogy to the Fourier transformation of the continuous intensity pattern leading to the autocorrelation function $\gamma(x)$ (see Eq. 6 and 7), the intensities of the discrete reflections from stacked samples can be used to construct a Fourer series according to

$$P(x) = \sum_{n} I_t(n/d) \cos (2\pi nx/d) \tag{15}$$

where $P(x)$ is the one-dimensional Patterson function. However, the interpretation is less straightforward than that of $\gamma(x)$, since $P(x)$ is in fact the latter convoluted with the lattice function (which ideally is a series of delta functions spaced at regular distances d, the unit cell size). This can lead to an overlap between neighboring autocorrelation functions sampled at the lattice points d, 2d, 3d, etc., unless d can be made larger than twice the membrane thickness by swelling (see Fig. 4). It was also suggested that the $\gamma(x)$ function can be extracted if the lattice function is not well developed, for instance by keeping the number of repeat units small within any one stack [43]. This approach, however, seems not really productive, since it counteracts the purpose of forming large regular stacks, i.e. to achieve high resolution. Moody [44] has suggested an

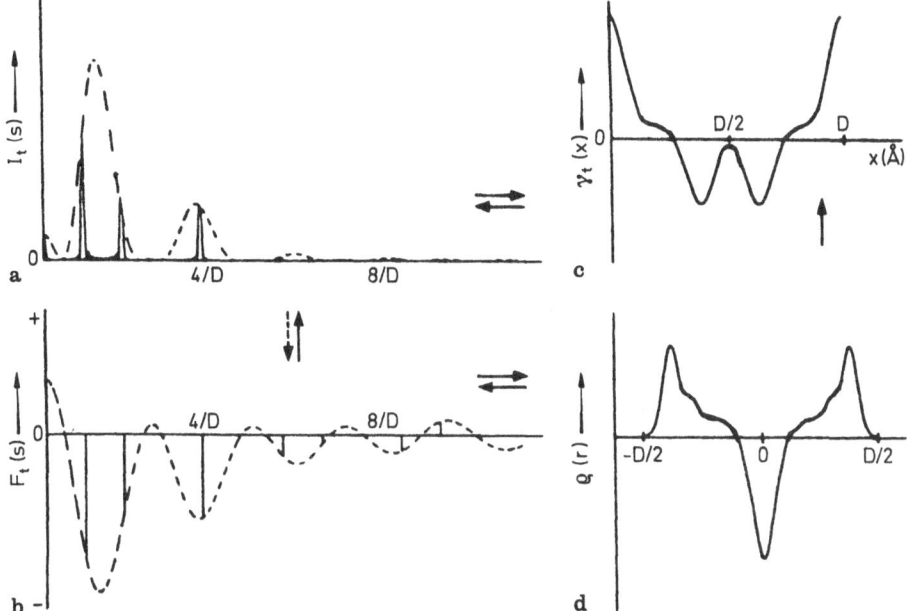

Fig. 4a–d. Scheme of diffraction patterns and Fourier transforms arising from a lamellar stack with centrosymmetrical unit cell. (**a**) The discrete maxima at the reciprocal spacings s = n/d sample the continuous scattering function of one lamella, which would be observed without stacking (dotted line). (**b**) The structure factor or amplitude function. (**c**) The Patterson function obtained by Fourier transformation of $I_t(s)$. (**d**) The electron density profile obtained by Fourier transformation of $F_t(s)$. Adapted from Ref. 3, with permission

alternative method to obtain $\gamma(x)$ from $P(x)$, but again this either requires the measurement of a swelling series, or is of low resolution.

Finally, a note on disorder of the membrane stacks and on attempts to correct for it in the analysis of diffraction data. Generally, two kinds of disorder are being discussed in crystal structure: Disorder of the first kind refers to displacements of the structural elements (for example the one-dimensional unit cell of a membrane stack) from the ideal positions prescribed by the periodic lattice. The effect on the diffraction pattern is indistinguishable from that of thermal vibrations and may, therefore, be expressed as a Debye-Waller temperature factor so that the structure factor, expressed as a cosine series, includes a Gaussian term, according to

$$F_n = \sum_j b_j \cos (2\pi n r_j/d) \cdot \exp (-n^2 B/4d^2) \tag{16}$$

where b_j is the scattering amplitude of the atom at coordinate r_j, and B is the exponent in the temperature factor adjusted to get agreement with the intensities of the observed reflections. Disorder of the first kind is often referred to frozen-in thermal displacements. This type of disorder leads to a progressive decay in the intensities of a series of reflections but not to broadening.

Disorder of the second kind, originally characterized by Zernike and Prins [45] and by Kratky [46] in the discussion of the structure of liquids, refers to a situation,

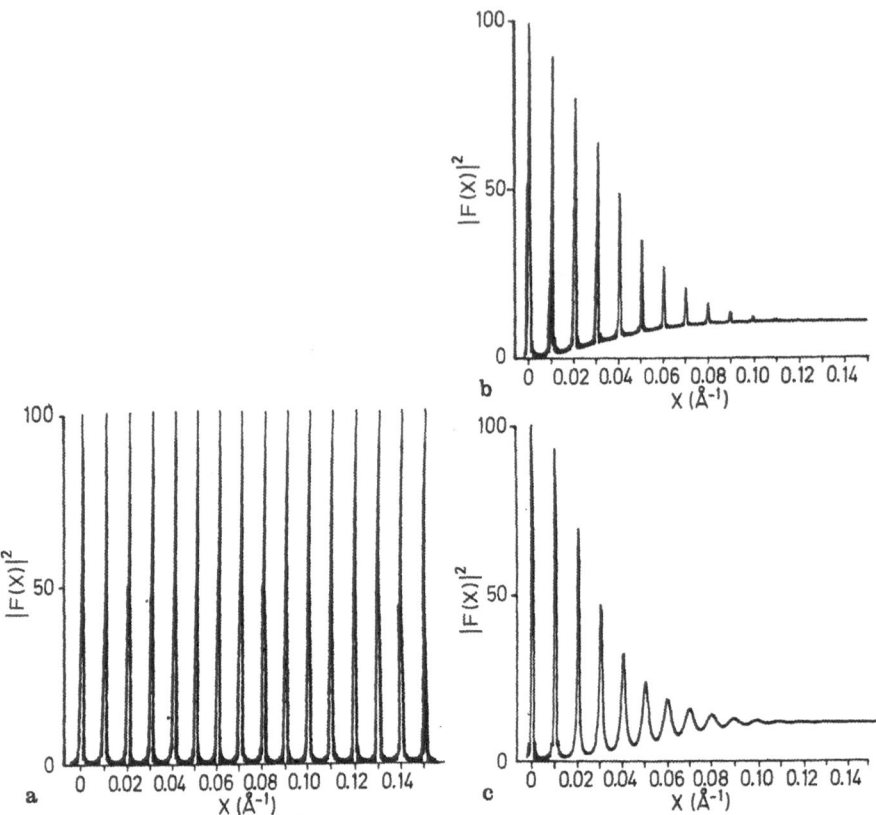

Fig. 5a–c. Diffraction patterns of lamellar stacks with perfect stacking order (**a**), with disorder of the first kind (**b**), and disorder of the second kind (**c**). For details of the model parameters, see Ref. 4. (From Ref. 4, with permission)

where each lattice point (or the width of the unit cell) varies only in relation to its nearest neighbors rather than to an ideally periodic lattice. Thus the absolute magnitudes of the displacements from an ideally periodic lattice increase with the square root of the distance. The effect of this type of disorder is both diminuition and broadening of the reflections. The two types of effects are shown for the case of point-scatters in Fig. 5.

The significance of disorder has been largely neglected in membrane studies. Worthington, in a recent work [47] has taken up again this problem and proposed a solution based upon quantitative analysis of the line broadening [47]. It is necessary to observe, however, that such an analysis has also to take into account all kinds of instrumental broadening as it may arise from finite beam geometry and departures from monochromasy [4,48].

3.3 Wide-Angle Diffraction

In most cases, the wide-angle diffraction pattern of membranes or phospholipid model systems shows more or less well-defined peaks which relate to the mutual packing of the hydrocarbon chains, for instance within the bilayer. In samples of biological membranes, where the lipids are in a "fluid" state, this signal is broad and diffuse, centered around $4.5 \, \text{Å}^{-1}$. Sharper peaks are observed with lipids below the gel-to-liquid-crystal transition, often around $4.0\text{--}4.2 \, \text{Å}^{-1}$; there, the peak positions can be used to infer the crystallographic subcell dimensions and the mutual positions of the chains [49, 50]. In cases where the long chain axes are positioned within a defined lattice, but the chains are still free to rotate about this axis, hexagonal packing is frequently observed, and the nearest neighbor distance a can be readily calculated from the diffraction spacing s, according to a $= s \cdot (2/\sqrt{3})$.

Of course, if only one single diffraction maximum is observed, a strict assignment of a certain lattice type cannot be made. The assignment of triclinic, ortho-rhombic, monoclinic or hexagonal lattices, as they are frequently found in lipids, requires the indexing of several reflections, which in practice, however, are often rather obscure. The analysis can be aided considerably by the comparison to the diffraction from single crystals [51].

4 Results Obtained with Synchrotron Radiation

Most of the diffraction work with synchrotron on membranes done up to now can be grouped into one of the following two categories: Firstly, studies on thermotropic phase transitions of phospholipid model systems. These investigations are aimed at the kinetics and structural mechanisms of the transitions, and are beginning to close the gap between the wealth of existing data from spectroscopic techniques, with their inherent time resolution and the detailed structural data from static X-ray diffraction.

The second domain, where significant progress has been achieved is that of time-resolved studies on membrane processes during active transport. So far this type of studies has been restricted to the calcium transport system of sarcoplasmic reticulum membranes, but the work shows the great potential of such studies also for other functional membrane species, and shall, therefore, be reviewed here in some detail.

4.1 Phospholipid Phase Transitions

Of the large number of phospholipid phases that have been identified under different environmental conditions (i.e. type of phospholipid, degree of hydration, temperature, pH, ionic strength) there are three major types of phases which are most relevant to "physiological" conditions. A schematic representation of the structures in these phases is given in Fig. 6. According to their symmetries they are called "lamellar" (L), "hexagonal" (H) and "cubic" (C) phases (for general reviews on lipid polymorphism, see Refs. [52, 53]). Lamellar phases, furthermore, show an important thermotropism between gel and liquid-crystalline forms, commonly denoted by the subscripts β and α. So far, only the lamellar $\beta \leftrightarrows \alpha$ transitions, with the so-called pretransition preceding the main transitions, and the transition between lamellar and hexagonal phases L $\leftrightarrows$ H have been studied by time-resolved diffraction methods.

Peter Laggner

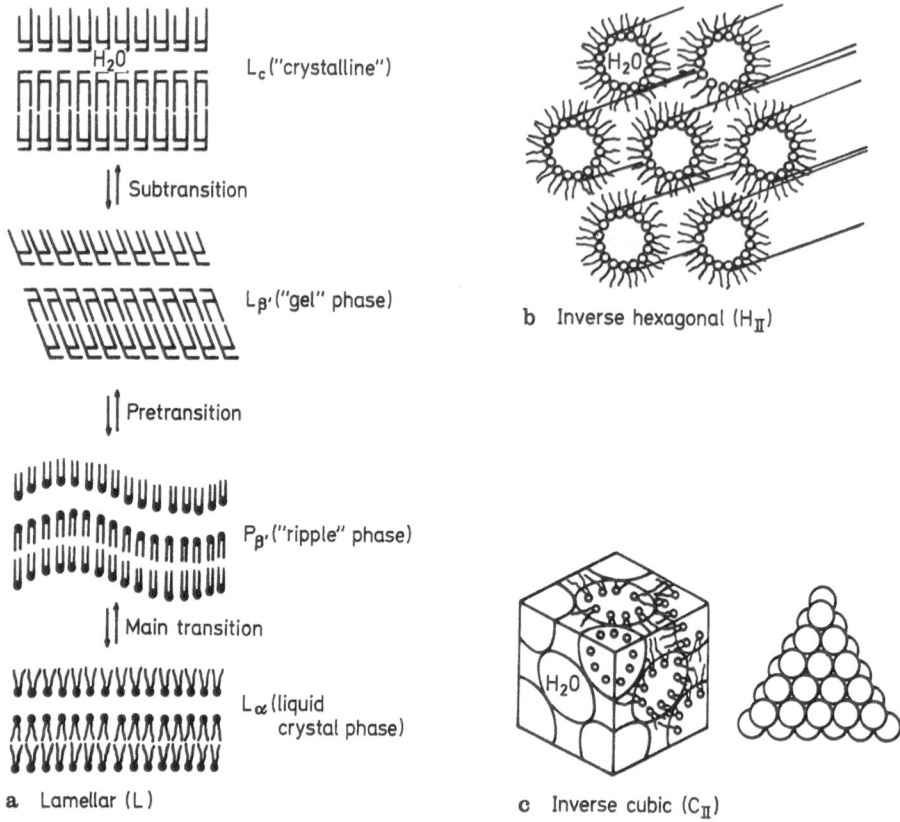

Fig. 6. Schematic depiction of various thermotropic phospholipid phases

4.1.1 Temperature-Jump Studies

Since the aforementioned transitions can be most easily effected by temperature changes, practically all time-resolved studies in this field have so far been performed by using different T-jump techniques. The limiting factor in the time-resolution has been the time required to heat (or cool) the sample through the transition. Apart from the specific heat conductivity of the sample and the latent heat of transitions, this depends on the geometry of the sample cell and of the irradiated volume. Furthermore it depends on the rate by which the temperature of the heating (or cooling) elements in contact with the sample can be changed and on the heat conductivities of the materials involved. For cylindrical capillaries heated by air flow (the method used in Caffrey's studies; for details, see Ref. [54]), the characteristic times can be calculated [66]. It was shown that the relevant time limits are, for this arrangement, approximately $0.\overset{.}{8}$ s for an $L_\beta \rightarrow L_\alpha$ transition, and 0.5 s for the $L_\alpha \rightarrow H_{II}$ transition. With other sample geometries, and heating media not homogeneously surrounding the sample, the calculation becomes difficult so that direct measurement of the sample temperature in the vicinity of the X-ray beam has to be performed which, however, is also affected by errors due to the finite response times of temperature sensors. Owing to the unavoidable temperature gradient across the irradiated sample,

188

the time resolution of such experiments can hardly be pushed below about 0.5 s, with todays instrumentation. In view of the fact that the primary intensities and detectors now available are sufficiently powerful to perform measurements with good statistics within the order of one millisecond, it is clear that future efforts in T-jump cell design will determine progress in this field. Alternatively, pressure jump or rapid mixing methods may become important as trigger methods for the transitions.

4.1.1.1 Lamellar Phase Transitions of Phospholipid Multilayer Dispersions

Fully hydrated samples of dipalmitoylphosphatidylcholine (DPPC, 1,2-Dipalmitoyl-sn-glycero-3-phosphocholine), the lipid used in the first studies with synchrotron radition [54, 55] show a series of three well-defined transitions in the temperature range between 15 and 45 °C which are termed sub- (around 17 °C), pre- (around 33 °C) and main transition (41,4 °C): for a detailed investigation of the static X-ray diffraction behaviour of this system, see Refs. [56, 57]. The underlying structures are schematically depicted in Figure 6a. The transition parameters, temperature and enthalpies, are constant only above a limiting value of full hydration (25 moles H_2O per mol DPPC, i.e. approximately 38 weight-% H_2O), and are sensitive also to the ionic environment.

Although the equilibrium properties of these transitions have been extensively studied and are well documented relatively little information is available on kinetic and mechanistic aspects and the data appear to depend both on the nature of the experiment (among the methods used are T-jump, p-jump, dielectric relaxation, ultrasonic relaxation) and the detection method (e.g. turbidity, label fluorescence). For the main transition, characteristic time constants between 1 ms and 1 s have generally been reported [58, 59] and were assigned to a sequence of events from fast, noncooperative motional changes of individual molecules to much slower cooperative processes, such as nucleation, growth and fusion of clusters, and changes in solvation. Both the pre- and the subtransition of DPPC are much slower and show a strong hysteresis. Their transition half-times are generally strongly temperature dependent and vary from minutes to many hours within a few degrees of the equilibrium transition temperature [60-62]. Particularly in the cooling direction, the pretransition appears to proceed via metastable states [61]. All these kinetic features call for a better analysis in terms of the structures involved, and this is certainly a potentially very fruitful area for time-resolved X-ray diffraction studies.

A representative result of a time-resolved temperature jump experiment on a fully hydrated sample of DPPC is shown in Fig. 7a. The experiment was performed at DORIS II (operating at 3.7 GeV, 4 bunch mode, 70 mA beam current) with the X-33 camera of the European Molecular Biology Laboratory, Hamburg Outstation, using a four-quadrant position sensitive detector [64] at a sample-to-detector distance of 130 cm. Heating of the sample was performed by perfusing the brass sample holder with thermostat fluid through a 4-valve ·servo-system [65]. The temperature jump was from 10 to 47 °C, of which the section between 30 and 47° is shown in the graph. Scattering patterns were recorded in time slices of 0.25 s.

The first two orders of the lamellar lattice are clearly resolved throughout the experiment, and in the low temperature L'_β phase even the third order can be discovered above the noise. The peak intensities in the first order maxima are in the order of 10^4 counts. This demonstrates that the time resolution of such an

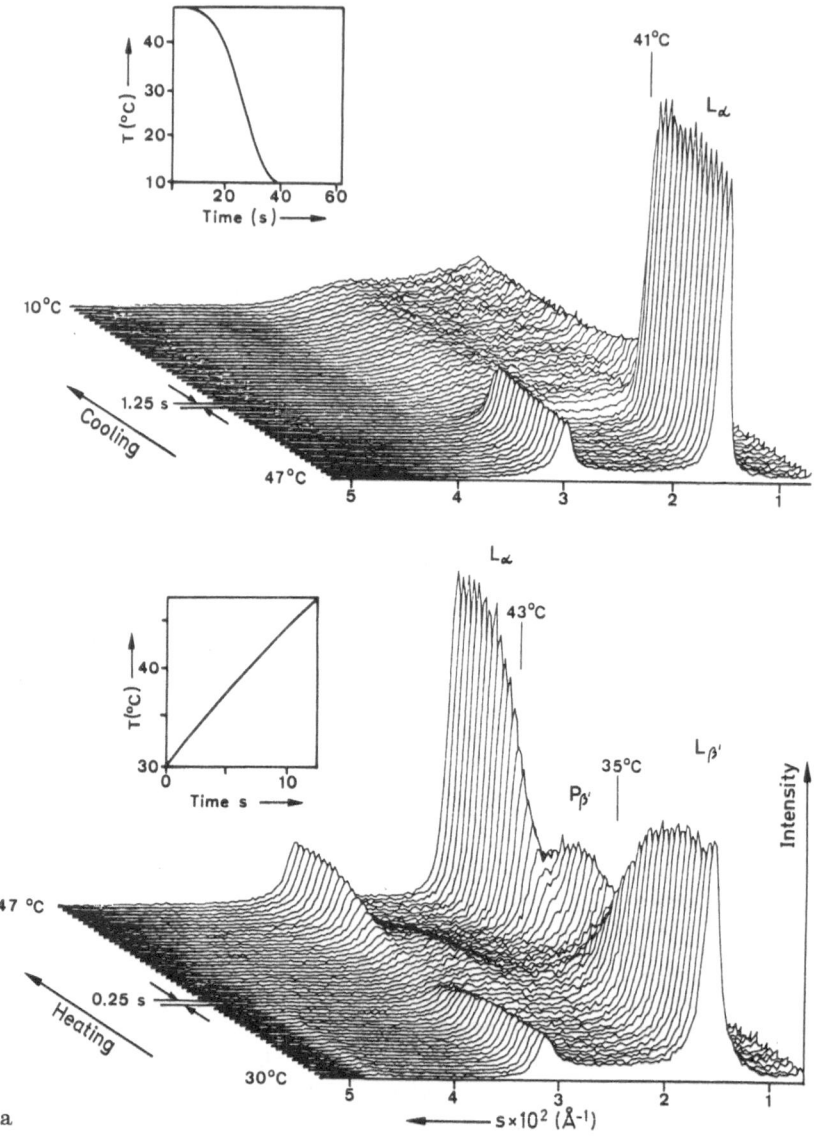

a

experiment could be extended into the millisecond region, considering merely the available intensities.

Both the three-dimensional intensity plot and the contour plot show quite clearly that the two transition, the pretransition centered around 35 °C and the main transition at about 42°, appear not isothermal under these conditions. For the intrinsically faster main transition the patterns indicate a half-width of about 2 s (or 3 °C) where the peaks of the P'_β and L_α phases coexist. This width is considerably larger than that expected from precise isothermal diffraction experiments [63] and is

190

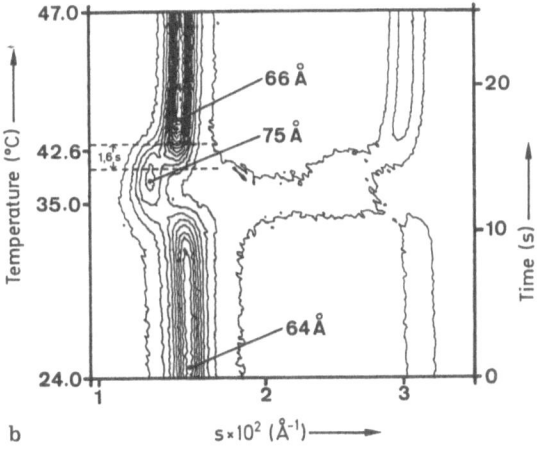

b

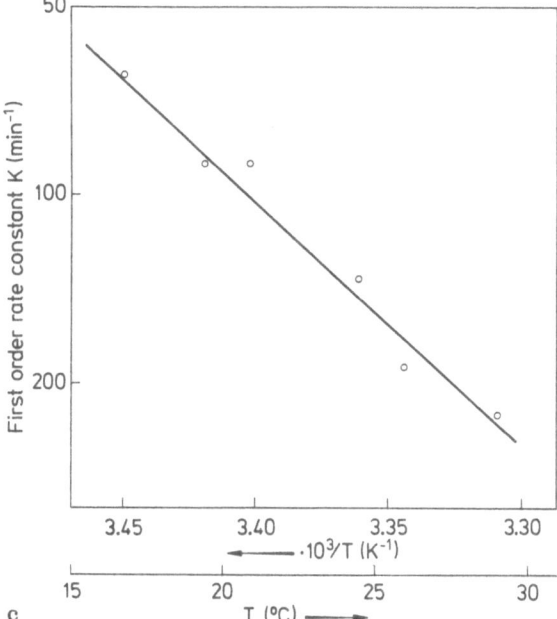

c

Fig. 7a. Time-resolved small-angle diffraction from a multi-lamellar aqueous dispersion ($c_{lip} \sim 0.2$) of dipalmitoyl-phosphatidylcholine (DPPC) during a temperature-jump and -drop experiment. The respective temperature courses are shown in the inserts. b Contour-line plot of the intensities obtained in the heating-experiment. c Temperature-dependence of the recovery rates of the L'_β phase, on the equilibration temperature, upon cooling from 37 °C. (From Ref. 74, with permission)

limitation is caused to a large extent by the heat conductivity of the aqueous sample. in good agreement with theoretical estimates for the range of thermal variations throughout the sample under these heating conditions. Thus, the effective time resolution of this experiment is equal to or slightly less than 2 s, a rather disappointing value if compared to the possibilities given by the intensities and the detector. This Quite similar experiences were made with an air-flow heating device as reported by Caffrey [54, 66].

While the main transition occurs close to the expected temperature and is merely broadened, the pretransition temperature (35 °C) lies by about 2 °C higher than the value found at slow heating rates [60, 61, 67]. This is a reflection of the kinetic hindrance

191

of the pretransition involving the long-range rearrangement of a planar to a rippled bilayer structure [50] in addition to changes in solvation [57].

The most striking anomaly in the thermal transitions of DPPC, however, is the extremely slow reversibility of the pretransition. Upon heating through the pretransition to 38°, the first order peak of the P'_β phase develops to its final maximum within less than 15 minutes. However, in the reverse direction, the recovery of the ordered lamellar arrangement of the L'_β phase takes much longer. This rate is slow enough to be followed by time resolved experiments on a laboratory X-ray

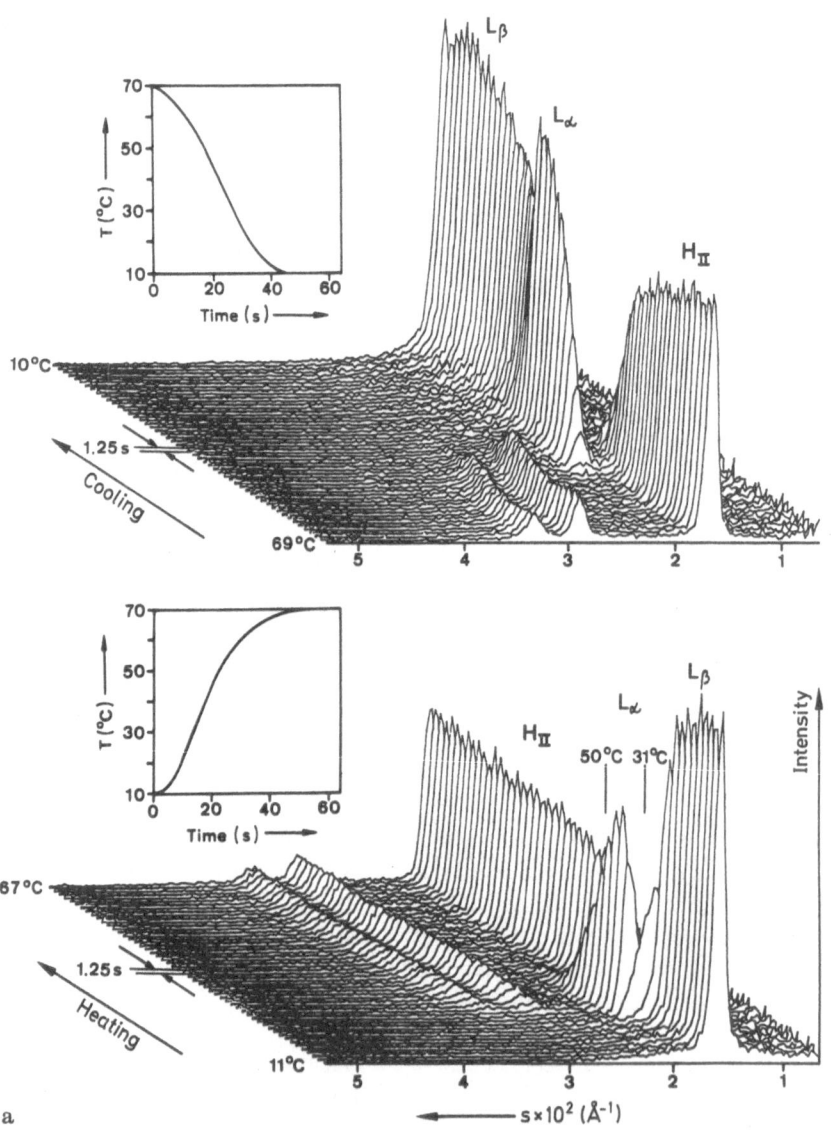

a

generator. Figure 7c shows the dependence of the recovery rates on the final quenching temperature. The rates show a negative temperature coefficient such that they grow exponentially with the difference between quenching and equilibrium transition temperature ($\sim 33\,^\circ C$) which indicates that the rates are governed by nucleation kinetics [68, 69]. It is also possible that the reformation of a partly ordered hydration shell requires time and low temperatures. A complete, kinetic analysis, however, must include also the short-time limits of the process and will require a broad overlap in time-ranges between synchrotron and coventional X-ray techniques.

4.1.1.2 b) Lamellar/Hexagonal Phase Transition

The question of intermediate structures around a phase transition has caught particular attention in the case of the lamellar-to-inverted hexagonal transition of ethanolamine phospholipids, since this involves a major topological change (see Fig. 6). Inverted micellar structures were proposed as intermediates on the basis of ^{31}P-NMR and electron microscopic results [70, 71], and also rationalized in a theoretical mechanism [72, 73]. On the other hand, first results of time-resolved X-ray diffraction

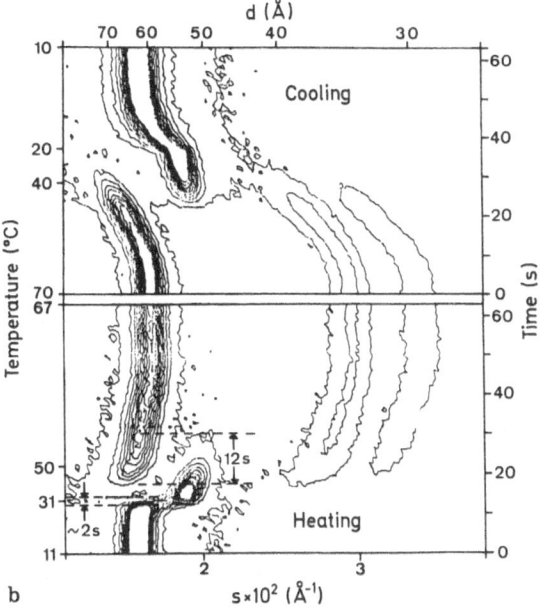

Fig. 8a. Time-resolved small-angle diffraction from a dispersion of a hexagonal-phase forming lipid (1-hexadecyl-2-oleoyl-phosphatidylethanolamine, HOPE) in the presence of excess water ($c_{lip} \sim 0.2$), during a heating- and cooling experiment. The temperature course are shown in the inserts. **b** contour-line plots of the intensities. (From Ref. 74, with permission)

studies by Caffrey [66] were interpreted in terms of a triggered growth mechanism involving no appreciable amounts of intermediates. As these latter results are described in more detail in the chapter by Caffrey, we will concentrate here on the more recent results of our own [74].

Figure 8 shows a pair of heating and cooling scans of fully hydrated 1-hexadecyl-2-oleoyl-phosphatidylethanolamine (HOPE), in the presence of excess water, between 10 and 69 °C. The experiment resolves very clearly the two lamellar phases L_β and L_α at low temperatures and the hexagonal phase at high temperatures. In the heating scan, the $L_\beta \rightarrow L_\alpha$ transition appears centered around 31 °C and is as sharp as can be expected for an isothermal transition under the conditions of an external heating experiment: the maximum range of coexistence is about 2 s, or 5 °C, reflecting again the limit of time resolution given by the inhomogeneous sample temperature. Very notably, however, the lamellar-to-hexagonal transition is less sharp and shows a broader range of phase coexistence: while the onset of H_{II} formation appears at 45 °C and coincides with the initial decay of the L_α phase, the latter coexists with growing amounts of hexagonal phase for more than 12 seconds. Allowing for an instrumental time-smearing of about 2 s leaves still about 10 s for coexistence, spanning the temperature range of 50–60 °C. According to the phase rule such a coexistence cannot be in equilibrium, with the given system of 2 components, 3 phases (excess water plus two lipid phases), and with (atmospheric) pressure freely chosen as one independent variable.

The cooling behaviour is very similar: again the three phases appear within the time-scale of the experiment. The reciprocal decay and growth of H_{II} and L_α phases, respectively, is similar to the heating experiment, and spans the same period of time (with similar heating and cooling rates). Notably, however, there appears a structural hysteresis: the inflection of the three H_{II}-maxima towards smaller angles during in the coexistence region is considerably more pronounced upon cooling. The last coexisting hexagonal structures have a first order spacing (the $10\bar{1}0$ reflection) of 73 Å while on heating, the respective value is 68 Å. This indicates that the first cylindrical tubes formed upon heating are smaller in diameter than the last ones that disappear upon cooling.

The above example is only one out of a series of similar experiments with various lipid system showing the lamellar-hexagonal transition. In combination they allow a number of important conclusions regarding the mechanism and kinetics of this processes:

1) The $L_\alpha \rightarrow H_I$ transition is a two-state process, i.e. it proceeds directly from one structure to the other. It neither involves detectable amounts of intermediates such as spherical micelles or other particles which would generate a broad continuous scattering background in the transition range, nor does it involve the loss of long range order before or within the transition.
2) The process is readily reversible, with a minor but noticeable hysteresis in the limiting lattice parameters.
3) L_α and H_{II} phases coexist over relatively broad temperature intervals, both in heating and cooling scans, with a third phase of excess water. This indicates that equilibration is slower than the time constant in such experiments, which is approximately 2 s.

It should be noted that the transition rates are expected to be temperature dependent. Therefore it is not surprising that shorter transition times have been observed with faster heating and cooling rates as used e. g. in the work of Caffrey [66)] and Ranck

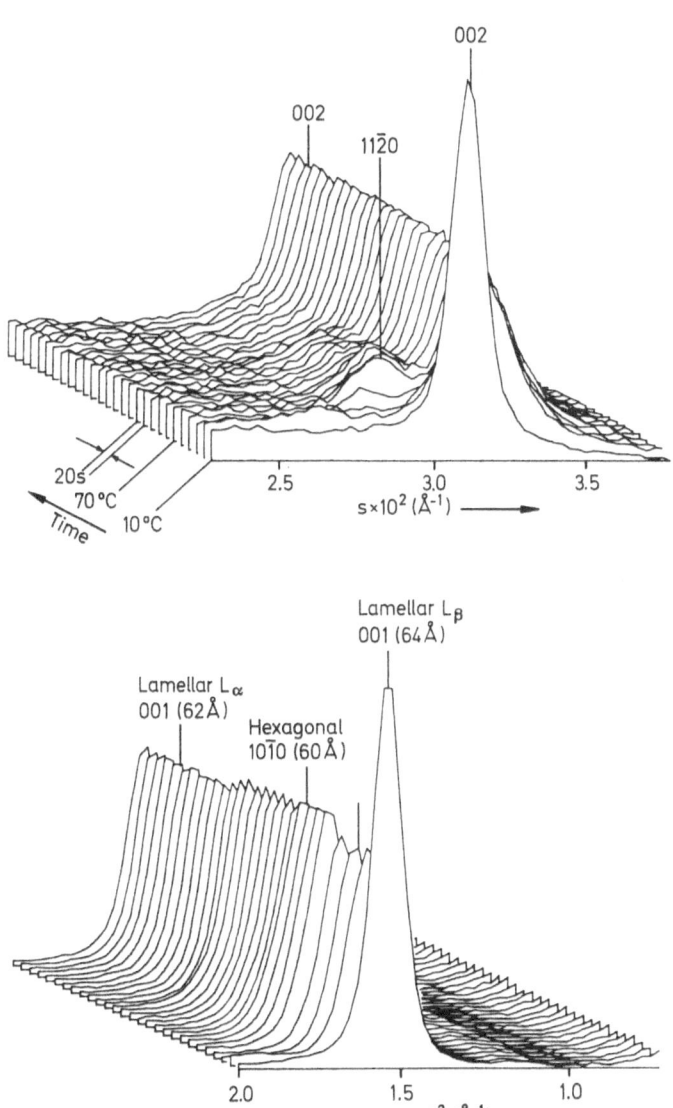

Fig. 9. Time-resolved small-angle diffraction from a mixture of 1-palmitoyl-2-oleoyl-phosphatidylcholine and 1-hexadecyl-2-oleoyl-phosphatidylethanolamine. Two minutes after mixing of the aqueous dispersions ($c_{lip} \sim 0.2$) the temperature was raised from 10 to 70 °C, where the latter lipid species forms a hexagonal H_{II} phase, while the former stays lamellar. Fusion of the two phases and reformation of a lamellar phase is evident from the decay of the hexagonal $11\overline{2}0$ reflection. (From Ref. 74, with permission)

et al. [75]. Furthermore, in natural lipid mixtures, which have to be considered as multi-component systems, coexistence between lamellar and H_{II} phases can obviously be an equilibrium phenomenon over wide temperature spans. In the light of the results obtained with pure lipids, this is probably the situation given in phosphatidyl-ethanolamine from egg yolk which we previously interpreted as a single-lipid-phase system [55].

4.1.1.3 Fusion of Lipid Phases

Fusion of membranes is a very important biological event which, in its detailed mechanisms, is still poorly understood [76, 77]. Also here, time-resolved X-ray diffraction offers attractive possibilities. This shall be illustrated by the first study that has recently been undertaken in this direction [74].

If two phospholipid species, one favouring lamellar structures and the other tending to form hexagonal phases are mixed, their two profoundly different X-ray small-angle diffraction patterns are very well suited to follow the ensuing changes with time, and thus to directly monitor the process of fusion. In particular the $(11\bar{2}0)$ reflection of the hexagonal phase, appearing at a spacing of $\sqrt{3} . s_1$, where s_1 is the prominent $(10\bar{1}0)$ reflection, is well suited for this purpose, since it lies well separated from any lamellar phase reflections. In the experiment shown in Fig. 9 the system starts from two separate lamellar phases (at low temperature); heating through the $L_\alpha \rightarrow H$ transition of the ethanolamine lipid leads to a situation where the hexagonal phase (as clearly evident by the $10\bar{2}0$ reflection) coexists with the lamellar L_α phase of the choline lipid. Fusion of the two phases proceeds with a half-time of about 3 minutes and leads to the expected lamellar phase of the equimolar mixture.

The important conclusion to be drawn is that the process has a clear two-state character, i.e. the structures go directly from one into the other, without involving noticeable amounts of irregular intermediary structures: once the local concentration of the choline lipid exceeds a critical value, the entire domain transforms from a well-ordered hexagonal to the lamellar state, without loosing long-range order as evident from the relatively constant width of the lamellar (002) reflection which is rather sensitive to lattice distortions.

Clearly, this experiment demonstrates the high potential of synchrotron radiation studies in this important field of membrane biophysics, and a wealth of very relevant results can be expected from future improvements in the design of the experiments, particularly from isothermal rapid mixing experiments, where the time resolution can be enhanced probably by two orders of magnitude.

4.2 Time-Resolved Diffraction Studies on Membranes During Active Transport Processes

A very attractive candidate for structural studies on biological membrane are vesicles of sarcoplasmic reticulum from striated muscle. It is characterized by a high degree of homogeneity, both from the compositional and from the functional point of view [78, 79]: more than 90% of the total protein is constituted by the Ca^{2+}-

ATPase (MW 10^5) which is responsible for the active transport function of Ca^{2+} ions across the membrane; about 40% of the total membrane dry mass is lipid, mainly phosphatidylcholine, in a typical bilayer arrangement, as has been confirmed by static X-ray small-angle diffraction of stacked vesicle preparations [12,13].

By a combination of X-ray and neutron scattering methods involving contrast matching techniques through selective deuteration, the separate profile structures of the lipid bilayer and the Ca^{2+}-ATPase molecule within the membrane have been determined to low (~ 10 Å) resolution [80]. A representative diffraction pattern obtained with synchrotron radiation is shown in Fig. 10.

The kinetics of Ca^{2+}-transport, as studied by spectrophotometric techniques, show a fast and a slow phase; the latter, lying in the range of seconds, can be identified with the translocation of Ca^{2+} across the membrane [81]. Synchronous triggering of the ensemble of Ca^{2+}-ATPase molecules within a oriented multilayer of membranes can be achieved by flash photolysis of caged ATP [82]. The time-scale of the effective synchronization of the ensemble depends on the duration of the UV-light flash required to produce a sufficient quantity of ATP and is ultimately limited to the millisecond range due to the kinetics of the dark-reaction of the photolytic process.

The results of a typical time-resolved X-ray experiment on a stack of sarcoplasmic reticulum membranes is shown in Fig. 11. In frame a, the intensity function of the lamellar stack obtained within 0.5 s before the UV flash, is shown; it displays the typical appearance of the first four orders of the lamellar diffraction pattern with $d \sim 200$ Å. Immediately after the 0.25 s photolysis flash, the intensity recorded

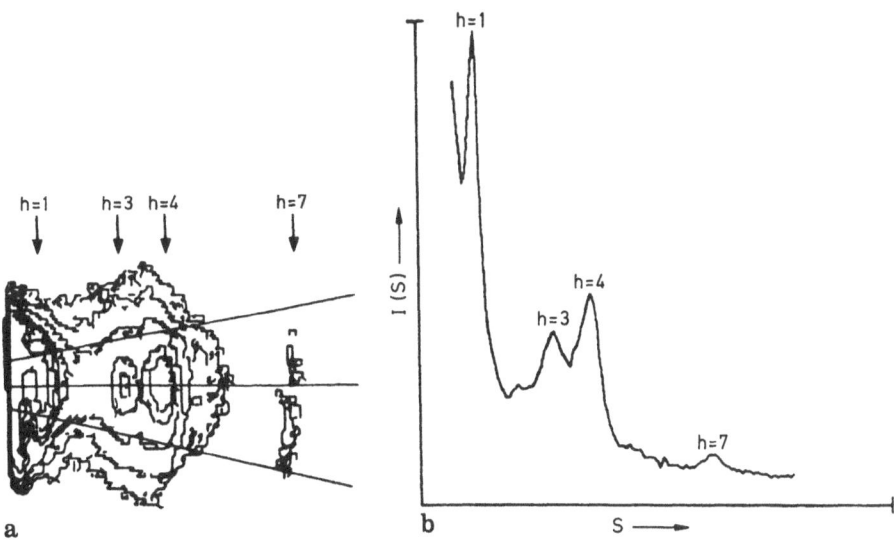

Fig. 10a and b. Small-angle diffraction pattern from an oriented, stacked sample of sarcoplasmic reticulum vesicles, obtained in 1 s with synchrotron radiation at Stanford Synchrotron Radiation Laboratory. The peaks correspond to the first seven orders of a lamellar diffraction pattern with a lattice periodicity of ~ 200 Å and exhibit the significant effects of lattice disorder of the second kind. The intensity distribution in (b) was obtained by integration of the two-dimensional pattern (a) between the limits indicated in (a). (From Ref. 85, with permission)

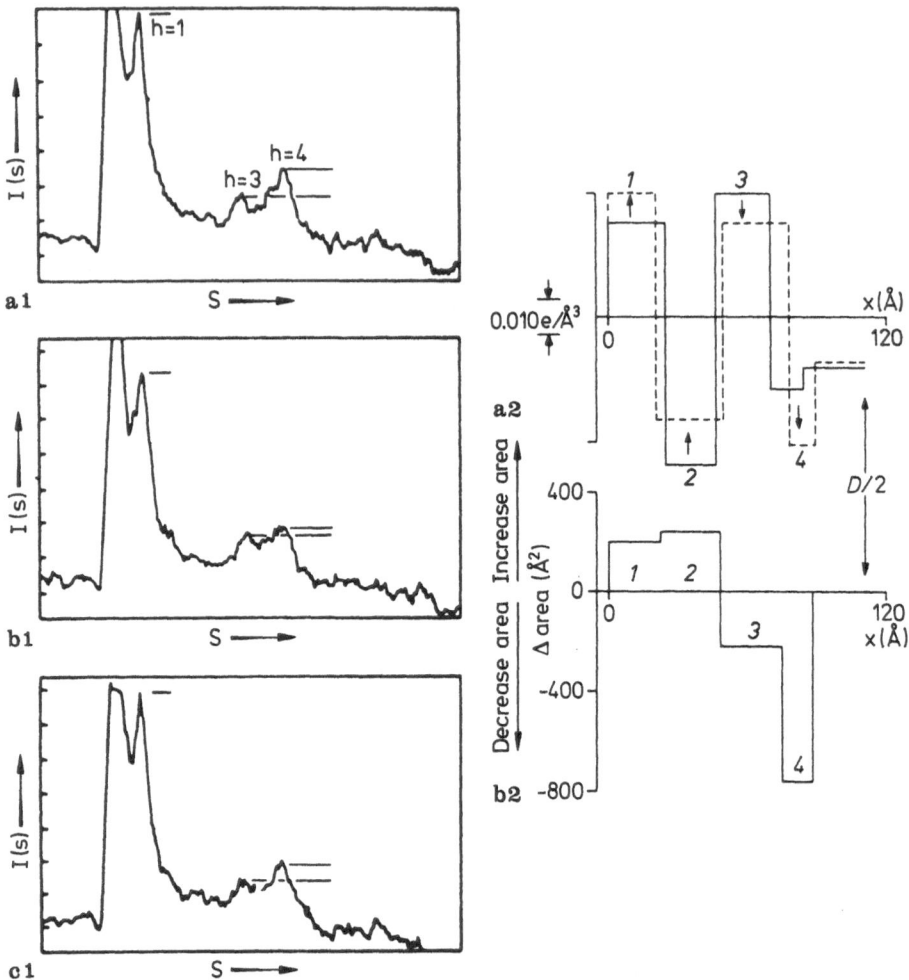

Fig. 11 a–c. *Left*: Small-angle diffraction patterns of sarcoplasmic reticulum membranes before (**a1**) and immediately after flash photolysis (0.25-s UV flash) of caged ATP (**b1**) with a 0.5-s exposure; pattern (**c1**) shows the result of a 0.5-s exposure 1 min or more after the UV flash.
Right: Refined step-function models (**a2**). The solid line corresponds to the membrane profile before photolysis of caged ATP and the dashed line to that immediately after. (**b2**) The differences of the electron density levels in regions *1–4* of the two profiles shown in **a**; these differences correspond to the protein profile and indicate changes in protein structure upon phosphorylation of the ATPase. (From Ref. 85, with permission)

over a period of 0.5 s shows significant differences in the relative intensities of the lamellar diffraction orders h = 1, 3, 4, which reverse to the original situation within about 1 min. The analysis of the results indicates a redistribution of electron densities within the unit cell while the multilayer lattice periodicity stays constant. The electron density profiles calculated from the data by Fourier analysis (see Chapt. 3.2) show this redistribution to involve about 8 % of the mass of the ATPase, which is moved from the extravesicular surface to the lipid hydrocarbon core and intravesicular surface regions of the membrane profile within 0.2–0.5 s after triggering of the Ca^{2+}

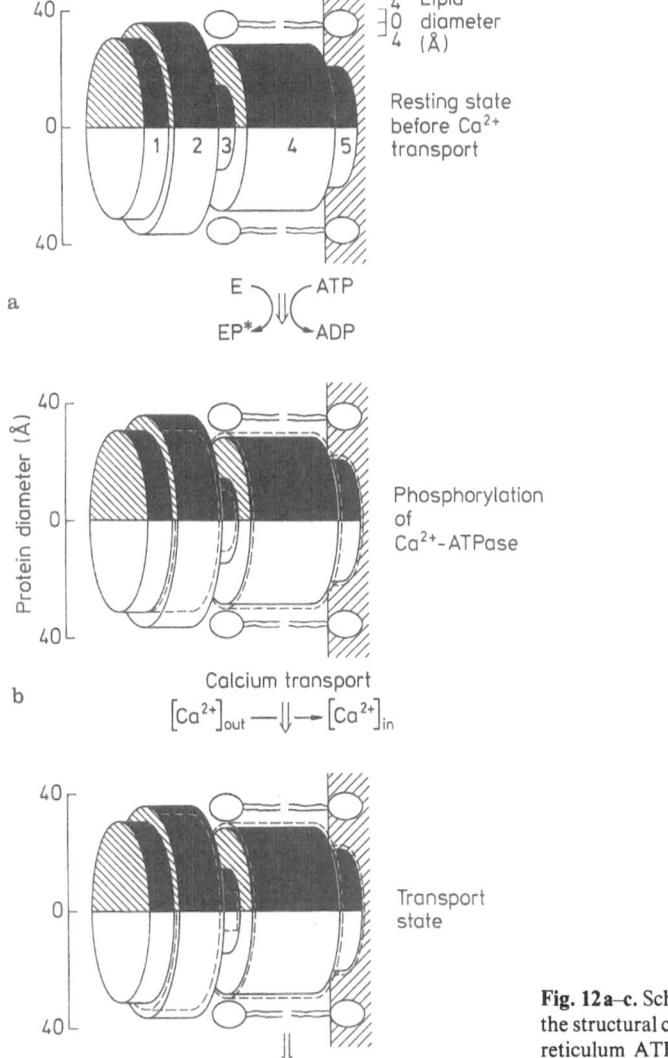

Fig. 12a–c. Schematical representation of the structural changes of the sarcoplasmic reticulum ATPase. (From Ref. 85, with permission)

pump cycle (at 7–8 °C). In terms of a low resolution structural model this can be represented as shown in Fig. 12.

The example of this study on a functional membrane system shows the present possibilities in this field. To those used to viewing biological systems at atomic resolution this may seem a rather modest progress, but this would be neglecting the inherent problems posed by the complex nature of membrane structure. In view of the numerous and largely hypothetical proposals for transport mechanisms from other physico-chemical and biochemical sources, however, results like those obtained on the Ca^{2+}-ATPase system gain strong importance.

An avenue which is of high potential for further progress seems to be the study of detergent-solubilized ATPase under different conditions resembling those of individual, transient steps of the transport cycle. A crucial question for such studies is the time required for a single exposure to obtain a continuous small-angle scattering curve with sufficient precision to allow an analysis in terms of structural parameters, such as the radius of gyration. Preliminary experiments in that direction [83] have shown that a stopped-flow experiment with a 1 % solution of ATPase can provide statistically significant results in the low-angle Guinier region within about 0.1 s.

5 Concluding Remarks

Membrane-associated processes, both in the lipid and protein domains, bear many kinetic aspects which require a cinematographic approach for their full, structural understanding. While spectroscopic data on many different natural and model membranes are plentiful, such is not the case with time-resolved X-ray data. This field, therefore, carries great prospects for synchrotron radiation studies. It was the aim of this review to outline the specific problems posed by the very nature of biological membranes to structure analysis, and the theoretical and present experimental feasibilities. The application of synchrotron radiation has entered this field only within the past four years, and therefore the available experience is still limited. Nevertheless, particularly in the field of lipid polymorphism, the work published so far has already provided essential new results, and the continuing efforts promise to be of central importance to an understanding of membrane functions such as fusion and transitory structures. Greater difficulties are involved, naturally, in the investigation of functional, biological membranes, but also there the few approaches taken so far have shown that this will be not only a worthwhile but necessary area of future activities.

6 Acknowledgements

The author wishes to thank Drs. Lohner and Müller for their help with the experiments described here, and the scientific and technical staff of the Hamburg Outstation at DESY, of the European Molecular Biology Laboratory for helping to operate their facilities. The author's own work benefited from grants no. 5264 and 2473 of the Österreichischer Fonds zur Förderung der Wissenschaftlichen Forschung and the Österreichische Nationalbank, respectively.

7 References

1. Worthington CR (1973) X-ray diffraction studies on biological membranes, in: Current Topics in Bioenergetics, Vol. V (eds) Sanadi DR, Packer L, p. 1, New York, Academic Press
2. Shipley GG (1973) Recent X-ray diffraction studies of biological membranes and membrane components, in: Biological Membranes, Vol. II (eds) Chapman D, Wallach DFH, p. 1, London, Academic Press
3. Franks NP, Levine YK (1981) Low-angle X-ray diffraction, in: Membrane Spectroscopy (ed) Grell E, p. 437, Springer, Berlin

4. Blaurock AE (1982) Biochim. Biophys. Acta *650*: 176
5. Blaurock AE (1975) J. Mol. Biol. *93*: 139
6. Henderson R (1975) ibid. *93*: 123
7. Kirschner DA, Kaspar DLD (1977) Diffraction studies of molecular organization, in: Myelin (ed) Morele P, p. 154, New York, Plenum Press
8. Worthington CR (1973) Exp. Eye Res. *17*: 487
9. Chabre M (1975) Biochim. Biophys. Acta *382*: 322
10. Kreuz W (1972) Angew. Chem. (Int. Ed.) *11*: 551
11. Finean JB, Coleman R, Kuntton S, Limbrick AR, Thompson JE (1968) H. Gen. Physiol. *51*: 19 s
12. Dupont Y, Harrison SC, Hasselbach W (1973) Nature *244*: 555
13. Herbette L, Marquardt J, Scarpa A, Blasie JK (1977) Biophys. J. *20*: 245
14. Thomson JE, Coleman R, Finean JB (1968) Biochim. Biophys. Acta *150*: 405
15. Limbrick AR, Fineman JB (1970) J. Cell. Sci. *7*: 373
16. Wilkins MHF, Blaurock AE, Engelman DM (1971) Nature New Biol. *230*: 72
17. Luzzati V (1968) X-ray diffraction studies of lipid-water systems, in: Biological Membranes, Vol. I (ed) Chapman D, p. 71, New York, Academic Press
18. Luzzati V, Tardieu A (1974) Ann. Rev. Phys. Chem. 79
19. Hauser H, Pascher I, Pearson RH, Sundell S (1981) Biochim. Biophys. Acta *650*: 21
20. Kratky O, Glatter O (1982) Small-Angle X-Ray Scattering, London, Academic Press
21. Kratky O, Laggner P (1987) X-ray small angle scattering, in: Meyers RA (ed) Encyclopedia of Physical Science and Technology, Vol. 14, p. 693, New York, Academic Press
22. Sardet C, Tardieu A, Luzzati V (1976) J. Mol. Biol. *105*: 383
23. Beverley Osborne H, Sardet C, Michel-Villaz M, Chabre M (1978) ibid. *123*: 177
24. Le Maire M, Møller JV, Tardieu A (1981) ibid. *150*: 273
25. Pachence JM, Edelman IS, Schoenborn BP (1987) J. Biol. Chem. *262*: 702
26. Deisenhofer J, Epp O, Mike K, Huber R, Michel M (1984) J. Mol. Biol. *180*: 385
27. Sherwood D (1976) Crystals, X-Rays and Proteins, London, Longman
28. Alexander LE (1969) X-Ray Diffraction Methods in Polymer Science, New York, Wiley-Interscience
29. Kratky O, Porod G (1948) Acta Phys. Austriaca *2*: 133
30. Porod G (1948) ibid. *2*: 255
31. Hosemann R, Bagchi SN (1962) Direct Analysis of Diffraction by Matter, Amsterdam, North-Holland
32. Bradaczek H, Luger P (1978) Acta Cryst. *A34*: 681
33. Pape EH (1974) Biophys. J. *14*: 284
34. Pape EH, Kreuz W (1978) J. Appl. Cryst. *11*: 421
35. Glatter O, Hainisch B (1984) ibid. *17*: 435
36. Glatter O (1981) ibid. *14*: 101
37. King GI (1975) Acta Cryst. *A31*: 130
38. Mitsui T (1978) Adv. Biophys. *10*: 97
39. Blaurock AE (1971) J. Mol. Biol. *56*: 35
40. Sayre D (1952) Acta Cryst. *5*: 843
41. Franks NP (1976) J. Mol. Biol. *100*: 345
42. Franks NP, Arunchalam T, Caspi E (1978) Nature *276*: 530
43. Lesslauer W, Blasie JK (1972) Biophys. J. *12*: 175
44. Moody MF (1974) ibid. *14*: 697
45. Zernike F, Prins JA (1927) Z. Physik *41*: 184
46. Kratky O (1933) Physik Z. *34*: 482
47. Worthington CR (1986) Biophys. J. *49*: 98
48. Blaurock AE, Nelander JC (1976) J. Mol. Biol. *103*: 421
49. Tardieu A, Luzzati V, Reman RC (1973) ibid. *75*: 711
50. Janiak MJ, Small DM, Shipley GG (1976) Biochemistry *15*: 4575
51. Abrahamsson S, Dahlén B, Löfgren H, Pascher I (1978) Progr. Chem. Fats other Lipids *16*: 125
52. Cullis P, De Kruijff B (1979) Biochim. Biophys. Acta *559*: 393
53. De Kruijff B, Cullis P, Verkleij A, Hope MJ, Van Echteld CJA, Taraschi TF, Van Moogevast P, Killian JA, Rietveld A, Van Der Steen ATM Modulation of lipid polymorphism by lipid-

protein interactions, in: "Progress in Protein-Lipid Interactions" Vol. I Watts A, De Pont JJHHM (eds) 1985 p. 89, Amsterdam, Elsevier

54. Caffrey M, Bilderback DM (1984) Biophys. J. *45*: 627
55. Laggner P FEBS Advanced Course "Structure and Dynamics of Membrane Lipids" April 1984, Zeist, Holland; Symp. on "New Methods in X-Ray Absorption, Scattering and Diffraction for Applications in Structural Biology", Aug. 1984, Bristol U.K. Chance B, Bartunik HD (eds) (1986) Academic Press, London
56. Inoko Y, Mitsui T (1978) J. Phys. Soc. Japan *6*: 1918
57. Ruocco MJ, Shipley GG (1982) Biochim. Biophys. Acta *691*: 309
58. Tsong TY, Kanehisa MI (1977) Biochemistry *16*: 2674
59. Gruenwald B (1982) Biochim. Biophys. Acta *687*: 71
60. Lentz RB, Freire E, Biltonen RL (1978) Biochemistry *17*: 4475
61. Cho KC, Choy CL, Young K (1981) Biochim. Biophys. Acta *663*: 14
62. Stumpel J, Eibl H, Niksch A (1983) ibid. *727*: 246
63. Gottlieb MH, Eanes ED (1974) Biophys. J. *14*: 335
64. Boulin C, Gabriel A, Hendrix J (1984) Research Report EMBL 1983, p. 135, EMBL Heidelberg
65. Bordas J, Mandelkow E In: Fast Methods in Physical Biochemistry and Cell Biology, Sha'afi RI, Fernandez SM (eds) (1983) p. 137, Elsevier, Amsterdam
66. Caffrey M (1985) Biochemistry *24*: 4826
67. Rand RP, Chapman D, Larsson K (1975) Biophys. J. *15*: 1117
68. Kashchiev D (1984) Crystal Res. Technol. *19*: 1413
69. Wunderlich B (1976) Macromolecular Physics, Vol. 2, Crystal Nucleation, Growth, Annealing, p. 7, New York, Academic Press
70. Dr Kruijff B, Verkleij AJ, Van Echteld CJA, Gerritsen WC, Mombers C, Noordam PC, De Gier J (1979) Biochim. Biophys. Acta *555*: 200
71. Hui SW, Stewart TP, Boni LT (1983) Chem. Phys. Lipids *33*: 113
72. Siegel CP (1984) Biophys. J. *45*: 399
73. Siegel DP (1986) ibid *49*: 1155
74. Laggner P, Lohner K to appear in Chem. Phys. Lipids
75. Ranck J-L, Letellier L, Shechter E, Krop B, Pernod P, Tardieu A (1984) Biochemistry *23*: 4955
76. Lucy JA Biomembrane fusion, in: Biological Membranes, Vol 4 Chapman D (ed) (1982) p. 367, Academic Press, London
77. Siegel DP Membrane-membrane interactions in lamellar-to-inverted hexagonal phase transitions, in: Membrane Fusion Sowers AE (ed) Plenum, New York; in the press
78. Ebashi S, Endo M, Ohtsuki IQ (1969) Rev. Biophys. *2*: 351
79. Meissner G, Conner GE, Fleischer S (1973) Biochim. Biophys. Acta *298*: 246
80. Herbette LG, De Floor P, Fleischer S, Pascalini D, Scarpa A, Blasie JK (1985) ibid. *817*: 103
81. Pierce D, Scarpa A, Trentham DR, Tapp MR, Blasie JK (1983) Biophys. J. *44*: 365
82. McCray JA, Herbette LG, Kihava T, Trentham DR (1980) Proc. Natl. Acad. Sci. USA *77*: 7237
83. Laggner P, Ludi H, Hasselbach W in preparation
84. Laggner P, Glatter O, Müller K, Kratky O, Kostner G, Holasek A (1977) Eur. J. Biochem. *77*: 165
85. Blasie JK, Herbette LG, Pascolini D, Skita V, Pierce DH, Scarpa A (1985) Biophys. J. *48*: 9

Synchrotron X-Ray Scattering Studies of the Chromatin Fibre Structure

Zehra Sayers

European Molecular Biology Laboratory Hamburg Outstation EMBL, c/o DESY, Notkestraße 85, 2000 Hamburg 52, FRG

Table of Contents

Small angle X-ray scattering and diffraction methods can provide information about chromatin structure at the levels of nucleosomes, chromatin fibre and whole nuclei. Making use of the high flux and point focusing achieved on synchrotron radiation cameras it has been possible to carry out systematic static and kinetic investigations of chromatin structure with minimum damage or perturbation on fresh samples. The results provide a strong evidence for a preformed 3-D zigzag structure at low ionic strength which folds instantaneously into the "30 nm filament" upon increasing the ionic strength. The extent of folding at a given ionic strength depends on the nature of the cations and can be influenced by other substances which bind to DNA. Models of chromatin fibre structure are discussed in the context of various experimental results.

Topics in Current Chemistry, Vol. 145
© Springer-Verlag, Berlin Heidelberg 1988

1 Introduction

In eukaryotic organisms the genetic processes of DNA replication and RNA synthesis are isolated by the nuclear envelope from translation which occurs on the ribosomes in the cytoplasm. In the nucleus, DNA forms a compact complex with various proteins, known as chromatin. Packing of chromatin into the nucleus is achieved through several orders of folding as illustrated schematically in Fig. 1.

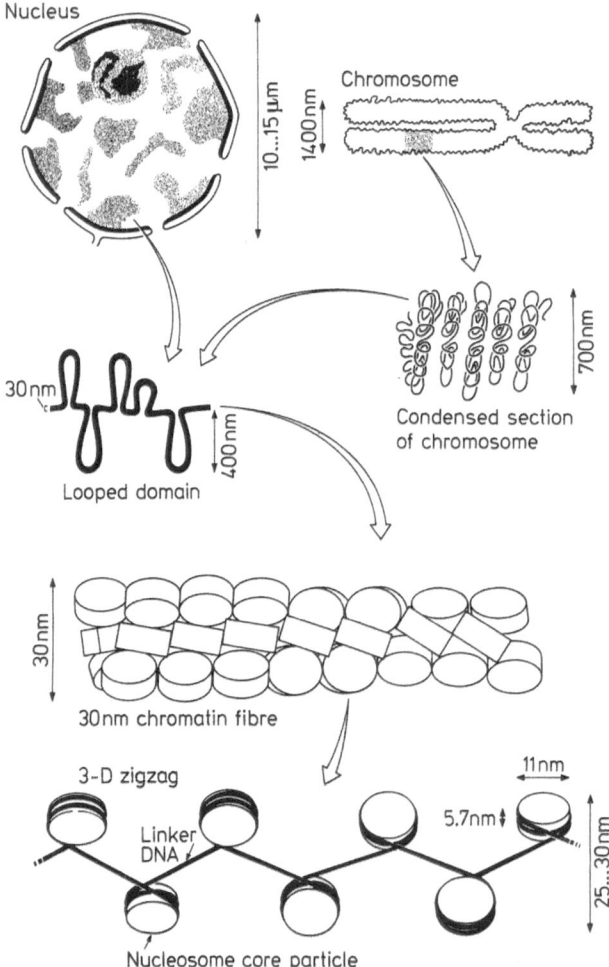

Fig. 1. Hierarchy of chromatin folding: in the nucleus, during interphase, chromosomes are spread in a diffuse form and the nucleoli appear as dense structures. During metaphase chromosomes which consist of condensed sections of chromatin are formed. Chromatin in interphase nuclei and in the condensed sections of chromosomes contains looped domains which are formed by the 30 nm chromatin fibre. In vitro it is possible to further unfold the 30 nm fibre to a 3-D zig-zag structure revealing the nucleosomes that consist of histone H1, the core particle and the linker DNA

Changes occur in the chromatin folding pattern throughout the cell cycle. While during metaphase chromosomes form unique and compact structures, during interphase the chromosomal structure is reorganized and extended into a diffuse form. Although the degree of compaction of the latter is lower, it still allows the packing of the total length of DNA (about 1 m) into a nucleus of about 10 μm in diameter. The volume of one meter of DNA fibre with a diameter of 2 nm would, however, occupy only about 0.6 % of the total nuclear volume, and DNA-binding proteins would represent another 1 % of the total nuclear volume.

In the hierarchy of chromatin folding the largest structures are the chromosomes seen in metaphase cells. These are about 5 μm in length and 1400 nm in width and consist of condensed sections of chromatin which are 700 nm in thickness. The condensed sections, in turn, result from the compaction of looped domains [1] which are present in interphase nuclei as well. In the latter, chromatin is also organized in less compact regions which appear to be correlated with gene expression and DNA replication [2,3]. The structure which folds to form the looped domains is known as the "30 nm chromatin fibre" and represents the lowest form of folding that has been observed by electron microscopy in cell nuclei [4,5]. In vitro at low ionic strength, the 30 nm fibre further unfolds to give a "beads-on-a-string" appearance in electron micrographs [6,7]. The "beads" of the fibre (i.e. the nucleosome core particles) are connected by the linker DNA.

1.1 The Nucleosome Core Particle

Prolonged enzymatic digestion of chromatin gives the nucleosome core particle: about 146 base pairs (bp) of DNA coiled around a globular protein octamer consisting of two of each of the histones H2A, H2B, H3 and H4. The histones are relatively small basic proteins and are among the most conserved of all proteins [8,9,10]. This strongly suggests that the conserved regions are involved in specific interactions. Both the slightly lysine-rich histones H2A and H2B and the arginine-rich histones H3 and H4 are composed of two domains: the amino-terminal region where the net positive charge is clustered, and the globular carboxyl-terminal region, that is rich in the highly conserved hydrophobic residues. NMR studies indicate that these hydrophobic regions are involved in histone complex formation [11,12]. Enzymatic digestion experiments suggest that histones with amino terminal regions cleaved retain the structural elements necessary for protein-protein association as well as the ability to organize DNA into a nucleoprotein complex resembling chromatin core particles [13,14]. The histone octamer aggregate, free of DNA, has been isolated in 2 M NaCl from both cross-linked [15,16] and non-cross-linked chromatin [17,18]. The radius of gyration of the histone octamer measured by neutron scattering is 3.0–3.3 nm [19,20]. Burlingame et al. [21,22] have proposed a prolate ellipsoid with dimensions $11.0 \times 7.0 \times 6.5$ nm^3 for the shape of the histone octamer in about 5 M ammonium sulfate on the basis of crystallographic data extending to 0.33 nm resolution.

The structure of the nucleosome core particle has been determined by X-ray crystallography at a resolution of 0.7 nm [23] and 1.5 nm [24] and by neutron diffraction in conjunction with contrast matching at 2.5 nm resolution [25]. The core particle can be described as a flat disc of diameter about 11 nm and thickness 5.7 nm, with 1.8 turns of DNA of pitch 2.8 nm coiled around the octamer. The X-ray crystallographic analysis [23] suggests that the right-handed B-DNA superhelix on the outside contains several sharp bends and makes several interactions with the histone core. The central turn of superhelix and H3-H4 tetramer have dyad symmetry but the H2A and H2B dimers show deviations.

1.2 The Nucleosome

Mild enzymatic digestion of chromatin yields two types of particles: chromatosomes and nucleosomes. Chromatosomes are metastable particles consisting of 165 bp of DNA, the core histone octamer and the histone H1 [26]. The H1 histones, H1, H1°, H5 and their variants, constitute a group of lysine-rich proteins of which there are several different but closely related varieties. H1 is found in all cells except yeasts, H1° is found in high amounts in some quiescent cells and H5 is only found in avian erythrocytes [27, 28, 29]. H1 is evolutionarily the most variable class of histones. The basic amino and carbonyl termini constitute the highly variable regions whereas the globular central region has a conserved sequence [10].

The nucleosomes [30], the basic repeating units of chromatin structure, consist of about 150 bp of DNA, the core histone octamer, the histone H1 and a length of linker DNA usually in the range 30 to the 50 bp. The length of linker DNA is not only variable within one nucleus but is also species dependent [31]. Chromatin in avian or amphibian erythrocyte nuclei has 210 bp nucleosomal repeat, whereas in nuclei of yeast or neuronal cells the repeat is 163 bp [32, 33, 34, 35].

The radius of gyration of the nucleosome is 4.0 to 4.5 nm as determined by neutron scattering [36, 37, 38] and X-ray scattering [39]. The model proposed for the nucleosome structure based on the histone octamer crystal structure [22] differs from that resulting from the crystal structure studies of the nucleosome core particle [23]. This controversy [40, 41, 42, 43] will be resolved when higher resolution structures become available.

1.3 Higher Orders of Chromatin Folding

Higher orders of chromatin folding i.e., the modes of packing of nucleosomes, can be visualised by electron microscopy. These are the "beads-on-a-string" or the "10 nm nucleofilament", and the 30 nm diameter chromatin fibre [44, 45, 7, 46].

The "beads-on-a-string" appearance of nucleosomes in electron micrographs has been observed only at very low ionic strength or when the structure has been stretched out [47, 7, 46]. In this form nucleosomes display a more or less regular zig-zag structure with linker DNA extending in between.

Fairly uniform fibres or threads of nucleosomes, "10 nm nucleofilament", have also been reported for the low ionic strength chromatin structure [44].

In interphase nuclei, metaphase chromosomes or in solutions of isolated chromatin fragments at ionic strength above 40 mM monovalent salt, the 30 nm fibre is observed. This structure is generally described as resulting from the condensation of the "10 nm nucleofilament". Electron microscopy [7, 46] and chemical cross-linking experiments [48, 49] indicate that histone H1 molecules are involved in the folding of nucleosomes into the 30 nm fibre.

Comparison of the structure of intact chromatin with that of H1-depleted chromatin shows that the presence of H1 prevents unraveling of the DNA from the nucleosomes at very low ionic strength and in chromatin with H1, DNA enters and leaves the nucleosome on the same side. Based on these observations it is suggested that H1 is located at the site where DNA enters and leaves the nucleosome.

Arrays of neighbouring H1 molecules can be cross-linked with bis-imidoesters and as the ionic strength of the chromatin solution is increased from 0 to 20 mM NaCl and above, the largest product of cross-linking changes from $(H1)_4$ to $(H1)_6$ oligomers. The lack of any strong dependence of the H1 cross-linking pattern on ionic strength has been interpreted in the framework of the solenoid model (for chromatin fibre structure) as an indication that the major H1-H1 interactions are lateral i.e. the H1-H1 interactions would be stronger between the molecules in one turn of the solenoid than between successive turns.

Enzymatic digestion and chromatin reconstitution experiments indicate that the globular domain of H1 can interact with the core histones but the end domains of both H1 and the core histones are necessary for the correct positioning of H1 with respect to the nucleosome [50, 26, 51, 52].

The exact location of the H1/H5 histones and the organization of linker DNA in the higher order structure of chromatin remain, however, unknown.

Table 1 summarizes some of the techniques that have been used to investigate the higher order structure of chromatin. There is substantial agreement that the 30 nm chromatin fibre has a structure with about 6 nucleosomes/11 nm and that the nucleosomes are arranged with their flat faces oriented approximately parallel to the fibre axis. There are, however, still significant differences between the models proposed for the detailed structure of the chromatin fibre which will be discussed separately.

Among various methods which can be used for structural investigations on chromatin X-ray diffraction and scattering play a central role. The resolution range (3 to 100 nm) of these methods makes it possible to obtain information about the chromatin structure at several levels: nucleosomes, chromatin fibre and whole nuclei, as well as about the changes in the higher order structure with ionic strength. In the following sections the use of synchrotron X-ray solution scattering in studies of chromatin folding will be presented and some recent results will be discussed in the context of the existing models of the fibre structure.

2 Methods

The principle of a small angle X-ray scattering experiment using synchrotron radiation (SR) is illustrated in Fig. 2. The optical system selects X-rays with a wavelength of 0.15 nm and a narrow band-width $(\Delta\lambda/\lambda)$ $5 \cdot 10^{-3}$. This beam is focused on the detector with an adequate cross section at the sample position. The incident beam intensity I_0, which follows the slow decay of the current in the storage

Table 1. Survey of the physical methods used in studies of chromatin fibre structure

Source of Nuclei/Chromatin

Method	Chicken erythrocytes	Rat liver	Thymus	Sperm	Others
X-ray scattering and diffraction	Langmore and Schutt, [71] Langmore and Paulson, [72] Perez-Grau et al., [79] Greulich et al., [39,82] Lasters et al., [96] Widom and Klug, [76] Williams et al., [74] Bordas et al., [61,83] Notbohm, [73,97] Widom, [84] Koch et al., [62,63]	Sperling and Tardieu, [86] Sperling and Klug, [87] Brust and Harbers, [98] Notbohm and Harbers, [99] Langmore and Paulson, [72] Nothbohm, [73] Koch et al., [62]	Azorin et al., [100] Hollandt et al., [101] Damaschun et al., [102] Notbohm and Harbers, [99] Koch et al., [62]	Azorin et al., [100] Langmore and Paulson, [72] Widom et al., [77] Williams et al., [74]	*HeLa and Lymphocytes* Langmore and Paulson, [72] *Yeast* Koch et al., [62]
Neutron scattering	Suau et al., [103] Notbohm, [73]	Bram et al., [104] Baudy and Bram, [105] Notbohm, [73]	Bram et al., [104,106] Baudy and Bram, [105,107]		*HeLa* Imai et al., [38]
Electron Microscopy	Worcel et al., [90] Azorin et al., [108] Subirana et al., [93,109] Woodcock et al., [91] Williams et al., [74] Widom, [84]	Finch and Klug, [44] Thoma and Koller, [6] Dubochet and Noll, [110] Thoma et al., [7] Thoma and Koller, [46] Dixon and Burkholder, [111]	Allan et al., [52]	Subirana et al., [93,109] Widom et al., [77] Williams et al., [74]	*Yeast* Rattner et al., [112]
Hydrodynamic methods: Sedimentation Viscosity, Flow Birefringence	Bates et al., [113] Allan et al., [114] Ausio et al., [81] Muyldermans et al., [115] Harrington [67a]	Butler and Thomas [116] Thomas and Butler, [117] Böttger et al., [118] Brust, [119,120]	Rees et al., [121] Hollandt et al., [101] Allan et al., [52]		*Erlich ascites Cell* Kubista et al., [122] *Neuronal Cells* Pearson et al., [123] Allan et al., [114]
Electric Dichroism and Birefringence	Fulmer and Fasman, [124] McGhee et al., [89,125]	Marion et al., [128a] McGhee et al., [89]	Mandel and Fasman, [132] Crothers et al., [133]	McGhee et al., [89]	*Neuronal, glial* Allan et al., [114]

			HeLa, CHO
Photochemical Dichroism, Circular Dichroism	Houssier et al., [126] Houssier et al., [127] Allan et al., [114] Marion et al., [128] Sen and Crothers, [85]	Marion, [129] Marion et al., [130] Brust, [131]	Houssier et al., [134] Lee et al., [135] Yabuki et al., [136] Mitra et al., [137]	McGhee et al., [89]
Elastic/Quasi elastic Light scattering	Ramsay-Shaw and Schmitz, [138,139] Schmitz and Ramsay-Shaw, [140] Gordon et al., [141] Ausio et al., [80]	Marion et al., [142] Roche et al., [143]		
Fluorescence Anisotropy Decay	Ashikawa et al., [144]		Ashikawa et al., [145]	

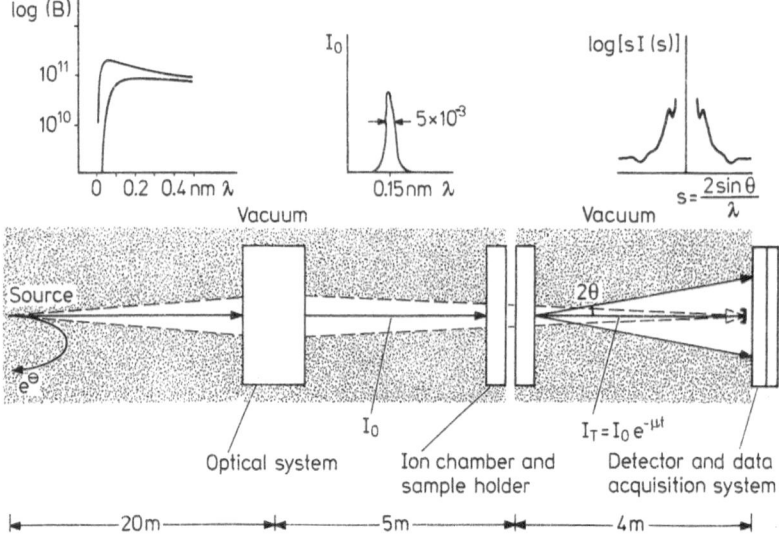

Fig. 2. The layout of synchrotron X-ray small angle scattering measurements. The optical system selects X-rays with a narrow band-width from the continous wavelength distribution of synchrotron radiation. Intensity of the primary beam at the sample I_0 is monitored by an ion chamber. I_T is the transmitted beam, and, $I(s)$, the scattered intensity is recorded with a position sensitive detector. Scattering patterns of fibres are presented as $\text{Log}(sI(s))$ vs s plots where $s = 2 \sin \theta / \lambda$ is the scattering vector. 2θ is the scattering angle and λ is the wavelength.

ring, is monitored using an ion chamber. I_T is the intensity of the beam transmitted through the sample and $I_T = I_0 e^{-\mu t}$, where the factor $e^{-\mu t}$ represents the absorbance of a solution of thickness t. $I(s)$ is the scattered intensity which depends on the scattering vector s defined as $s = 2 \sin \theta / \lambda$, where 2θ is the scattering angle and λ the wavelength. For fibres, the scattering pattern is usually presented as $\log(sI(s))$ vs s and within the range of 2θ used, the approximations $\sin \theta = \theta$ and $\cos \theta = 1$ are valid.

Before the recorded patterns can be interpreted, background scattering is subtracted and they are normalised according to the following equation:

$$I(s) = \frac{e^{\mu t}}{cD(s)} \left[\frac{I_x(s)}{I_0} - \frac{I_B(s)}{I_0} \right] \tag{1}$$

where $I_x(s)$ is the scattered intensity of the sample solution and $I_B(s)$ is that of the corresponding buffer solution. In experiments on chromatin the absorption of the buffer and the sample solution is the same and can be neglected, however when one deals with concentrated salt solutions absorption factors need to be taken into account. $D(s)$ refers to the detector response which is determined separately by exposing the detector to a homogeneous source of radiation (usually a ^{55}Fe source) and c is the concentration of the sample.

2.1 Synchrotron Radiation

The characteristics of synchrotron radiation have been described in several books and reviews (e.g. [53]).

A light source is characterized by its spectral brilliance:

$$B = \frac{\text{number of photons/s}}{(\text{mrad})^2 \, (\text{mm}^2) \, 0.1\% \, (\Delta\lambda/\lambda)} \qquad (2)$$

where $(\text{mrad})^2$ refers to the horizontal and vertical divergence, i.e., to the solid angle of emission of a source of unit area, $(\text{mm})^2$ to the horizontal (σ_x) and vertical (σ_y) source size and $(\Delta\lambda/\lambda)$ to the relative band-width. The high spectral brilliance of SR coupled with wavelength tunability, illustrated in the insert in Fig. 2, and the pulsed time structure make storage rings a unique source for X-ray experiments. The good collimation of SR offers the possibility of using crystal monochromators and total reflection mirrors to achieve a high flux of monochromatic photons at the sample.

The high flux compensates for the poor scattering power and the short life time of the biological samples. Since rapid experiments are possible, systematic investigations more akin to those say on a conventional spectrophotometer can be performed. High intensity also allows time-resolved studies of structural intermediates of systems in solution.

The narrow band-width obtained with crystal monochromators allows the resolution of bands in the solution scattering patterns which would not be possible with neutrons, where high fluxes can only be achieved by using a large bandpass $(\Delta\lambda/\lambda \sim 0.08)$.

2.2 Instrumentation and Data Acquisition

The experiments on chromatin solutions were performed on the X33 camera in HASYLAB on the storage ring DORIS of the Deutsches Elektronen Synchrotron (DESY) in Hamburg. The double focusing monochromator-mirror cameras and the detector and data acquisition systems developed for small angle scattering and diffraction experiments on biological samples using SR are discussed in various publications [54, 55, 56, 57, 58, 59, 60].

Data were recorded using a quadrant detector; a proportional gas chamber with delay line readout (Hendrix, Gabriel and Boulin, unpublished).

The optical system of X33 yields a near point focus which dispenses with the need for correction for beam size, wavelength distribution etc. With various sample-detector distances the range of s-values from $0.01 \, \text{nm}^{-1}$ to $0.5 \, \text{nm}^{-1}$ can be covered, and studies of higher order structures are thus possible.

The thermostatized multicompartment sample cell used in the experiments is shown in Fig. 3.

2.3 Biochemical Methods

Most of the results given in the following sections were obtained using nuclei and chromatin fragments isolated from chicken erythrocytes (CE). Rat liver nuclei and chromatin, calf thymus and yeast nuclei were also used in some experiments for

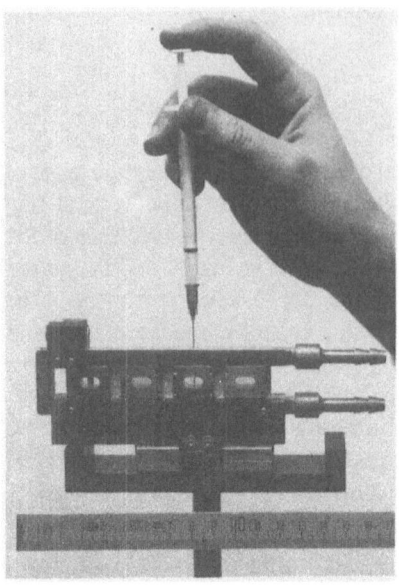

Fig. 3. The thermostatized multicompartment cell used in chromatin experiments. Each compartment has a volume of 100 μl and a thickness of 1 mm. The windows of the cells consist of thin (25 μm) mica sheets. A compartment can be filled from the top using a syringe fitted with a hypodemic needle. Temperature is monitored by means of a sensor placed in the first compartment

comparison. The preparation methods are [61,62,63] described in detail elsewhere. Briefly CE nuclei (at $A_{260} = 100$) in 150 mM NaCl, 10 mM Tris. HCl, 0.5 mM $CaCl_2$ and 0.5 mM phenylmethylsulfonyl fluoride (PMSF), pH 7.5, were digested for 1 min at 37 °C with Micrococcal Nuclease (15 Boehringer units/mg DNA). Digestion was terminated by adding ethylene diaminetetraacetic acid (EDTA) to 2 mM and cooling on ice. The nuclear pellet was washed once in the above digestion buffer without Ca^{2+}, but supplemented with 1 mM EDTA. Washed nuclei were resuspended in 5 mM Tris. HCl, 1 mM EDTA and 0.5 mM PMSF, pH 7.5, (TE buffer) which facilitates the release of chromatin fragments. The fragment solution, clarified by a final centrifugation, was dialysed extensively against TE buffer and was used without further fractionation. Chromatin fragments had a length distribution of 50–150 nucleosomes and the presence of the full histone complement was monitored by electrophoresis. Samples of nuclei and chromatin fragments refered to as nuclei-EDTA and chromatin-EDTA were kept in TE buffer. TE buffer or buffers below 20 mM NaCl are refered to as low ionic strength buffers and modifications in the ionic conditions are given when necessary. Chromatin concentration was determined by measuring the absorbance at 260 nm and a concentration of 1 mg chromatin/ml, or 0.52 mg DNA/ml corresponds to $A_{260} = 10.4$.

Radiation damage on chromatin samples was monitored and within the exposure times (typically 3 minutes per sample) no significant effect on the scattering pattern was observed.

2.4 Theory and Data Interpretation

Small angle X-ray scattering from a solution of macromolecules results from inhomogeneities in the electron density due to the macromolecules dispersed in the uniform electron density, ϱ_0, of the solvent. The scattering pattern of the system is thus determined by the excess electron density of the solute $\varrho(r)$

$$\varrho(r) = (\varrho_p - \varrho_0) \, \varrho_c(r) + \varrho_s(r)$$
$$= \bar{\varrho}\varrho_c(r) + \varrho_s(r) \tag{3}$$

ϱ_p is the average electron density of the particle, and $\bar{\varrho}$ is thus the average electron density of the particle above the level of the solvent, i.e. the contrast. $\varrho_c(r)$ is a dimensionless function which is independent of the contrast and has a value of 1 inside the boundaries of the particle and zero elsewhere. It thus describes the volume of the particle. $\varrho_s(r)$ is also independent of the contrast and describes the fluctuations of electron density inside the particle above and below its mean value ϱ_p [64].

For a solution of chromatin ϱ_0 would represent the average electron density of the buffer, about 330 e/nm^3 for pure water, ϱ_p would be the weighted average electron density of the DNA and the histones in the fibre and has a value of 484 e/nm^3 for the nucleosome in pure water. $\varrho_c(r)$ depends on the shape of the fibre and $\varrho_s(r)$ represents the deviations from the average electron density of the fibre due to the presence of regions containing the linker DNA and the regions with the nucleosome core particles. Calculations show that the excess scattering mass of the linker DNA is about $7 \cdot 10^3$ electrons against $4 \cdot 10^4$ electrons for the nucleosome. The features of the scattering patterns at low angles are thus dominated by the contribution of the nucleosome [62].

In ideal solutions all solute particles are (assumed to be) identical and randomly positioned and oriented in the solvent. The scattering pattern thus contains information only about the spherically averaged structure of the solute, which is described by a distance probability function, p(r). p(r) is the spherically averaged autocorrelation function of $\varrho(r)$ and $r^2p(r)$ is the probability of finding a point inside the particle at a distance between r and r + dr from any other point inside the particle.

As illustrated in Fig. 4 p(r) for a globular particle (e.g. a protein) has two main regions:
a) A region of sharp fluctuations due to neighbouring atom pairs (0.1 nm $\leq$ r $\leq$ 0.5 nm) and of damped oscillations corresponding to structural domains (e.g., α-helix in proteins) for which 0.5 nm $\leq$ r $\leq$ 1 nm.
b) A smooth region corresponding to larger intramolecular vectors. Beyond a certain value of r (r = D_{max}), p(r) vanishes and all vectors correspond to the solvent.

The scattering intensity and p(r) are related by a Hankel transformation:

$$I(s) = 4\pi \int_0^\infty r^2 p(r) \, \frac{\sin(2\pi sr)}{2\pi sr} \, dr \tag{4}$$

As a result, the scattering curve also contains two regions: a) a region at small angle containing mainly information about the long range organization of the particle (e.g., its shape); b) a large angle region where the internal structure of the particle, i.e., deviations from ϱ_p, dominates. The specific features of the scattering pattern are determined by the symmetries of the particles, i.e. the fact that certain distances occur with greater frequency. This can be best understood when the

213

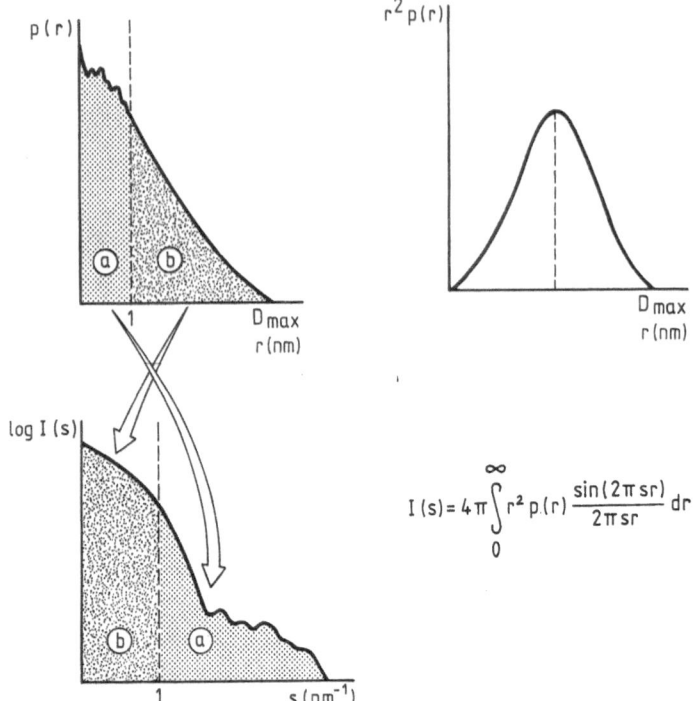

Fig. 4. Correspondance between the structural features and the scattering patterns of globular molecules. The distance distribution function p(r) contains information about neighbouring atoms, Region a, and structural domains, Region b. These regions appear in reversed order in the scattering patterns. The probability of finding a point inside the particle ($r^2 p(r)$) vanishes beyond the largest distance in the particle. The equation shows the transformation relating the distance distribution function and the scattered intensity

structure is described in terms of spherical harmonics [65]. Experimental conditions however, impose limitations on the data evaluation since I(s) for very small angles can only be obtained by extrapolation and the signal to noise ratio at large angles is poor.

The scattered intensity is also obtained by Fourier transforming Eq. (3) and multiplying with the complex conjugate:

$$I(s) = \bar{\varrho}^2 I_c(s) + \bar{\varrho} I_{cs}(s) + I_s(s) \tag{5}$$

The three functions $I_c(s)$, $I_s(s)$ and $I_{cs}(s)$ are the basic scattering functions or the characteristic functions corresponding to the shape (I_c), the internal structure (I_s) and the convolution of the two (I_{cs}) [64].

At small angles a simplified form of the Debye formula [66] for scattering from randomly oriented systems can be used to describe the scattering pattern because the absorption factors for the incident and the scattered beam are assumed to be the same and multiple scattering can be neglected. Then,

$$I(s) = \sum_m \sum_n f_m(s) \, f_n(s) \, \frac{\sin (2\pi r_{mn} s)}{2\pi r_{mn} s} \tag{6}$$

214

Where f_m is the form factor for the mth atom and the summations extend over all atoms (N_0). The r_{mn} are the distances between pairs of atoms m and n.

The simplest interpretation of the scattering curves for globular particles is based on the Guinier approximation [67]. At very low angles, the scattering curve can be described by the following equation:

$$I(s) = I(0) \exp\left(-4\pi^2 Rg^2 s^2/3\right) \tag{7}$$

$I(0)$ is the extrapolated zero angle intensity and R_g is the radius of gyration of the particle. $I(0)$ is proportional to the molecular weight of the particle. Thus, it can easily be seen that even small amounts of larger non-specific aggregates will strongly distort the scattering pattern at low angles. For particles which have nearly homogeneous scattering density, i.e., in which $\varrho_s(r)$ can be neglected, interpretation usually proceeds using simple models in a best fit calculation.

In the case of rod-like particles, fibres in solution, the plot log $(sI(s))$ vs s^2 is used [68]. The radius of gyration of the cross-section (R_x) for the particle is obtained from the equation:

$$sI(s) = I(0)_x \exp\left(-4\pi^2 R_x^2 s^2/2\right) \tag{8}$$

and $\lim_{s \to 0} (sI(s))$ is proportional to the mass per unit length (M/L). Since chromatin solutions are polydisperse with respect to the length of the fragments but homogeneous as far as the cross sections are concerned the formula is readily applicable.

The systems discussed so far were assumed to be dilute and the interparticle interactions were neglected. However, in concentrated systems the correlation in orientation of the particles cannot be neglected, e.g., the fibres are roughly parallel to each other with solvent gaps in between. The particles no longer scatter independently and the small angle scattering patterns display interference effects. The influence of a finite set of structural units whose orientations have a spread is observed as a decrease of the scattered intensity at the origin and the appearance of interference bands at higher angles [69, 70]. These features appear when, for instance, studying scattering by gels or intact nuclei.

3 Small Angle Scattering Patterns of Chromatin

3.1 Scattering Patterns from Cell Nuclei

The 30 nm chromatin fibre in intact nuclei of several types of cells has been investigated by small angle scattering [61, 62, 71, 72, 73, 74]. The results which help to gain an insight into the packing of chromatin in vivo can be used for comparison with the patterns of isolated chromatin fibres in solution. Features of the scattering patterns of CE nuclei-EDTA (a) and nuclei in TE buffer with 100 mM NaCl (b) are illustrated in Fig. 5.

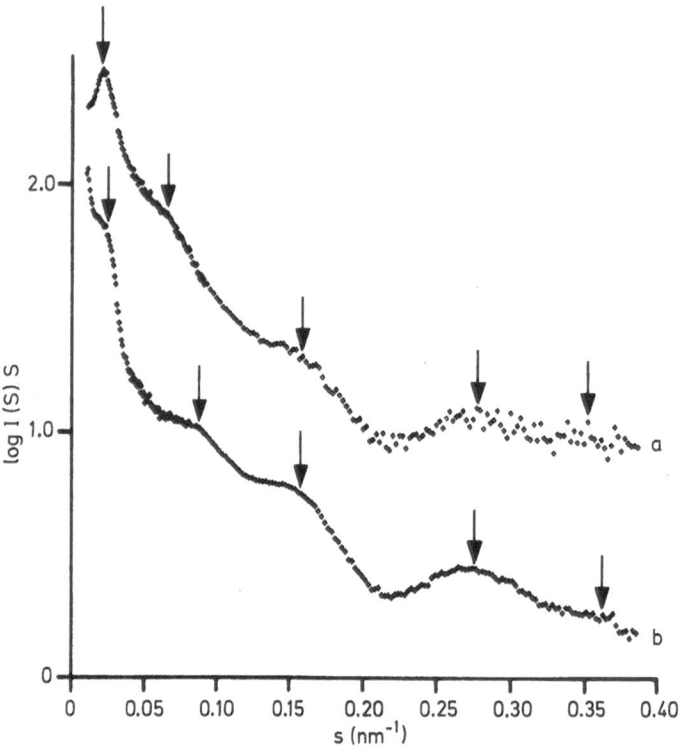

Fig. 5. Scattering patterns of CE nuclei. **a** Scattering pattern of nuclei-EDTA. Bands indicated by arrows are the 0.025 nm^{-1} interfibre interference, the 0.06 nm^{-1} internucleosomal interference and the 0.156, 0.27 and 0.36 nm^{-1} bands which are due to the contributions from the nucleosome core particles. **b** Scattering pattern of nuclei at high ionic strength (150 mM NaCl). The 0.05 nm^{-1} band is no longer visible but a weak band at 0.045 nm^{-1} is detected. Arrows indicate the interfibre interference and the intranucleosomal bands as well as the 0.083 nm^{-1} and 0.15 nm^{-1} bands which result from the close approach of nucleosomes in the condensed chromatin fibre (Fig. 3a) from Bordas et al., 1986a)

Notice that the scattering bands in the patterns of nuclei are superimposed on a high background which makes the determination of their exact position difficult. Most patterns have an interference band near s = 0.025 nm^{-1}. Correlating the scattering patterns with electron micrographs of whole cell nuclei shows that this band results from the side by side packing of the 30 nm fibres in the nucleus. The position of the band is dependent on the cell type; ranging from (40 nm)$^{-1}$ in chicken erythrocyte nuclei to (32 nm)$^{-1}$ in mouse lymphocyte nuclei [72].

A theoretical analysis of the scattering by an array of fibres [69] indicates that the intensity and position of the interference band are determined by the parameter γ, which is the ratio of the centre-to-centre distance between fibres to the fibre diameter. The variation in the position of the interference band from different types of nuclei under similar ionic conditions can thus be attributed to differences in the chromatin fibre diameter [72].

With decreasing ionic strength the band shifts to smaller s-values and its intensity increases probably because the increased repulsion between fibres results in a larger value of the centre-to-centre distance [72, 73].

The interfibre interference band is not observed in sperm cell nuclei [72] or when the chromatin in CE nuclei is condensed in 10 mM Tris · HCl, pH 7.5, with 3–4 mM MgCl$_2'$ [61]. Sperm cell nuclei display densely packed condensed chromatin in electron micrographs [75] and it is suggested that the interference band moves to larger s-values or disappears because of the loss of contrast between fibres.

The main features of the low ionic strength CE nuclei-EDTA patterns, besides the interence peak, are: a broad and strong band at 0.06 nm^{-1} which is absent in the high salt patterns, a prominent band at 0.156 nm^{-1} and bands at 0.27 and 0.36 nm^{-1}. In high salt conditions one observes a weak band at about 0.045 nm^{-1}, a band at 0.083 nm^{-1}, a prominent band at 0.15 nm^{-1} and the bands at 0.275 and 0.37 nm^{-1}. The trough observed at 0.22 nm^{-1} is present in all the patterns independently of the ionic strength, and is attributed to the mononucleosome transform.

The band at about s = 0.045 nm^{-1} in the high salt patterns [61, 74] would correspond to the equatorial first side maximum of the fibre transform, if the 30 nm fibre is considered as a solid cyclinder. Its position would depend on the fibre diameter. A variation in the position of this band in nuclei patterns from different cell types has been reported and correlated with the differences in fibre diameter due to the differences in the length of linker DNA [74].

As discussed below scattering patterns of (partially) oriented chromatin fibres and chromatin gels can be used to determine the origin of the various bands observed from nuclei.

3.2 Scattering Patterns from Oriented Chromatin Fibres, Concentrated Solutions and Gels

Patterns from (partially) oriented condensed chromatin fibres allow to identify the origin of some of the bands seen in the patterns of intact nuclei. Some insight into the behaviour of chromatin in nuclei under different ionic conditions can also be gained by studies on gels obtained by concentrating isolated chromatin fragments.

On the (partially) oriented fibre patterns the bands at 0.025, 0.175, 0.27 and 0.37 nm^{-1} appear mainly as equatorial whereas the 0.09 nm^{-1} band is meridional [76, 77]. Observation of bands as opposed to reflections in these patterns indicate the limited degree of order in the structure.

The 0.045 nm^{-1} band attributed to the fibre transform in the nuclei patterns [72, 61, 74] is absent in the oriented fibre patterns. This could be due to the aggregated state of chromatin in the oriented gels; Bordas et al. [61] report that the 0.045 nm^{-1} band is lost at ionic strengths higher than 100 mM NaCl (or >5 mM MgCl$_2$) where the chromatin fibres in solution begin to precipitate.

The appearance of 0.025 nm^{-1} band mainly as equatorial verifies that it arises from the lateral packing of the fibres. Parallel to the observations on whole nuclei the interference band is absent in the patterns of fibres prepared from sperm chromatin. This feature is attributed to the loss of contrast due to close packing [77].

Unoriented gels obtained from chromatin-EDTA fibres and chromatin fibres condensed in the presence of 100 mM NaCl also contain the interfibre interference band, as can be seen from Fig. 6 (a) and (b). However when condensation is

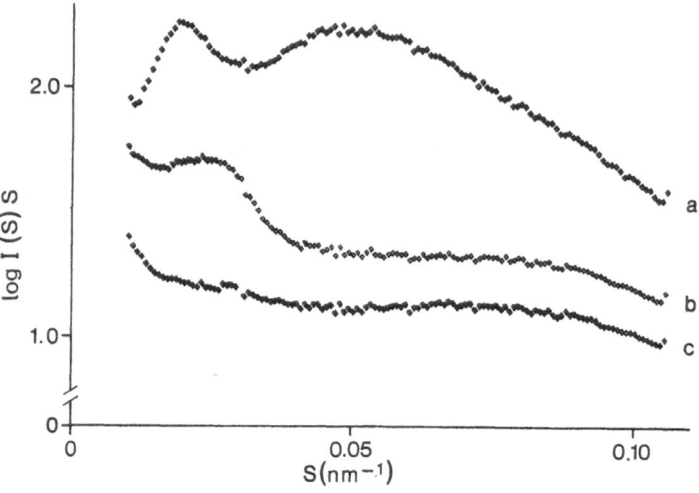

Fig. 6. Scattering patterns illustrating the interfibre interference band in chromatin gels. **a** chromatin gel in TE buffer. **b** chromatin gel condensed in the presence of 100 mM NaCl and **c** chromatin gel condensed in the presence of 3 mM $MgCl_2$. Note that the interfibre interference band is not observed in the $MgCl_2$ patterns (Fig. 4 from Bordas et al., 1986a)

induced by 3 mM $MgCl_2$, figure 6 (c), the 0.025 nm^{-1} band is no longer visible in accordance with the observations on intact nuclei. Studies on chromatin gels isolated together with non-histone nuclear scaffold proteins indicate that at least during the initial stages of condensation the effect of say Na^+ cannot be replaced by Mg^{2+} and vice versa [78]. The effects of mono- and multivalent cations on the condensation of chromatin are different and will be discussed in the ·following section.

Figure 7 illustrates the features of scattering patterns, obtained in the region $0.01 < s < 0.1$ nm^{-1} from low ionic strength chromatin solutions at 3 (a), 12 (b) and 16 (c) mg DNA/ml. With increasing concentration an interfibre interference band develops at very low angles. This interference band cannot be made to move beyond 0.033 nm^{-1} even for the most concentrated uncondensed chromatin preparations [79, 61]. The appearance of the interfibre interference band indicates that the uncondensed chromatin fibres can be packed in a similar fashion to the condensed fibres in the nuclei and that the effective diameter of the uncondensed fibre is comparable to that of the condensed one. The prominent $s = 0.05$ nm^{-1} band is present in all patterns and, as discussed in the next section, arises from the interference of internucleosomal distances in the extended fibre at low ionic strength.

The 0.27 and 0.37 nm^{-1} bands arise from the internal contrast of the nucleosome core particle; the former has contributions from both the DNA and the protein organization whereas the latter is due to the separation between turns of DNA in the core particle. The equatorial 0.175 nm^{-1} and the meridional 0.09 nm^{-1} band, on the other hand, are due to the organization of the nucleosomes in the condensed fibre. The kinetics of the condensation process in gels have been studied [61]. Even in very compact gels the condensation is fast (half-time about 5 s) independently of the use of mono-or divalent cations. Most, if not all, of the lag time appears to be due to the dilution and diffusion of the condensing agent. As the condensation proceeds, the interfibre interference peak shifts to smaller angles and vanishes, the 0.05 nm^{-1} band shifts to larger angles and looses intensity.

If the intensity changes in different regions of the scattering curves are compared, it is seen that the transition from uncondensed to condensed structure proceeds through a series of intermediate states.

3.3 Scattering Patterns from Chromatin Fragments in Solution

Some of the modifications in the structure of chromatin which are dependent on ionic conditions can be studied on isolated fragments in solution. The fragments prepared by mild endonuclease digestion of nuclei have been distinguished according to their solubility in NaCl solutions [80,81,73]. Soluble chromatin (S) refers to the fragments that remain in solution at 150 mM NaCl concentration whereas insoluble chromatin (I) is already precipitated above 10 mM salt. Using very low ionic strength buffers containing EDTA (e.g. TE buffer) fragments can be further extracted from nuclei after the S and I fractions, these constitute the EDTA-extractable chromatin (E). The three fractions, I, S and E, appear to be biochemically identical and it is suggested that the lower solubility of some fractions is due to a larger proportion of the linker DNA being in association with the nucleosome core [81]. The chromatin fragments used in the experiments below are extracted from digested nuclei in low ionic strength buffers containing EDTA as described in Sect. 2.3.

Solution scattering patterns of chicken erythrocyte chromatin fragments in buffers of increasing ionic strength are shown in Fig. 8 (a–c). A detailed description of these patterns is given in [61]. Comparison between Figs. 8 and 5 reveals that in the region $s > 0.03 \text{ nm}^{-1}$ the features of the solution scattering patterns are very similar to those obtained from whole nuclei. The only band which is observed in nuclei patterns and is absent in the solution patterns is the 0.025 nm^{-1} interfibre interference maximum.

Since the maximum dimensions of the core particle are $11 \cdot 11 \cdot 5.7 \text{ nm}^3$ features in the region $s < 0.09 \text{ nm}^{-1}$ $((11 \text{ nm})^{-1})$ are mainly associated with the distribution of the nucleosomes in the fibre. The region of featureless decay $(s < 0.025 \text{ nm}^{-1})$ in the scattering patterns can be used to determine the radius of gyration of the cross section (R_x) and the extrapolated forward scattering intensity $(I(0)_x)$ which is proportional to the mass per unit length (M/L).

In TE buffer, R_x is about 10.0 nm for chicken erythrocyte chromatin and 7.5 nm for rat liver chromatin. These values would yield equivalent solid cylinder diameters of 28 nm and 21 nm respectively. Under these conditions, the mass/unit length for chicken erythrocyte chromatin determined relative to F-actin is $21.6 \pm 2.0 \times 10^3$ Daltons/11 nm corresponding to 0.8 ± 0.1 nucleosomes/11 nm [82,62].

A prominent and specific feature of the scattering patterns at low ionic strength ($<20 \text{ mM}$ NaCl), as seen in Fig. 8 (a), is the band at 0.05 nm^{-1}. This band disappears at higher ionic strength. Its presence indicates that significant contrast exists in the uncondensed fibre and that even at low resolution the structure cannot be regarded as a solid cylinder. The band corresponds to an interference due to an

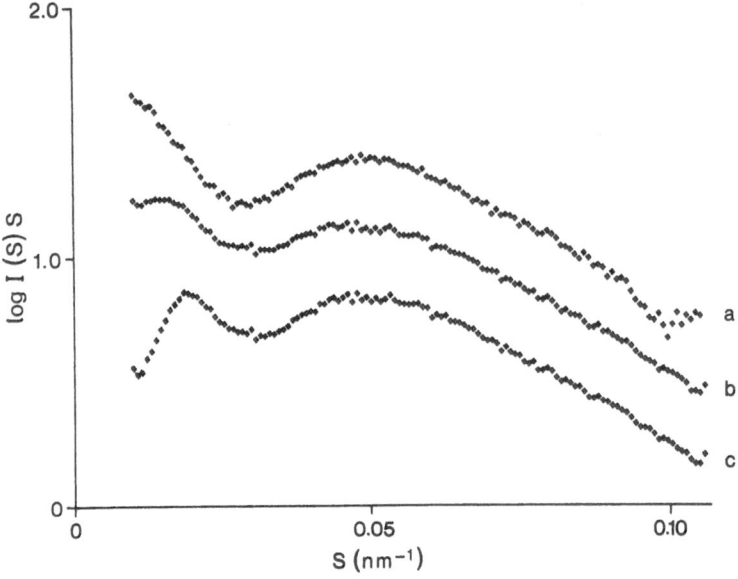

Fig. 7. Appearance of the interfibre interference band in the scattering patterns of chromatin-EDTA with increasing chromatin concentration. **a** 3 mg DNA/ml, **b** 12 mg DNA/ml and **c** 16 mg DNA/ml. Development of the interfibre interference band at high chromatin concentrations indicates that the uncondensed and the condensed fibre have comparable diameters (after figure 6 from Bordas et al., 1986a)

average internucleososomal distance of about 20 nm. This assignment is verified by experiments where the enzymatic digestion of chromatin fragments is monitored by X-ray acattering. During micrococcal nuclease digestion a chromatin solution contains, at any given time, fragments of different lengths and the scattering of such a polydisperse system can be expressed in terms of the scattering of nucleosomes (the total number of which remains constant) and the interference which is a function of the (decreasing) number of links and the average distance between the nucleosomes in the fragments. As a first approximation the interference function can be obtained by dividing the scattering pattern of the undigested sample by that of the fully digested one. The amplitude of the maxima of the interference function is a direct measure of the number of internucleosomal links and the average internucleosomal distance can be determined from the position of the maxima and minima. Digestion with micrococcal nuclease results in a progressive decrease in intensity of the 0.05 nm^{-1} band. From the interference function an average internucleosomal distance of 23 nm is obtained [63].

Model calculations based on a helix-like structure with 9.3 nm radius, a pitch of 3.3 nm and with about 2.9 nucleosomes per turn where the internucleosomal distance is about 20 nm have provided good agreement with the experimental data for uncondensed chromatin [83].

An alternative approach is to consider the extended fibre as a convolution of a wormlike chain with spherically averaged nucleosomes. At the resolution of solution scattering experiments the features of the patterns are dominated by the contribution

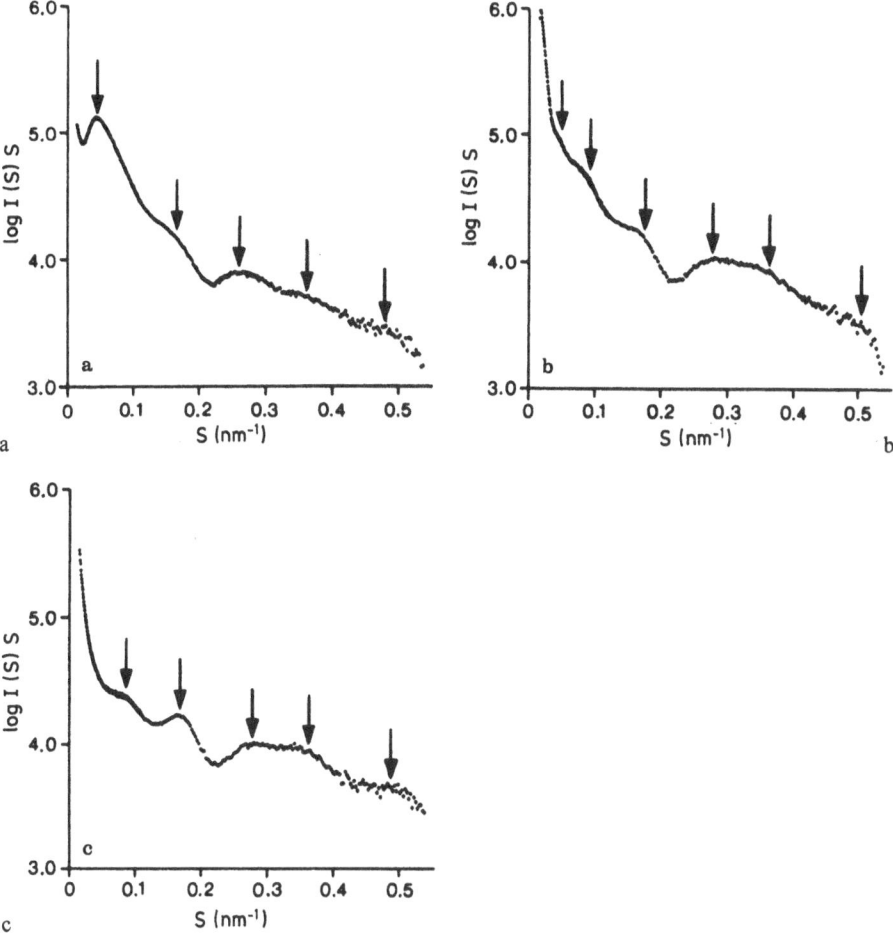

Fig. 8 a–c. Scattering patterns from a solution of chromatin fragments at 6.2 mg DNA/ml. **a** chromatin-EDTA, **b** chromatin fragments in TE buffer supplemented with 5 mM Mg^{2+} (similar patterns are obtained when ionic strength is raised with 75–100 mM NaCl) and **c** chromatin fragments in TE buffer supplemented with 8 mM Mg^{2+}. At this concentration of the cation chromatin begins to precipitate slowly however, due to the data acquisition with synchrotron radiation, patterns can be recorded before precipitation. Arrows indicate in **a** the 0.05 nm^{-1} band, the shoulder at 0.145 nm^{-1} and the high angle bands at 0.27, 0.365 and 0.47 nm^{-1}. **b** the weak band near 0.04 nm^{-1}, the bands developing at 0.075 and 0.16 nm^{-1} and the high angle bands. **c** the bands at 0.09 and 0.17 nm^{-1} characteristic of the condensed state of chromatin and the high angle bands (after figure 2 from Bordas et al., 1986 a)

of the nucleosomes since the linker DNA represents only about 20% of the excess scattering mass, and at low ionic strength, as a first approximation the correlation between the orientation of nucleosomes can be neglected. Model calculations in which the segment length is constrained in the range 18–22 nm and the angle between the segments is larger than 60° with no constraints on the dihedral angles involving three successive segments have been carried out. The average of 100 conformations of a wormlike chain with 70 segments obtained by a Monte Carlo calculation is sufficient to reproduce the features observed experimentally [62].

The 0.05 nm^{-1} band can be used to monitor changes in the unfolded chromatin structure. Upon removal of H1/H5 histones the band is displaced to lower s-values (i.e. the average distance between successive nucleosomes increases) showing the requirement for H1/H5 for maintaining the structural scaffolding of the low ionic strength fibre. Similarly when ethidium bromide, which partly intercalates between the bases, is bound the structure of uncondensed chromatin fibre is further extended. In parallel to the shift in the position of the 0.05 nm^{-1} band the R_x and the M/L decrease. At 0.5 mM ethidium bromide with 3.5 mg DNA/ml, M/L is about 2.5 times less than in the "native" uncondensed chromatin, and $R_x = 3.5$ nm which corresponds to that expected from a fully extended fibre [61].

Increasing concentrations in the range 0–10 molecules/100 bp, of both Distamycin and Netropsin-antibiotics which bind in the minor groove of DNA without intercalation — lead to a broadening of the 0.05 nm^{-1} interference band, reflecting a broadening of the distribution of internucleosomal distances. A parallel decrease in the M/L is also observed. Above 20 molecules/100 bp denaturation takes place [62].

Increasing the ionic strength of chromatin solutions results in two apparently successive processes: condensation and aggregation of the fibres. Condensation can be monitored by the changes in the scattered intensity at very low angles, i.e., by changes in the M/L and by the presence and the position of the 0.05 nm^{-1} band. The fully condensed state of the fibre can be determined by the absence of the 0.05 nm^{-1} band and the appearance of the 0.09 nm^{-1} $((11 \text{ nm})^{-1})$ and 0.175 nm^{-1} $((5.7 \text{ nm})^{-1})$ bands which indicate close approach of the nucleosomes. Figure 8 (b) and (c) represent patterns of condensed chromatin and precipitating chromatin respectively.

The presence of aggregates in the solution can be inferred from the trends in the variation of R_x and $I(0)_x$, however, whenever possible, the onset of this process is monitored also by other techniques such as light scattering and solubility measurements [80] (Koch et al. unpublished).

With increasing ionic strength the 0.05 nm^{-1} band shifts to higher angles and decreases in intensity as expected from an interference band. The central scatter shows a continuous increase whereas R_x decreases up to a critical concentration of the cation used for condensation, beyond which it starts to increase due to aggregation. The change in the position of the 0.05 nm^{-1} band together with values of R_x and $I(0)_x$ for rat liver chromatin at NaCl concentration below the aggregation point are given in Fig. 9.

Although similar trends of changes are seen in the scattering patterns when chromatin condensation is induced by different cations, the efficiencies of mono- and multivalent cations in inducing condensation are quite different. Systematic studies on the effects of mono- and multivalent cations on chromatin condensation using X-ray scattering have been carried out [84, 62]. Results of these studies show that the efficiency of a cation for inducing condensation increases with its valence and the condensation results mainly from the neutralization of negative charges on the DNA in the chromatin. In the case of monovalent cations (Li$^+$, Na$^+$, K$^+$, Cs$^+$), at every bulk salt concentration the number of counterions bound to chromatin is independent of their nature and there are no major changes of hydration. Over the range 0–70 mM, M/L of chicken erythrocyte chromatin increases by a factor of up to 8, independently of the nature of the cation [62].

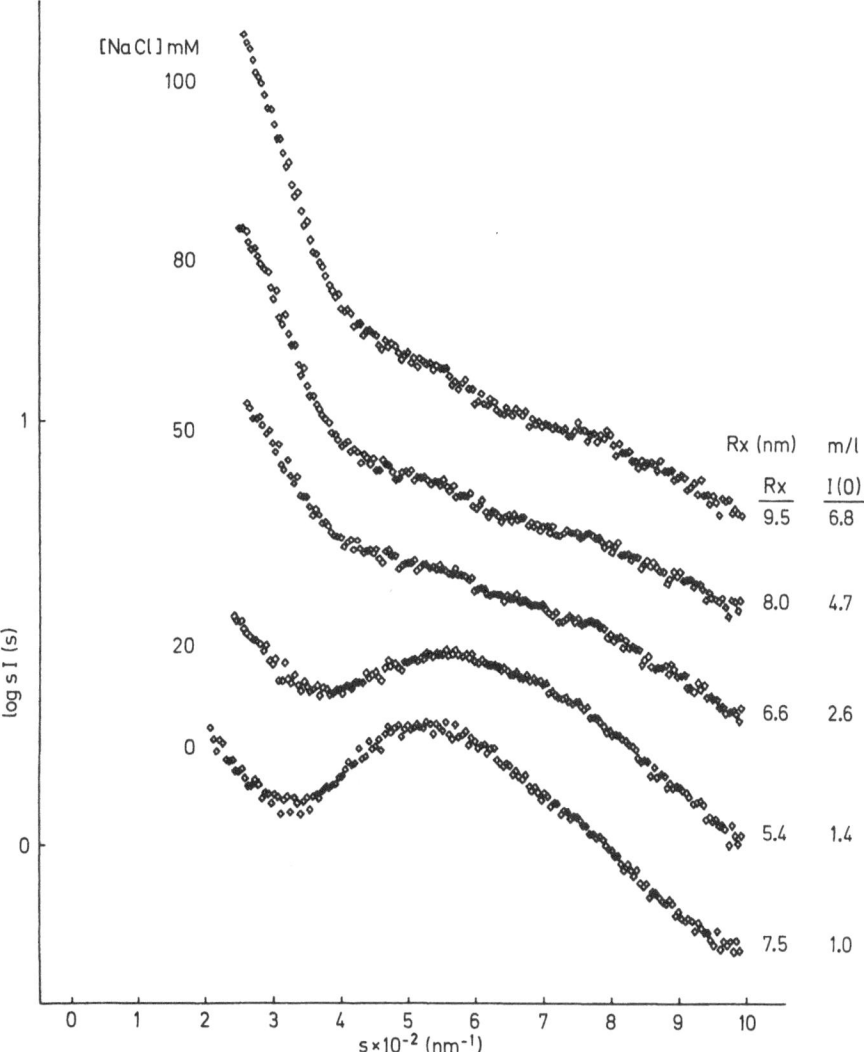

Fig. 9. Changes in the scattering pattern of rat liver chromatin fragments (3.5 mg DNA/ml) as a function of NaCl concentration and corresponding values of R_x and relative M/L. As the fibres condense the 0.05 nm^{-1} band shifts to larger s-values and R_x and $I(O)_x$ progressively increase (after figure 5 from Koch et al., 1987a)

In the case of divalent cations significant differences between various cations are observed. In all cases however, the condensation process, as monitored by the disappearance of the 0.05 nm^{-1} band appears to be completed before precipitation sets in. Precipitation occurs at concentrations of 3.5 mg DNA/ml and 2.5 mM MgCl$_2$, or at about 0.5 Mg^{2+}/bp. The M/L at a given salt concentration follows the expected sequence for the alkali earth chlorides, Ba^{2+} > Sr^{2+} > Mg^{2+} but Ca^{2+} is more efficient. Cu^{2+} and Cd^{2+} and the trivalent complex Co(NH$_3$)$_6^{3+}$ induce condensation at lower concentrations than Mg^{2+}, Co(NH$_3$)$_6^{3+}$ being more efficient than any of the

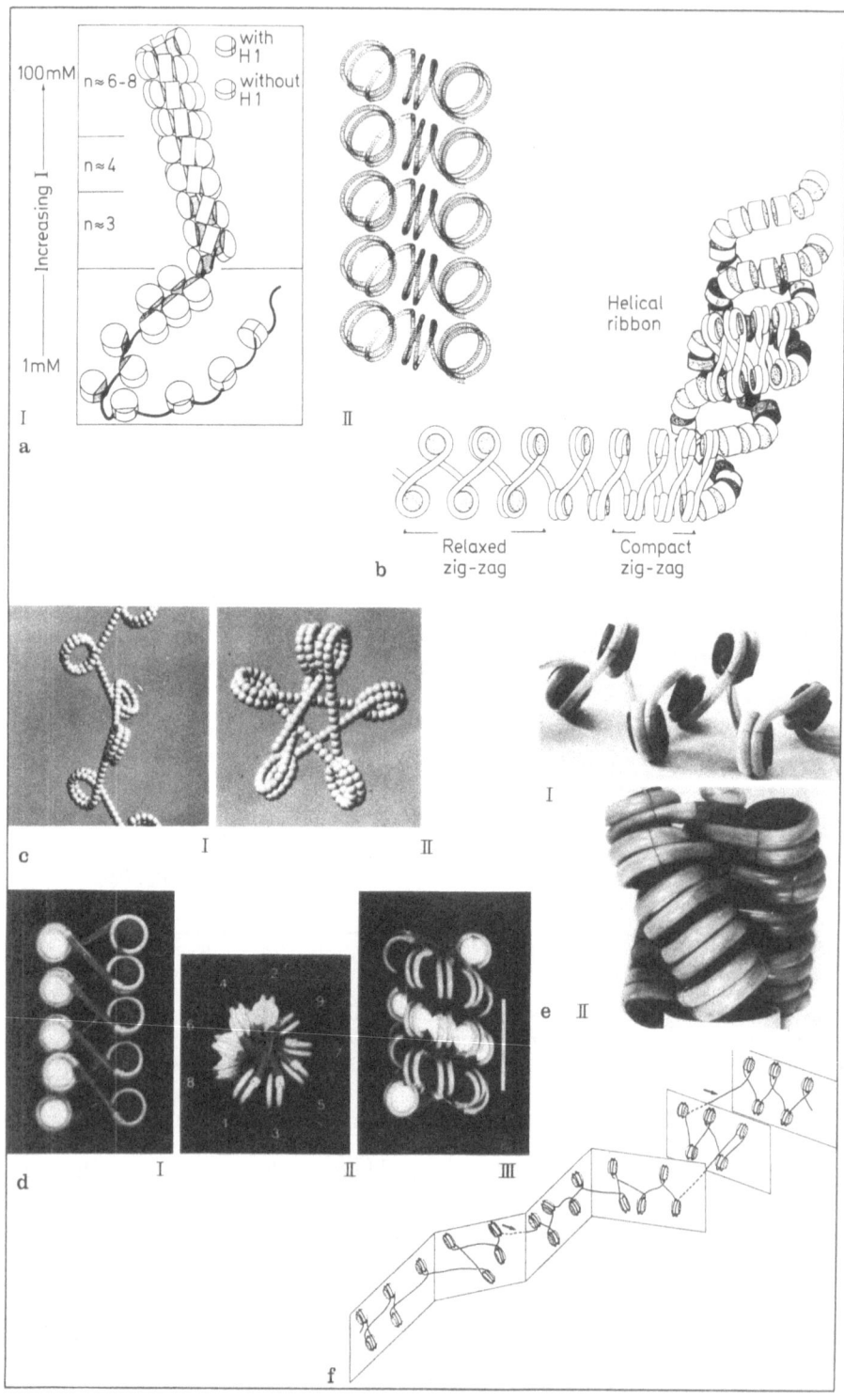

divalent cations. Effects on condensation of spermine ($+4$), spermidine ($+3$), Gd^{3+} and Cr^{3+} have also been investigated (Koch et al., unpublished). Again the condensation behaviour is similar to mono- and divalent cations and at about 0.6 mM concentration of the cation (at about 3.0 mg DNA/ml) precipitation sets in.

If the condensation takes place in the presence of more than one type of cation, e.g., in a buffer containing both NaCl and $MgCl_2$, trends of changes in the R_x and $I(0)_x$ are similar to the case of single cation, but the concentrations at which condensation and precipitation occur are different. It has been reported that the two types of cations act in an antagonistic manner [84]. It appears that NaCl delays the condensation induced by $MgCl_2$ but no longer affects the point of precipitation above 10 mM NaCl [84], Koch et al., unpublished). Electric dichroism studies, on the other hand, have been interpreted as the the the effects of two cations being synergistic rather than competitive [85]. Preliminary X-ray and light scattering studies (Koch et al., unpublished) show that the situation may be more complex.

4 Models of Higher Order folding of Chromatin

Studies of the condensed chromatin fibre structure and the condensation mechanism have resulted in basically two classes of models: models based on a helical arrangement of nucleosomes along the fibre and those based on a linear array of globular nucleosome clusters (superbeads) along the fibre. The first class includes the solenoid, twisted ribbon and crossed linker models whereas the latter are the superbead models and related layered structures. Schematic representations of some models are shown in Fig. 10.

In the original solenoid model [44] the low ionic strength structure of the chromatin fibre is described by a cylindrical 10 nm nucleofilament which shows no internal contrast for the nucleosome arrangement neither on the electron micrographs [44] nor in the small angle diffraction or scattering patterns [86, 87]. Condensation is achieved by the winding of the 10 nm nucleofilament into a solenoid, or contact helix, with about 6 nucleosomes per turn, a pitch of 11 nm and a helical diameter of 25–30 nm. Linker DNA in the condensed fibre has been modelled as kinked into the centre of the fibre [88] or in a helical path between the nucleosome cores leaving a central hole [89].

In the twisted ribbon models [90, 91] the low ionic strength structure of chromatin is described by a flat ribbon consisting of two parallel stacks of nucleosomes

◀ **Fig. 10a–f.** Schematic representation of models of higher order folding of chromatin. **a** I. The solenoid model from Thoma et al., [7], Figure 14. II. A possible path of DNA in the fibre from McGhee et al. [89], Figure 5, for clarity only three nucleosomes on the front surface of the fibre are shown. **b** The twisted ribbon model from Woodcock et al., [91], Fig. 8. For clarity, linker DNA is shown only in the first turn of the condensed structure. **c** The low ionic strength 3-D zigzag structure of chromatin (I) and the end-on view of a complete turn of the condensed fibre making contact with the first nucleosome in the next turn (II) from Bordas et al., [83], Fig. 5. For clarity the linker is shown fully stretched. **d** The longitudinal cross section (I), cross section (II) and side view (III) of the crossed linker model with DNA linking number -2 from Williams et al., [74], Fig. 2. **e** The extended triple nucleosome structure at low ionic strength (I) and the triple helix model at high ionic strength (II) from Makarov et al., [94], Fig. 2. **f** The layered structure from Subirana et al., [93], Fig. 10. The two planes at the right indicate how the nucleosomes would be arranged in the condensed fibre, proceeding to the left, the fibre unfolds as it would upon decreasing the ionic strength

225

Table 2. Observations resulting from studies of chromatin fibre structure compiled from the references given in table 1. Sources are noted only when values given differ from those generally reported

	X-ray diffraction and scattering	Electron microscopy
Bands		
0.02–0.03 nm⁻¹	– Interfibre interference (equatorial). – Observed in nuclei and gels. – Not observed in sperm nuclei and nuclei in Mg^{2+} buffers. – Exact position depends on species, ionic strength and in gels, on fibre concentration. – 1st side maximum of the fibre transform (model calculations).	
0.045–0.05 nm⁻¹	– Observed in nuclei and chromatin fragments in solution at 50–100 mM ionic strength. – Exact position species dependent.	
0.05 nm⁻¹	– Internucleosomal interference in the uncondensed fibre. – Observed in nuclei and chromatin fragments in solution at ionic strength below 20 mM.	
0.09 nm⁻¹	– Internucleosomal interference in the condensed fibre (meridional). – Observed in nuclei and chromatin fragments in solution at high ionic strength (100 mM salt) or in the presence of multivalent cations.	
0.145 nm⁻¹	– Observed in solutions of isolated nucleosomes and fragments.	
0.175 nm⁻¹	– Observed in nuclei and chromatin fragments in solution at 100 mM ionic strength. – Internucleosomal interference in the condensed fibre (equatorial).	
0.27 nm⁻¹	– Arises from intranucleosomal DNA and protein. – Observed in solutions of isolated nucleosomes, nuclei and chromatin fragments.	
Fibre diameter		– 30 nm — in intact nuclei. – 20–25 nm — rat liver chromatin fragments at 5 mM ionic strength. – 25 nm — rat liver chromatin fragments at 20 mM ionic strength. – 38 nm — thyone sperm chromatin fragments.
Mass per unit length		8.1 ± 1.0 nucleosomes/11 nm — necturus erythrocyte condensed chromatin fragments. 13.1 ± 2.0 nucleosomes/11 nm — thyone sperm chromatin fragments. 2.5 nucleosomes/11 nm — uncondensed chicken erythrocyte chromatin fragments.[91] 11.6 nucleosomes/11 nm — condensed chicken erythrocyte chromatin fragments.[91]
Other Observations		– Supranucleosomal particles and layered structures at high ionic strength. – Open zigzag form for uncondensed chromatin.
Nucleosome and linker DNA Orientation		*Electric dichroism and birefringence/photo electric dichroism/circular dichroism* – Nucleosome faces and linker DNA parallel to the uncondensed chromatin fibre axis – Nucleosome faces tilted 20–30° to the condensed chromatin fibre axis.

0.36 nm^{-1}	— Arises from the DNA wound around the histone octamer.	
0.47 nm^{-1}	— Observed in solutions of isolated nucleosomes, nuclei and chromatin fragments.	
	— Observed in solutions of isolated nucleosomes.	
Apparent Radius of gyration of the cross section	10 ± 1 nm (0.013 < s < 0.024 nm^{-1})	Uncondensed and condensed CE chromatin fragments.
	2.5 ± 0.05 nm (0.055 < s < 0.077 nm^{-1})	Uncondensed rat liver chromatin fragments. [86]
Mass per unit length	0.8 ± 0.1	Nucleosomes/11 nm uncondensed CE chromatin fragments.
	6.4 ± 0.8	Nucleosomes/11 nm condensed CE chromatin fragments.

Light scattering

— Low ionic strength fibre behaves as a flexible coil.
— Condensation is followed by aggregation and precipitation.

Sedimentation

— Sedimentation coefficient of hexanucleosomes or longer oligomers show a power-law dependence on the ionic strength. That of shorter oligomers remains constant above 25 mM ionic strength.

connected by the linker DNA in the form of an open zig-zag. In the condensed state the distance between alternating nucleosomes decreases by the collapse of the zigzag and the ribbon is wrapped on the surface of a cylinder to form a fibre in which linker DNA runs up and down the helical axis between the adjacent nucleosomes. In this model the helical pitch would be about 20 nm or more depending on the length and the orientation of the linker.

The crossed linker models are based on the nonsequential arrangement of nucleosomes in the condensed fibre. Nucleosome cores are connected by the linker DNA running across the central part of the fibre. Such arrangements of nucleosomes are discussed by Staynov [92].

In the model proposed by Bordas et al. [83] at low ionic strength the chromatin fibre already has a 3-D zig-zag superstructure with an outer diameter comparable to that of the condensed fibre. Upon condensation the distance between successive nucleosomes diminishes and the structure collapses in a way similar to the folding of an accordion, resulting in about a 10-fold decrease in the pitch and an 8-fold increase in the mass per unit length. In the semi-condensed state the structure would be very similar to that observed by Subirana et al. [93] by electron microscopy.

In the crossed linker model [74] the odd and even nucleosomes form a double helical structure with the linker DNA criss-crossing the condensed fibre transversely. The double helical pitch is about 26 nm and the fibre diameter would depend on the length of the linker. A model in which a repeating unit of a trinucleosome forms a 3-D zig-zag has also been proposed [94]. Twisting and compression of the zig-zag would result in a triple helix structure for the condensed fibre.

The superbead models and related layered structures [95, 75, 93] are based on the results of digestion experiments, the beaded appearance of chromatin fibres at intermediate ionic strength as well as the tendency of the 30 nm fibre to fragment into globular domains of variable size. In one of the models [93] at low ionic strength domains of 5–6 nucleosomes are arranged in zig-zag planes which, upon increasing the ionic strength, fold by a decrease in the distance between the planes. The path of the linker DNA cannot be modelled in a unique way directly from the electron micrographs.

In order to be acceptable models of chromatin structure need to be consistent with observations made by different methods as summarized in Table 2. Due to the lack of direct information on the exact location of histones H1/H5 and disposition of the linker DNA the results do not lead to a unique model however, they provide evidence to rule out some of the existing models.

The twisted ribbon model of Worcel et al. [90] and the superbead model of Zentgraf and Franke [75] can be ruled out on the basis of X-ray diffraction studies of oriented chromatin fibres [76]. The former model predicts the 0.175, 0.27 and 0.37 nm^{-1} band to be meridional whereas they appear as equatorial, and the latter predicts a meridional band at 0.03 to 0.02 nm^{-1} due to repeating supranucleosomal units along the condensed fibre which is not observed. The triple helix model of Makarov et al. [94] can also be ruled out since it would not give rise to the observed 0.175 and 0.09 nm^{-1} bands.

The twisted ribbon model of Woodcock et al. [91] is ruled out because it would in contradiction with experiment, give rise to a very negative value of the electric dichroism of condensed chromatin since all DNA would be oriented parallel to the fibre axis.

The double helical model of Williams et al. [74] predicts a helical pitch which is not detected by X-ray diffraction or scattering and other evidence given by the authors in support of this model, e.g. optical diffraction patterns from electron micrographs, need further clarification.

Currently the solenoid model [44] and the model of Bordas et al. [83] provide descriptions for the structure of the 30 nm fibre which are consistent with most of the observations. The parameters of the solenoid and of the model of Bordas et al. are similar. The two models differ in the path of the linker DNA: in the former the linker connects the adjacent nucleosomes but in the latter nucleosomes which are not packed adjacently are connected.

Observations on chromatin fragments from different types of eukaryotic cell nuclei thus indicate that at low ionic strength fibres are predisposed with an irregular helix-like superstructure in which the linker DNA is extended. The increased flexibility of the linker at increasing ionic strength facilitates the rapid transition to the condensed form. The condensed fibre is an irregular structure which can accommodate different lengths of linker DNA. Here the nucleosomes are stacked radially with their flat face at about 30° to the fibre axis. Details of the condensed structure and the condensation mechanism can be determined only when the exact path of the linker DNA and the position of histones H1/H5 are known. It should be stressed that in any of the existing models both of these crucial features are speculative.

The high flux on SR small angle X-ray scattering instruments has made it possible to carry out systematic static and kinetic investigations of chromatin structure in various conditions with minimum damage or perturbation on fresh samples. The point focusing achieved in SR cameras facilitated resolving features like the $0.05 \, nm^{-1}$ band unambiguously. These results provide strong evidence for a preformed 3-D zig-zag structure at low ionic strength which folds instantaneously into the "30 nm filament" upon increasing the ionic strength. The extent of folding at a given ionic strength depends on the nature of the cations and can be influenced by other substances which bind to DNA.

Complementary use, in future, of small angle solution synchrotron X-ray scattering methods with electron microscopy could contribute to the clarification of the path of the linker DNA. Thus further experiments seeking a universal model for the chromatin fibre structure need to be undertaken although the existence of such a universal structure is questionable. More importantly future investigations need to be aimed at understanding what relevance the structure of the fibre has to the control of DNA transcription and replication.

5 Acknowledgement

I am grateful to M. H. J. Koch for constant support and critical reading of the manuscript. Thanks are due to A. M. Michon, E. Dorrington and R. Kläring for their contribution to the experiments and to S. Mottram for her patient help with the literature survey. I also wish to thank the colleagues who have provided me with the originals for Fig. 10.

6 References

1. Paulson JR In: Harris R (ed) (1981) Electron microscopy of proteins, Vol. 3, Academic Press, London
2. Weisbrod S (1982) Nature *297*: 289
3. Yaniv M, Cereghini S (1986) CRC Crit. Rev. Biochem. *21(1)*: 1
4. Davies HG (1968) J. Cell Sci. *3*: 129
5. Olins AL, Olins DE (1974) Science *183*: 330
6. Thoma F, Koller T (1977) Cell *12*: 101
7. Thoma F, Koller Th, Klug A (1979) J. Cell Biol. *83*: 403
8. Isenberg I (1979) Ann. Rev. Biochem. *48*: 159
9. Von Holt C, Strickland WN, Brandt WF, Strickland MW (1979) FEBS Lett. *100*: 201
10. Wu RS, Panusz HT, Hatch CL, Bonner WM (1986) CRC Crit. Rev. Biochem. *20(2)*: 201
11. Moss T, Cary PD, Abercrombie BD, Crane-Robinson C, Bradbury EM (1976) Eur. J. Biochem. *71*: 337
12. Bohm L, Hayashi H, Cary PD, Moss T, Crane-Robinson C, Bradbury EM (1977) Eur. J. Biochem. *77*: 487
13. Whitlock JP Jr., Simpson RT (1977) J. Biol. Chem. *252*: 6516
14. Whitlock JP Jr., Stein A (1978) J. Biol. Chem. *253*: 3857
15. Thomas JO, Kornberg RD (1975a) Proc. Natl. Acad. Sci. USA *72*: 2626
16. Thomas JO, Kornberg RD (1975b) FEBS Lett. *58*: 353
17. Weintraub H, Palter K, Van Lente F (1975) Cell *9*: 409
18. Eickbush TH, Moudrianakis EN (1978) Biochemistry *17*: 4955
19. Braddock GW, Baldwin JP, Bradbury EM (1981) Biopolymers *20*: 327
20. Uberbacher EC, Harp JM, Wilkinson-Singley E, Bunnick GJ (1986) Science *232*: 1247
21. Burlingame RW, Love WE, Moudrianakis EN (1984) Science *223*: 413
22. Burlingame RW, Love WE, Wang B-C, Hamlin R, Yuong N-H, Moudrianakis EN (1985) Science *228*: 546
23. Richmond TJ, Finch JT, Rushton B, Rhodes D, Klug A (1984) Nature *311*: 532
24. Uberbacher EC, Bunnick GJ (1985a) J. Biomol. Struct. Dyn. *2*: 1033
25. Bentley GA, Lewit-Bentley A, Finch JT, Podjarny AD, Roth M (1984) J. Mol. Biol. *176*: 55
26. Simpson RT (1978) Biochemistry *17*: 5524
27. Neelin JM, Callahan PX, Lamb DC, Murray K (1964) Can. J. Biochem. *42*: 1743
28. Hnilica LS (1972) The structure and function of histones, CRC Press Cleveland, Ohio
29. Smith BJ, Johns EW (1980) FEBS Lett. *110*: 25
30. Kornberg RD (1974) Science *184*: 868
31. Kornberg RD (1977) Ann. Rev. Biochem. *46*: 931
32. Morris NR (1976) Cell *9*: 627
33. Lohr D, Corden J, Tatchell K, Kovacic RT, Van Holde KE (1977) Proc. Natl. Acad. Sci. USA *74*: 79
34. Thomas JO, Thompson RJ (1977) Cell. *10*: 633
35. Weintraub H (1978) Nucleic Acids Res. *5*: 1179
36. Pardon JF, Worcester DL, Wooley JC, Cotter RJ, Lilley DHJ, Richards BW (1977) Nucleic Acids Res. *4*: 3199
37. Suau P, Kneale GG, Braddock GW, Baldwin JP, Bradbury EM (1977) Nucleic Acids Res. *4*: 3769
38. Imai BS, Yau P, Baldwin JP, Ibel K, May RP, Bradbury EM (1986) J. Biol. Chem. *261*: 8784
39. Greulich KO, Ausio J, Eisenberg H (1985) J. Mol. Biol. *186*: 167
40. Klug A, Finch JT, Richmond TJ (1985) Science *229*: 1109
41. Moudrianakis EN, Love WE, Wang BC, Xuong NG, Burlingame RW (1985a) Science *229*: 1110
42. Moudrianakis EN, Love WE, Burlingame RW (1985b) Science *229*: 1113
43. Uberbacher EC, Bunick GJ (1985b) Science *229*: 1112
44. Finch JT, Klug A (1976) Proc. Natl. Acad. Sci. USA *73*: 1879
45. Olins AL, Olins DE (1979) J. Cell Biol. *81*: 260
46. Thoma F, Koller Th (1981) J. Mol. Biol. *149*: 709
47. Ris H, Kubai DF (1970) Ann. Rev. Genet. *4*: 263

48. Thomas JO, Khabaza AJA (1980) Eur. J. Biochem. *112*: 501
49. Lennard AC, Thomas JO (1985) EMBO J. *4(13A)*: 3455
50. Noll M, Kornberg RD (1977) J. Mol. Biol. *109*: 393
51. Allan J, Hartman PG, Crane-Robinson C, Avilles FX (1980) Nature *288*: 675
52. Allan J, Mitchell T, Nerina H, Bohm L, Crane-Robinson C (1986) J. Mol. Biol. *187*: 591
53. Koch ·EE, Eastman DE, Farge Y In: Koch EE (ed) (1983) Handbook on Synchrotron Radiation North Holland Publishing Co. Amsterdam, New York, Oxford, vol. 1a, pp. 1–65
54. Hendrix J, Koch MHJ, Bordas J (1979) J. Appl. Cryst. *12*: 467
55. Bordas J, Koch MHJ, Clout PN, Dorrington E, Boulin C, Gabriel A (1980) J. Phys. E: Sci. Instrum. *13*: 938
56. Boulin C, Dainton D, Dorrington E, Elsner G, Gabriel A, Bordas J, Koch MHJ (1982) Nucl. Instrum. and Methods *201*: 209
57. Gabriel A, Dauvergne F (1982) Nucl. Instrum. and Methods *201*: 223
58. Hendrix J, Fuerst H, Hartfiel B, Dainton D (1982) Nucl. Instrum. and Methods *201*: 139
59. Koch MJH, Bordas J (1983) Nucl. Instrum. and Methods *208*: 461
60. Boulin C, Kempf R, Koch MHJ, McLaughlin S (1986) Nucl. Instrum. and Methods *A249*: 399
61. Bordas J, Perez-Grau L, Koch MHJ, Vega MC, Nave C (1986a) Eur. J. Biophys. *13*: 157
62. Koch MHJ, Vega MC, Sayers Z, Michon AM (1987a) Eur. Biophys. J. *14*: 307
63. Koch MHJ, Sayers Z, Vega MC, Michon AM (1987b) Eur. Biophys. J. *14*: in press
64. Stuhrmann HB, Kirste RG (1965) Z. physik. Chemie, Frankfurt *46*: 247
65. Stuhrmann HB In: Glatter O, Kratky O (eds) 1982) "Small angle X-ray scattering", Academic Press, London, pp. 197–213
66. Debye P (1915) Ann. Phys. *46*: 809
67. Guinier A, Fournet G (1955) Small Angle Scattering of X-rays. Wiley, New York
67a. Harrington R (1985) Biochemistry *24*: 2011
68. Porod G In Glatter O, Kratky O (eds) (1982) Small Angle X-ray Scattering, Academic Press, London, pp. 17–52
69. Oster G, Riley DP (1952) Acta Cryst. *5*: 272
70. Vainshtein BK (1966) Diffraction of X-rays by chain molecules, Elsevier Publishing Company, New York
71. Langmore JP, Schutt C (1980) Nature *288*: 620
72. Langmore JP, Paulson JR (1983) J. Cell Biol. *96*: 1120
73. Notbohm H (1986a) Eur. Biophys. J. *13*: 367
74. Williams SP, Athey BD, Muglia LJ, Schappe RS, Gough AH, Langmore JP (1986) Biophys. J. *49*: 233
75. Zentgraf H, Franke WW (1984) J. Cell Biol. *99*: 272
76. Widom J, Klug A (1985) Cell *43*: 207
77. Widom J, Finch JT, Thomas JO (1985) EMBO J. *4(12)*: 3189
78. Staron K (1985) Biochim. Biophys. Acta *825*: 289
79. Perez-Grau L, Bordas J, Koch MHJ (1984) Nucleic Acids Res. *12(6)* 2987
80. Ausio J, Borochov N, Seger D, Eisenberg H (1984) J. Mol. Biol. *177*: 373
81. Ausio J, Sasi R, Fasman G (1986) Biochemistry *25*: 1981
82. Greulich KO, Wachtel E, Ausio J, Seger D, Eisenberg H (1987) J. Mol. Biol. *193*: 709
83. Bordas J, Perez-Grau L, Koch MHJ, Nave C, Vega MC (1986b) Eur. J. Biophys. *13*: 175
84. Widom J (1986) J. Mol. Biol. *190*: 411
85. Sen D, Crothers DM (1986) Biochemistry *25*: 1495
86. Sperling L, Tardieu A (1976) FEBS Lett. *64(1)*: 89
87. Sperling L, Klug A (1977) J. Mol. Biol. *112*: 253
88. Butler PJG (1984) EMB J. *3*: 2599
89. McGhee JD, Nickol JM, Felsenfeld G, Rau D (1983) Cell *33*: 831
90. Worcel AS, Strogatz S, Riley D (1981) Proc. Natl. Acad. Sci. USA *78*: 1461
91. Woodcock CLF, Frado LLY, Rattner JB (1984) J. Cell Biol. *99*: 42
92. Staynov DZ (1983) Int. J. Biol. Macromol. *5*: 3
93. Subirana JA, Munoz-Guerra S, Aymami J, Radermacher M, Frank J (1985) Chromosoma *91*: 377
94. Makarov V, Dimitrov S, Smirnov V, Pashev I (1985) FEBS Lett. *181(2)*: 357
95. Renz M, Nehls P, Hozier J (1977) Proc. Natl. Acad. Sci. USA *74*: 1879

96. Lasters I, Wyns L, Muyldermans S, Baldwin JP, Poland GA, Nave C (1985) Eur. J. Biochem. *151*: 283
97. Notbohm H (1986 b) Int. J. Biol. Macromol. *8*: 114
98. Brust R, Harbers E (1981) Eur. J. Biochem. *117*: 609
99. Notbohm H, Harbers E (1981) Int. J. Biol. Macromol. *3*: 311·
100. Azorin F, Martinez AB, Subirana JA (1980) Int. J. Biol. Macromol. *2*: 81
101. Hollandt H, Notbohm H, Riedel F, Harbers E (1979) Nucleic Acids. Res. *6(5)*: 2017
102. Damaschun H, Damaschun G, Pospelov VA, Vorob'ev VI (1980) Molec. Biol. Rep. *6*: 185
103. Suau P, Bradbury EM, Baldwin JP (1979) Eur. J. Biochem. *97*: 593
104. Bram S, Baudy P, Lepault J, Hermann D (1977) Nucleic Acids Res. *4(7)*: 2275
105. Baudy P, Bram S (1979) Nucleic Acids Res. *6(4)*: 1721
106. Bram S, Butler-Brown G, Baudy P, Ibel K (1975) Proc. Natl. Acad. Sci. USA *72*: 1043
107. Baudy P, Bram S (1978) Nucleic Acids Res. *5(10)*: 3697
108. Azorin F, Perez-Grau L, Subirana JA (1982) Chromosoma *85*: 251
109. Subirana JA, Munoz-Guerra S, Radermacher M, Frank J (1983) J. Biomol. Struct. Dyn. *1*: 705
110. Dubochet J, Noll M (1978) Science *202*: 280
111. Dixon DK, Burkholder GD (1985) Eur. J. Cell Biol. *36*: 315
112. Rattner JB, Saunders C, Davie JR, Hamkalo BA (1982) J. Cell Biol. *93*: 217
113. Bates DL, Butler PJG, Pearson EC, Thomas JO (1981) Eur. J. Biochem. *119*: 464
114. Allan J, Rau DC, Harborne N, Gould H (1984) J. Cell Biol. *98*: 1320
115. Muyldermans S, Lasters I, Hamers R, Wyns L (1985) Eur. J. Biochem *150*: 441
116. Butler PJG, Thomas JO (1980) J. Mol. Biol. *140*: 505
117. Thomas JO, Butler PJG (1980) J. Mol. Biol. *144*: 89
118. Böttger M, Karawajew L, Fenske H, Grade K, Lindigkeit R (1981) Molec. Biol. Rep. *7*: 231
119. Brust R (1985) Molec. Biol. Rep. *10*: 231
120. Brust R (1986 b) Z. Naturforsch. *41c*: 917
121. Rees AW, Debuysere MS, Lewis EA (1974) Biochim. Biophys. Acta *361*: 97
122. Kubista M, Hard T, Nielsen PE, Norden B (1985) Biochemistry *24*: 6336
123. Pearson EC, Butler PJG, Thomas JO (1983) EMBO J. *2*: 1367
124. Fulmer AW, Fasman GD (1979) Biopolymers *18*: 2875
125. McGhee JD, Rau D, Charney E, Felsenfeld G (1980) Cell *22*: 87
126. Houssier C, Lasters I, Muyldermans S, Wyns L (1981 a) Nucleic Acids Res. *9(21)*: 5763
127. Houssier C, Lasters I, Muyldermans S, Wyns L (1981 b) Int. J. Biol. Macromol. *3*: 370
128. Marion C, Martinage A, Tirard A, Roux B, Daune M, Mazen A (1985 a) J. Mol. Biol. *186*: 367
128 a. Marion C, Bezot P, Hesse-Bezot C, Roux B, Bernengo JC (1981) Eur. J. Biochem. *120*: 169
129. Marion C (1984) J. Biomol. Struct. Dyn. *2(2)*: 303–317
130. Marion C, Roche J, Roux B, Gorka C (1985 b) Biochemistry *24*: 6328
131. Brust R (1986 a) Z Naturforsch. *41c*: 910
132. Mandel R, Fasman GD (1976) Nucleic Acids Res. *3*: 1839
133. Crothers DM, Dattagupta N, Hogan M, Klevan L, Lee KS (1978) Biochemistry *17(21)*: 4525
134. Houssier C, Hacha R, de Paw-Gillet MC, Pieczynski JL, Fredericq E (1981 c) Int. J. Biol. Macromol. *3*: 59
135. Lee KS, Mandelkern M, Crothers DM (1981) Biochemistry *20*: 1438
136. Yabuki H, Dattagupta N, Crothers DM (1982) Biochemistry *21*: 5015
137. Mitra S, Sen D, Crothers DM (1984) Nature *308*: 247
138. Ramsey-Shaw B, Schmitz KS (1976) Biochem. Biophys. Res. Commun. *73(2)*: 224
139. Ramsay-Shaw B, Schmitz KS In: Nicolini CA (ed) (1979) Chromatin structure and function, Plenum Publishing Corporation, New York
140. Schmitz KS, Ramsay-Shaw B (1977) Biopolymers *16*: 2619
141. Gordon VC, Knobler CM, Olins DE, Schumaker VN (1978) Proc. Natl. Acad. Sci. USA *75*: 660
142. Marion C, Hesse-Bezot C, Bezot P, Marion MJ, Roux B, Bernengo JC (1985 c) Biophysical Chem. *22*: 53
143. Roche J, Girardet JL, Gorka C, Lawrence JJ (1985) Nucleic Acids Res. *13(8)*: 2843
144. Ashikawa I, Knosita K Jr., Ikegami A, Nishimura Y, Tsuboi M (1984) Biochemistry *24*: 1291
145. Ashikawa I, Knosita K Jr., Ikegami A, Nishimura Y, Tsuboi M, Watanabe K, Iso K, Nakano T (1983) Biochemistry *22*: 6018

Author Index Volumes 101–145

Contents of Vols. 50–100 see Vol. 100
Author and Subject Index Vols. 26–50 see Vol. 50

The volume numbers are printed in italics

Alekseev, N. V., see Tandura, St. N.: *131*, 99–189 (1985).

Alpatova, N. M., Krishtalik, L. I., Pleskov, Y. V.: Electrochemistry of Solvated Electrons, *138*, 149–220 (1986).

Anders, A.: Laser Spectroscopy of Biomolecules, *126*, 23–49 (1984).

Armanino, C., see Forina, M.: *141*, 91–143 (1987).

Asami, M., see Mukaiyama, T.: *127*, 133–167 (1985).

Ashe, III, A. J.: The Group 5 Heterobenzenes Arsabenzene, Stibabenzene and Bismabenzene. *105*, 125–156 (1982).

Austel, V.: Features and Problems of Practical Drug Design, *114*, 7–19 (1983).

Badertscher, M., Welti, M., Portmann, P., and Pretsch, E.: Calculation of Interaction Energies in Host-Guest Systems. *136*, 17–80 (1986).

Baird, M. S.: Functionalised Cyclopropenes as Synthetic Intermediates, *144*, 137–209 (1987).

Balaban, A. T., Motoc, I., Bonchev, D., and Mekenyan, O.: Topilogical Indices for Structure-Activity Correlations, *114*, 21–55 (1983).

Baldwin, J. E., and Perlmutter, P.: Bridged, Capped and Fenced Porphyrins. *121*, 181–220 (1984).

Barkhash, V. A.: Contemporary Problems in Carbonium Ion Chemistry I. *116/117*, 1–265 (1984).

Barthel, J., Gores, H.-J., Schmeer, G., and Wachter, R.: Non-Aqueous Electrolyte Solutions in Chemistry and Modern Technology. *11*, 33–144 (1983).

Barron, L. D., and Vrbancich, J.: Natural Vibrational Raman Optical Activity. *123*, 151–182 (1984).

Bazin, D., Dexpert, H., and Lagarde, P.: Characterization of Heterogeneous Catalysts: The EXAFS Tool, *145*, 69–80 (1987).

Beckhaus, H.-D., see Rüchardt, Ch., *130*, 1–22 (1985).

Benfatto, M., see Bianconi, A.: *145*, 29–67 (1987).

Bestmann, H. J., Vostrowsky, O.: Selected Topics of the Witting Reaction in the Synthesis of Natural Products. *109*, 85–163 (1983).

Beyer, A., Karpfen, A., and Schuster, P.: Energy Surfaces of Hydrogen-Bonded Complexes in the Vapor Phase. *120*, 1–40 (1984).

Bianconi, A., Garcia, J., and Benfatto, M.: XANES in Condensed Systems, *145*, 29–67 (1987).

Binger, P., and Büch, H. M.: Cyclopropenes and Methylenecyclopropanes as Multifunctional Reagents in Transition Metal Catalyzed Reactions. *135*, 77–151 (1986).

Böhrer, I. M.: Evaluation Systems in Quantitative Thin-Layer Chromatography, *126*, 95–188 (1984).

Boekelheide, V.: Syntheses and Properties of the [2$_n$] Cyclophanes, *113*, 87–143 (1983).

Bonchev, D., see Balaban, A. T., *114*, 21–55 (1983).

Borgstedt, H. U.: Chemical Reactions in Alkali Metals *134*, 125–156 (1986).

Bourdin, E., see Fauchais, P.: *107*, 59–183 (1983).

Büch, H. M., see Binger, P.: *135*, 77–151 (1986).

Calabrese, G. S., and O'Connell, K. M.: Medical Applications of Electrochemical Sensors and Techniques. *143*, 49–78 (1987).

Cammann, K.: Ion-Selective Bulk Membranes as Models. *128*, 219–258 (1985).

Charton, M., and Motoc, I.: Introduction, *114*, 1–6 (1983).